Genetic Expression and Nutrition

The 50th Nestlé Nutrition Workshop, Genetic Expression and Nutrition, *was held in Montreux, Switzerland, 7–11 October 2001.*

Nestlé Nutrition Workshop Series
Pediatric Program Volume 50

Genetic Expression and Nutrition

Editors

Claude Bachmann *Clinical Chemistry, University of Lausanne, and Central Laboratory of Clinical Chemistry, Centre Hospitalier Universitaire Vaudois, Lausanne, Switzerland*

Berthold Koletzko *Division of Metabolic Diseases and Nutritional Medicine, Dr. von Hauner Children's Hospital, Kinderklinik and Kinderpoliklinik, University of Muenchen, Muenchen, Bavaria, Germany*

Lippincott Williams & Wilkins

Nestlé NUTRITION

Acquisitions Editor: Beth Barry
Developmental Editor: Maria McAvey
Production Editor: Melanie Bennitt
Manufacturing Manager: Colin Warnock
Compositor: Maryland Composition
Printer: Maple Press

Nestec Ltd., 55 Avenue Nestlé
CH-1800 Vevey, Switzerland
Lippincott Williams & Wilkins,
530 Walnut Street
Philadelphia, PA 19106 USA
LWW.com

Printed in the USA

Library of Congress Cataloging-in-Publication Data

Genetic expression and nutrition / editors, Claude Bachmann, Berthold Koletzko.
p. cm. — (Nestlé Nutrition workshop series. Pediatric Program ; vol. 50)
Includes bibliographical references and index.
ISBN 0-7817-4444-X
1. Metabolism, Inborn errors of—Nutritional aspects—Congresses. 2. Metabolism, Inborn errors of—Diet therapy—Congresses. 3. Gene expression—Congresses. I. Bachmann, C. (Claude) II. Koletzko, B. (Berthold) III. Nestlé Nutrition workshop series. Pediatric Program ; vol. 50.

RC627.8 .G465 2003
616.3′042—dc21

2002030170

Care has been taken to confirm the accuracy of the information presented and to describe generally accepted practices. However, the authors, editors, Nestec, and Lippincott Williams & Wilkins are not responsible for errors or omissions or for any consequences from application of the information in this book and make no warranty, expressed or implied, with respect to the currency, completeness, or accuracy of the contents of the publication. Application of this information in a particular situation remains the professional responsibility of the practitioner.

The authors, editors, Nestec, and Lippincott Williams & Wilkins have exerted every effort to ensure that drug selection and dosage set forth in this text are in accordance with current recommendations and practice at the time of publication. However, in view of ongoing research, changes in government regulations, and the constant flow of information relating to drug therapy and drug reactions, the reader is urged to check the package insert for each drug for any change in indications and dosage and for added warnings and precautions. This is particularly important when the recommended agent is a new or infrequently employed drug.

Some drugs and medical devices presented in this publication have Food and Drug Administration (FDA) clearance for limited use in restricted research settings. It is the responsibility of the health care provider to ascertain the FDA status of each drug or device planned for use in their clinical practice.

10 9 8 7 6 5 4 3 2 1

Preface

New areas of investigation and technological progress allow the testing of new and old hypotheses, raise questions, and stimulate alternative approaches. This can also be said of the rapid progress in the field of molecular biology, which has been significantly enhanced by the human genome project. However, with the mapping of the human genome it was realized at an early stage that the identification of the genetic cause of a phenotype is not the final objective of molecular research, but instead a powerful tool for gaining a better understanding of the pathophysiology leading to disease manifestation.

Although nuclear genes provide a similar potential for each nucleated cell, the actual utilization of the genetic potential varies between cells and over time, making the organism complex and conditionally able to adapt to different situations. Thus, there is a need to investigate the mechanisms of gene expression and regulation on the RNA, protein, and functional levels in cells, organs, and the whole organism, the impact of changing and complex environmental factors such as nutrition and also the potential differences between adult age and the period of growth and development.

In the past, morphological or biochemical markers have been primarily used for the understanding of normal or abnormal function in human physiology and pathophysiology. Progress in molecular biology studies starting at the gene level is now allowing the investigation of the impact of genetic factors involved in diseases. New findings tend to raise multiple new hypotheses and lead to question established models, and in this way may make their contribution to improved health and well-being.

The Program of the 50th Nestlé Nutrition Workshop held at Montreux, Switzerland, aimed at improving the understanding of nutritional physiology and biochemistry, by learning from observations in patients with genetic defects in their protein, lipid, and carbohydrate metabolism. It discussed nutritional aspects of genetic variation in populations and further included perspectives on recombinant modifications in food production, and gene therapy of inborn errors of metabolism. Distinguished experts presented data and interpretations, as well as their views on future developments, and they helped to avoid unfounded interpretations by their critical appraisal. The lively discussion with the very active involvement of a competent audience proved that the speakers had more than fulfilled expectations.

We sincerely thank the speakers, moderators, and participants for their constructive and challenging contributions. We are indebted to Mr. Ulrich Preysch and the enthusiastic staff of Nestlé Suisse SA, who hosted this most delightful event and provided impeccable logistic support, and to Dr. Anne-Lise Carrié-Fässler and Prof. Wolf Endres (Nestec Ltd, Vevey) for their excellent organizational support and helpful feedback during the planning of this workshop, which led to a most stimulating exchange of information and ideas.

Professor Claude Bachmann, Switzerland
Professor Berthold Koletzko, Germany

Foreword

These are the proceedings of a jubilee workshop—the 50th Nestlé Nutrition Workshop—since their introduction in 1980.

Knowledge about inherited metabolic disorders is growing fast, not only because of new technologies for the early detection of these diseases, but also because of the highly sophisticated differentiation of subtypes of well known defects and the development of sometimes successful treatments. In contrast, the link between our knowledge of genetic expression and our knowledge of general nutrition appeared to be rather poor. We therefore decided on the topic "Genetic Expression and Nutrition." The main objectives were to identify what insights can be gained from studying the natural course of inborn errors of metabolism, and the effects of dietary interventions in these diseases, and to see how these may be applied to people with normal metabolism or who are at risk for common diseases. At the end of this workshop we knew a great deal more about these issues, and we were in a better position to understand our present limitations in applying this knowledge practically.

I thank the two chairmen, Prof. Claude Bachmann and Prof. Berthold Koletzko, who are experts in this field, for putting the program together and inviting a group of speakers who are known to be opinion formers in the field of nutrition and genetics. Invited scientists from 25 countries contributed to the discussions that are published in the book. Mr. Ulrich Preysch's team from Nestlé Switzerland provided excellent logistical support and the participants enjoyed a taste of Swiss hospitality.

Dr. Anne-Lise Carrié Fässler from the Nutrition Strategic Business Division in Vevey, Switzerland, was responsible for the scientific coordination. Her excellent cooperation with the chairmen was essential for the success of this workshop.

PROFESSOR WOLF ENDRES, MD
Vice President
Nestec Ltd.
Vevey, Switzerland

Contents

Genetic Variation of Homocysteine Metabolism and Atherosclerosis 1
J. M. Scott

Minimum Protein Requirements in Infancy and Childhood: Insights from Patients with Protein-Restricted Diets 25
H. Ogier de Baulny, L. François, M. O. Grenèche, M. Giraud, and E. Depondt

How to Asses Amino Acid Requirements (With Particular Reference to Inborn Errors of Amino Acid Metabolism) 37
P. B. Pencharz and R. O. Ball

Long-Chain Omega-3 Fatty Acids are Conditionally Essential Substrates in Children: Effects on Visual Function in Children With Phenylketonuria 51
B. Koletzko, S. Beblo, A. Muntau-Heger, W. Mueller-Felber, and H. Reinhardt

Fatty Acid Regulation of Gene Expression: A Genomic Mechanism to Improve the Metabolic Syndrome 67
S. D. Clarke, C. Nelson, D. Gasperikova, J. DeMar, A. Lapillonne, and W. C. Heird

Genetic Disorders of Mitochondrial and Peroxisomal Fatty Acid Oxidation and Peroxisome Proliferator-Activated Receptors 85
R. J. A. Wanders

Oxidative Stress: Antioxidants in Degenerative Neurologic and Ophthalmologic Diseases 107
H. Sies, W. Stahl, and M. C. Polidori

Carbohydrate Therapy of Congenital Disorders of Glycosylation 123
T. Marquardt

Genetic Variability of Human Milk Oligosaccharides: Are There Biologic Consequences? 137
C. Kunz, L. Bode, and S. Rudloff

From Arginine to Creatine: Regional and Cellular Gene Expression of Enzymes and Transporters Linking Nitrogen and Energy Metabolism in Brain 153
C. Bachmann, H. Henry, and O. Braissant

Current Understanding of the Molecular and Cellular Basis of Celiac Disease 167
L. M. Sollid

Genetic Variations of Gliadins: New Approaches for the Treatment of Celiac Disease. 185
O. D. Anderson

Current Uses and Future Perspectives of Recombinant Modifications in Food Production 201
D. M. Bier

Diet as a Determinant of Drug Response. 219
D. G. Bailey

Diet, Gene Expression, and Apoptosis: Clues to Cancer Prevention? 239
H. Daniel and U. Wenzel

Gene Therapy: A Potential Solution for Nutrition-Dependent Disorders? 263
M. L. Batshaw, J. M. Wilson, X. Ye, and S. E. Raper

Concluding Discussion. 279
B. Koletzko

Subject Index 283

Contributing Authors

Olin D. Anderson
Western Regional Research Center
Agricultural Research Service
United States Department of Agriculture
Albany, California
USA

Claude Bachmann
Clinical Chemistry
University of Lausanne;
Central Laboratory of Clinical Chemistry
Centre Hospitalier Universitaire Vaudois (CHUV)
Lausanne, Switzerland

David G. Bailey
Lawson Health Research Institute
London Health Sciences Centre;
Departments of Medicine, Physiology, and Pharmacology
University of Western Ontario
London, Ontario
Canada

Mark L. Batshaw
Children's Research Institute
Children's National Medical Center
Department of Pediatrics
George Washington University School of Medicine and Health Sciences
Washington, DC
USA

Dennis M. Bier
Children's Nutrition Research Center, and Department of Pediatrics
Baylor College of Medicine
Houston, Texas
USA

Steven D. Clarke
Institute for Cellular and Molecular Biology
Division of Nutritional Sciences
University of Texas
Austin, Texas
USA

Hannelore Daniel
Department of Food and Nutrition Sciences
Molecular Nutrition Unit
Technical University of Munich
Freising-Weihenstephan, Germany

Berthold Koletzko
Department of Pediatrics
Ludwig Maximilians University;
Division of Metabolic Diseases and Nutritional Medicine
Dr. von Hauner Children's Hospital
Kinderklinik and Kinderpoliklinik
University of Muenchen
Muenchen, Bavaria
Germany

Clemens Kunz
Institute of Nutrition
University of Giessen
Giessen, Germany

Thorsten Marquardt
Klinik und Poliklinik für Kinderheilkunde
Münster, Germany

Helene Ogier de Baulny
Service de Neurologie
Maladies Métaboliques
Hôpital Robert Debré
Paris, France

Paul B. Pencharz
Department of Pediatrics
The Hospital for Sick Children;
Department of Nutritional Sciences
University of Toronto
Toronto, Ontario
Canada

John M. Scott
Department of Biochemistry
Trinity College Dublin
Dublin, Ireland

Helmut Sies
Institut für Physiologische. Chemie I
Heinrich-Heine Universität
Düsseldorf, Germany

Ludvig M. Sollid
Institute of Immunology
Rikshospitalet
University of Oslo
Oslo, Norway

Ronald J. A. Wanders
Department of Pediatrics
Academic Medical Center;
Department of Clinical Chemistry
University of Amsterdam
Amsterdam
The Netherlands

Session Chairmen

K. Baerlocher / *Switzerland*
D. Belli / *Switzerland*
B. Fowler / *Switzerland*
J.-M. Matthieu / *Switzerland*
B. Steinmann / *Switzerland*

Invited Attendees

Stephanie Williams / *Australia*
Mahabbat Ibrahimova / *Azerbaijan*
Chong Kim / *Brazil*
Stephen T.S. Lam / *China*
Xiaoping Luo / *China*
Ramzi El-Baroudy / *Egypt*
Cherif S. Erian / *Egypt*
Mohamed Fawzy Abdel Fattah / *Egypt*
Soad Wahba / *Egypt*
Reet Raukas / *Estonia*
Jean Rey / *France*
Hansjosef Bohles / *Germany*
Emmanuel Kanavakis / *Greece*
Christos Kostalos / *Greece*
Apurba Ghosh / *India*
Erijati Inrasanto / *Indonesia*
Subijanto Marto Sudarmo / *Indonesia*
Bambang Tridjaja Parwoto / *Indonesia*
Dedi Subardja / *Indonesia*
Massimo Agosti / *Italy*
Nicola Laforgia / *Italy*
Fabio Mosca / *Italy*
Yuichiro Yamashiro / *Japan*
Ki Sup Chung / *Korea*
Jae Ock. Park / *Korea*
Apisa Kushbakeeva / *Kyrgyzstan*
Majd Ariss-Timani / *Lebanon*
David Gonzales Flores / *Mexico*
Enrique Romero Velarde / *Mexico*
Edgar Vasquez Garibay / *Mexico*
Rebecca A. Ramirez / *Philippines*
Wojciech Cichy / *Poland*
Janusz Swietlinsky / *Poland*
Igor Kon / *Russia*
Fahad Alharbi / *Saudi Arabia*
Peter Baehler / *Switzerland*
Olivier Boulat / *Switzerland*
Christian P. Braegger / *Switzerland*
Guy Deleze / *Switzerland*
Sergio Fanconi / *Switzerland*
Ilse Kern / *Switzerland*
Jürg Lutschg / *Switzerland*
Jean-Léopold Micheli / *Switzerland*
Beatrice Mueller-Schenker / *Switzerland*
Jean-Marc Nuoffer / *Switzerland*
David H. Shmerling / *Switzerland*
Andrea Superti-Furga / *Switzerland*
René Tabin / *Switzerland*
François de Techtermann / *Switzerland*
Prasong Tienboon / *Thailand*
Peggy Borum / *USA*

Nestlé Representatives

Amani Hafez	Nestlé Egypt S.A.E., Cairo, Egypt
George Tadross	Nestlé Egypt S.A.E., Cairo, Egypt
Rana Bandyopadhyay	Nestlé India Ltd., Gurgaon, India

Yoichi Fukushima	Nestlé Japan Holding Ltd., Gokodori, Japan
Yoshiro Takada	Nestlé Japan Holding Ltd., Gokodori, Japan
Olga Netrebenko	Nestlé Food LLC, Moscow, Russia
Denis Barclay	Nestec Ltd., Vevey, Switzerland
Anne-Lise Carrié Fässler	Nestec Ltd., Vevey, Switzerland
Wolf Endres	Nestec Ltd., Vevey, Switzerland
Bianca-Maria Exl	Nestlé Suisse S.A., Vevey, Switzerland
Ulrich Preysch	Nestlé Suisse S.A., Vevey, Switzerland
Iskander Maged	Nestlé Middle East FZE, Dubai, United Arab Emirates
Linda Hsieh	Nestlé USA, Inc., Glendale, CA, USA

Nestlé Nutrition Workshop Series
Pediatric Program

Volume 50: Genetic Expression and Nutrition
Claude Bachmann and Berthold Koletzko, Editors, 320 pp., 2003.

Volume 49: Obesity in Childhood and Adolescence
Chunming Chen and William H. Dietz, Editors, 320 pp., 2002.

Volume 48: Public Health Issues in Infant and Child Nutrition
Robert E. Black and Kim Fleischer Michaelsen, Editors, 368 pp., 2002.

Volume 47 Supplement: Infant Formula: Closer to the Reference
Niels C. R. Räihä and Firmino F. Rubaltelli, Editors; 256 pp., 2002.

Volume 47: Nutrition and Growth
Reynaldo Martorell and Ferdinand Haschke, Editors; 448 pp., 2001.

Volume 46: Gastrointestinal Functions
Edgard E. Delvin and Michael J. Lentze, Editors; 352 pp., 2001.

Volume 45: Nutrition, Immunity, and Infection in Infants and Children
Robert M. Suskind and Kraisid Tontisirin, Editors; 464 pp., 2001.

Volume 44: Risk Assessment in the Food Chain of Children
Peter J. Aggett and Harry A. Kuiper, Editors; 304 pp., 2000.

Volume 43: Nutrition of the Very Low Birthweight Infant
Ekhard E. Ziegler, Alan Lucas, and Guido E. Moro, Editors; 288 pp., 1999.

Volume 42: Probiotics, Other Nutritional Factors, and Intestinal Microflora
Lars Å. Hanson and Robert H. Yolken, Editors; 320 pp., 1999.

Volume 41: Nutrition and Bone Development
Jean-Philippe Bonjour and Reginald C. Tsang, Editors; 304 pp., 1999.

Volume 40: Clinical Trials in Infant Nutrition
Jay A. Perman and Jean Rey, Editors; 304 pp., 1998.

Volume 39: Placental Function and Fetal Nutrition
Frederick C. Battaglia, Editor; 288 pp., 1997.

Volume 38: Diarrheal Disease
Michael Gracey and John A. Walker-Smith, Editors; 368 pp., 1997.

Volume 37: Feeding from Toddlers to Adolescence
Angel Ballabriga, Editor; 320 pp., 1996.

Volume 36: Long-Term Consequences of Early Feeding
John Boulton, Zvi Laron, and Jean Rey, Editors; 256 pp., 1996.

Volume 35: Diabetes
Richard M. Cowett, Editor; 320 pp., 1995.

Volume 34: Intestinal Immunology and Food Allergy
Alain L. de Weck and Hugh A. Sampson, Editors; 320 pp., 1995.

Volume 33: Protein Metabolism During Infancy
Niels C. R. Räihä, Editor; 264 pp., 1994.

Volume 32: Nutrition of the Low Birthweight Infant
Bernard L. Salle and Paul R. Swyer, Editors; 240 pp., 1993.

Volume 31: Birth Risks
J. David Baum, Editor; 256 pp., 1993.

Volume 30: Nutritional Anemias
Samuel J. Fomon and Stanley Zlotkin, Editors; 232 pp., 1992.

Volume 29: Nutrition of the Elderly
Hamish N. Munro and Günter Schlierf, Editors; 248 pp., 1992.

Volume 28: Polyunsaturated Fatty Acids in Human Nutrition
Umberto Bracco and Richard J. Deckelbaum, Editors; 256 pp., 1992.

Volume 27: For a Better Nutrition in the 21st Century
Peter Leathwood, Marc Horisberger, and W. Philip T. James, Editors; 272 pp., 1992.

Volume 26: Perinatology
Erich Saling, Editor; 208 pp., 1992.

Volume 25: Sugars in Nutrition
Michael Gracey, Norman Kretchmer, and Ettore Rossi, Editors; 304 pp., 1991.

Volume 24: Inborn Errors of Metabolism
Jürgen Schaub, François Van Hoof, and Henri L. Vis, Editors; 320 pp., 1991.

Volume 23: Trace Elements in Nutrition of Children—II
Ranjit Kumar Chandra, Editor; 248 pp., 1991.

Volume 22: History of Pediatrics 1850–1950
Buford L. Nichols, Jr., Angel Ballabriga, and Norman Kretchmer, Editors; 320 pp., 1991.

Volume 21: Rickets
Francis H. Glorieux, Editor; 304 pp., 1991.

Genetic Expression and Nutrition

Genetic Expression and Nutrition, edited by Claude Bachmann and Berthold Koletzko. Nestlé Nutrition Workshop Series, Pediatric Program, Vol. 50. Nestec Ltd., Vevey/Lippincott Williams & Wilkins, Philadelphia, © 2003.

Genetic Variation of Homocysteine Metabolism and Atherosclerosis

John M. Scott

Department of Biochemistry, Trinity College Dublin, Dublin, Ireland

HOW HOMOCYSTEINE ARISES IN CELLS

Homocysteine is an amino acid that is found in mammalian cells and in the plasma. It arises in the first instance from the catabolism of the essential dietary amino acid methionine, to which it is very similar in structure (Fig. 1). This process takes place mainly in the liver as a constituent of the so-called methylation cycle (Fig. 2).

Examination of the structures of methionine and homocysteine show them to be identical except for a conversion that has resulted in the removal of a methyl group ($-CH_3$) from the former (1). *In vivo,* this transfer of the methyl group of methionine has important consequences (2). In all cells, there are dozens of enzymes called methyltransferases. As the name suggests, these are involved in the donation (or transfer) of a methyl group from the activated form of methionine, s-adenosylmethionine (SAM). The products of such methyltransferase reactions are always s-adenosylhomocysteine (SAH) and the methylated version of the substrate for the reaction (3).

The substrates for the methyltransferases vary widely and encompass a wide range of compounds that have nothing to do with each other metabolically other than that they are all methylated using SAM as the methyl donor. A recent review suggests that there are 28 different methyltransferases, each with separate functions (4). Thus, the substrates can be any of the following: proteins, which will be methylated in their arginine or lysine side chains; phospholipids, which are also methylated with the conversion of phosphatidylethanolamine to phosphatidylcholine; DNA, which can be methylated in its cytosine residues; or the hormone DOPA, which is also methylated. The functions of these methylation reactions are as varied as their substrates. Such methylations can change protein function, alter membrane characteristics, down-regulate the transcription of DNA, or even contribute to a degradation pathway.

As can be seen from Fig. 2, these widely differing methyltransferase reactions are in fact part of the methylation cycle that exists in all cells. After donation of the methyl group, with the resultant formation of the methylated product and SAH, the latter is broken down immediately *in vivo* to produce adenosine and the amino acid homocysteine. The equilibrium for the hydrolase enzyme that carries out this reaction very much favors the back reaction and would be expected to result in the accumulation of a large concentration of SAH relative to the concentration of homocysteine or adenosine. However, it is known that *in vivo* the reaction is pulled to the right in the direction of the formation of homocysteine by the rapid and continuous metabolism of the component adenosine (Fig. 2). Thus under normal circumstances accumulation of SAH does not occur in the cell. This is important in that all of the methyltransferases with one exception are markedly inhibited by SAH. Methyltransferase activity is always a balance between the concentration of the methyl donor SAM and the concentration of the product SAH.

In some papers and reviews, the relative level of these two components is often referred to as "the methylation ratio." This conveys the concept that methyltransferase activity will be high with high SAM/SAH ratios and progressively inhibited as the ratio of SAM to SAH falls. It also conveys the concept that a high level of SAH can be compensated for to some extent by increasing the concentration of SAM. However, some view the ratio as being an absolute indicator of activity, which it is not. The same methylation ratio arrived at by different levels of SAM or SAH will not result in the same methyltransferase activity, an issue that becomes more important if non-physiologic levels are involved. However, it suffices to say that SAH is a potent regulator and that its accumulation in the cell is accompanied by impaired methyltransferase activity, which may be overcome by an increase in the level of SAM.

It is hardly by chance that methyltransferases are regulated by potent feedback inhibition. It may also be that such inhibition is graded with respect to different methyltransferases. If, for whatever reason, SAM levels became limiting, this

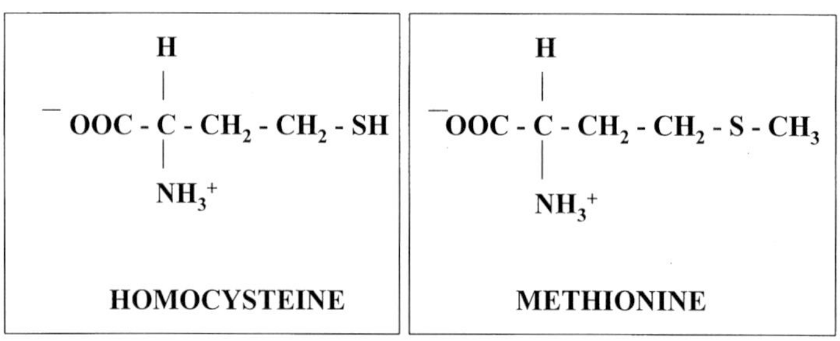

HOMOCYSTEINE - SH

HOMOCYSTEINE - S - S - HOMOCYSTEINE

HOMOCYSTEINE - S - S - CYSTEINE

HOMOCYSTEINE - S - S - GLUTATHIONE

HOMOCYSTEINE - S - S - PROTEINS

FIG. 1. Structures of methionine and homocysteine showing that *in vivo*, the latter is frequently attached to other thiols.

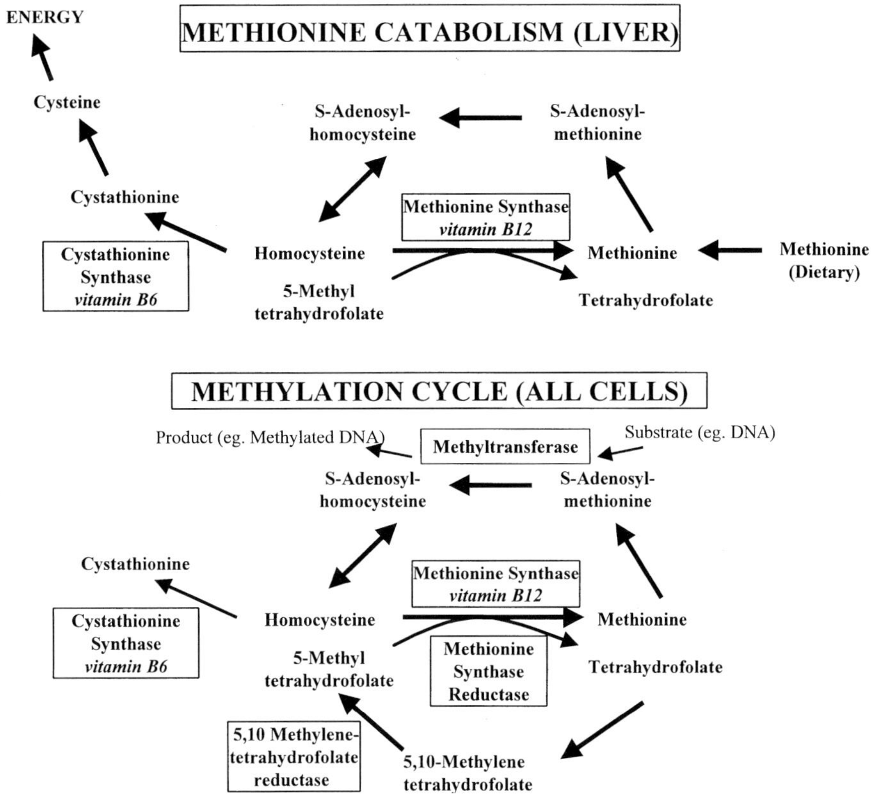

FIG. 2. Homocysteine arises in cells through methionine catabolism or through the methylation cycle. Its level is controlled by three enzymes, each of which requires one or more vitamins: cystathionine-β-synthase (vitamin B_6), methionine synthase (folate or vitamin B_{12}), and 5,10-methylenetetrahydrofolate reductase (folate and riboflavin).

would be accompanied by a reduction in the methylation cycle and an increased level of SAH. In those circumstances, the less important methyltransferase could be inhibited, allowing others the continued use of any available SAM. All of the above are important in understanding the potential way in which the accumulation of homocysteine—with the inevitable consequence of an accumulation of SAH and reduced methyltransferase activity—might impair cellular function in a large number of ways. As discussed later, it is also possible that homocysteine itself, having as it does a very reactive free thiol (−SH) group (Fig. 1), is toxic to cells in its own right.

THE FATE OF HOMOCYSTEINE IN CELLS

Once homocysteine has arisen in any cell, it has three possible fates. First, it can be remethylated back to methionine and thus conserved in the cell; second, it can

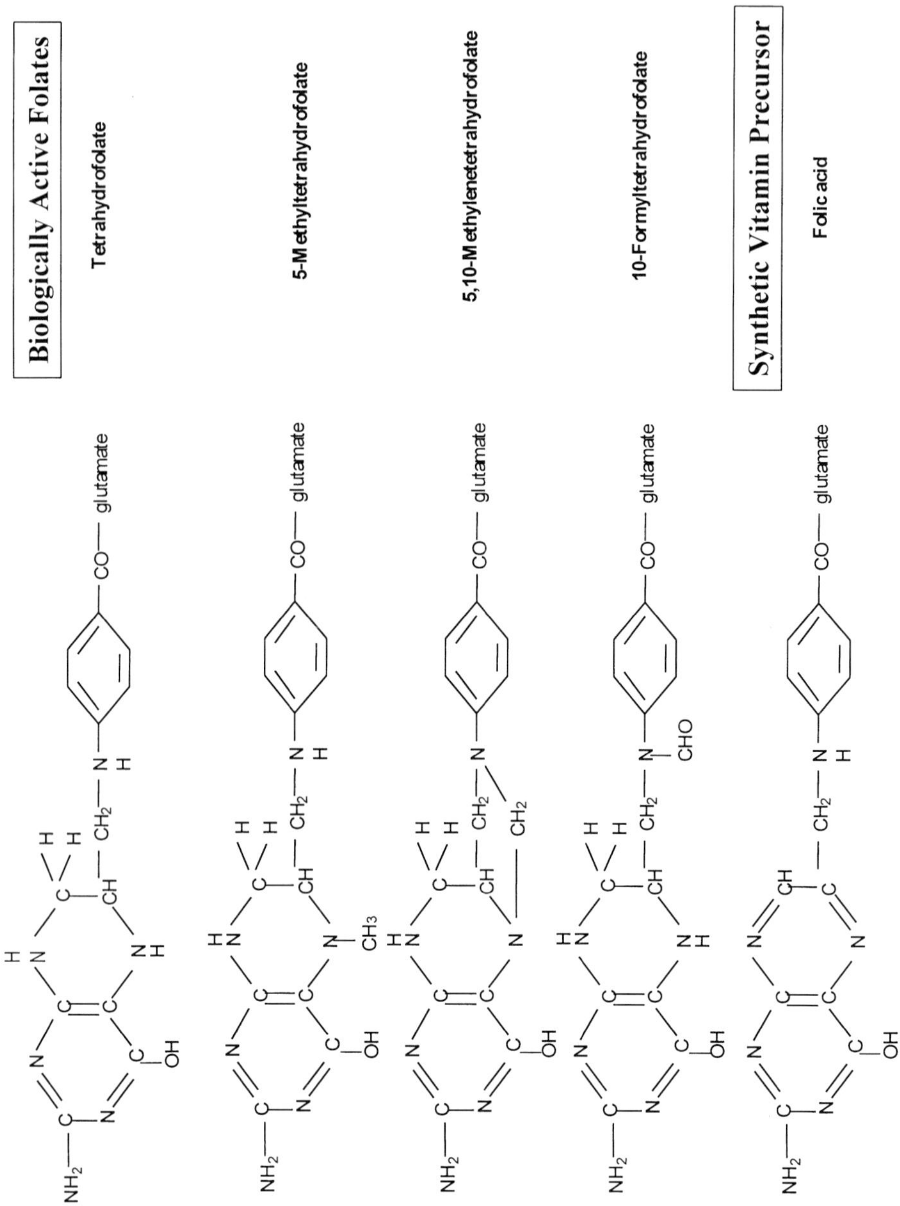

FIG. 3. Structure of the folates (natural forms) and folic acid (synthetic form).

be converted to cystathionine, which in most cells is the start of the transsulfuration pathway which sees the carbon sulfur skeleton of homocysteine degraded down to sulfate and pyruvate which can be used for energy; finally, homocysteine can be exported from the cell.

In most cells it would be anticipated that homocysteine is conserved by its conversion back to methionine, which in turn is activated with adenosine triphosphate (ATP) to make SAM and thus provide a methyl group for a further methyltransferase reaction. Thus, the methylation cycle in most cells regenerates methionine, using the enzyme methionine synthase to salvage homocysteine and to remethylate it back to methionine (3). This methyl group can only be provided while attached to the reduced form of the vitamin, folate, in the form of the cofactor 5-methyltetrahydrofolate (Fig.3). This in turn is synthesized from tetrahydrofolate using the enzyme 5,10-methylenetetrahydrofolate reductase (MTHFR).

The origin of this methyl group is in the carbon 3 (C3) of serine. Serine can be synthesized in most cells in three enzymatic steps from 3-phosphoglycerate, an intermediate in the glycolytic pathway, with the amino nitrogen being provided by glutamate. Both 3-phosphoglycerate and glutamate are readily available in all cells, implying that they have a ready supply of methyl groups. However, cells also require an adequate concentration of folate in the form of the methyl donor, 5-methyletrahydrofolate, and an adequate concentration of vitamin B_{12}, which is essential for the activity of methionine synthase.

METHIONINE CATABOLISM

As well as providing methyl groups, the methylation cycle also fulfills another function, namely the degradation of excess methionine.

Methionine is an essential amino acid that in most diets is provided in large amounts over and above the requirement. Thus after its utilization for protein biosynthesis there is usually a 60% or more excess over what the body needs. As with many other amino acids, this must be catabolized. In contrast to many of the other amino acids, however, transamination of methionine only occurs at nonphysiologically high concentrations of the amino acid and probably has no physiologic role as a pathway. Instead, methionine catabolism takes place mainly in the liver through its sequential conversion to SAM and SAH and then to homocysteine, followed by the transsulfuration pathway to cysteine, which is further converted to pyruvate and sulfate (3).

One can see in this pathway that there is a need for a methyl acceptor to remove the methyl group from the SAM. This role could not be fulfilled by any of the usual substrates for the methyltransferases because it would result in the degree of their methylation being dictated by the vagaries of how much methionine was in the diet. The alternative used by the liver, the major catabolic site, is to have a very high concentration of glycine-*N*-methyltransferase. This enzyme uses SAM to methylate glycine, converting it into sarcosine, which can be further catabolized in the mitochondria to yield carbon dioxide (CO_2) and energy. Thus, the liver uses a discretionary methyltransferase to methylate a varying amount of glycine as a means of catabolizing the excess dietary methionine.

HOMOCYSTEINE CATABOLISM

The homocysteine generated in this catabolic pathway can also be catabolized further in the liver by the transsulfuration pathway present in all animals. Excess homocysteine is seen to accumulate in the plasma during renal failure. However, this is not due to the loss of the organ's excretory function because very little homocysteine appears in the urine under normal circumstances.

An earlier suggestion was that the kidney must be a major site of homocysteine catabolism, and while this may be true in the rat, it now appears that the human kidney is responsible for the removal of very little, if not negligible, amounts of plasma homocysteine (5). Thus, the explanation for an up to 100% increase in plasma homocysteine in humans during renal failure remains unclear (6). It is only corrected to a minor degree by folic acid supplements and appears to be something intrinsic to an alteration in homocysteine metabolism rather than due to impaired activity of the methylation cycle (6).

HOMOCYSTEINE MAY BE CHEMICALLY TOXIC ITSELF

There is one further aspect of homocysteine metabolism to be discussed before considering whether its accumulation is likely to be detrimental to cells and thus cause cardiovascular disease. This relates to its chemistry. A look at the structure of methionine will show that it is quite chemically stable (Fig. 1). A similar examination of homocysteine shows that it has a free thiol (−SH) group. This group is highly reactive at the redox state that normally exists in cells and in the circulation; thus, under such circumstances homocysteine will tend to react with any free thiols present. This is most readily seen in human plasma, where perhaps as little as 1% to 4% of the homocysteine present is in the reduced form (7); most homocysteine is found attached to the thiol groups of plasma proteins such as albumin. An appreciable percentage is also found as a dimer of homocysteine with itself, namely homocystine. It can also be found to form a dimer with cysteine: cysteinylglycine and glutathione. While it is less clear what complexes may be formed by reduced homocysteine within cells, it seems very probable that with the reactivity of its thiol group at neutral pH similar complexes will occur.

HOW IS HOMOCYSTEINE TOXIC TO CELLS?

It is evident that homocysteine may be damaging to cells in a variety of ways. Its intracellular accumulation would certainly lead to accumulation of SAH and thus to the inhibition of intracellular methyltransferases, which could have various effects; it may have a direct toxic effect in cells; and its accumulation in the plasma—as a result, say, of its inadequate catabolism by the liver—could produce an externally delivered insult to cells.

It is hard to see how the major fraction of the plasma homocysteine, found bound as it is to protein and other thiols, could have a direct extracellular toxic effect. It may well be that at a very high plasma level, free homocysteine does reach toxic levels. However, at the levels reported to cause risk for cardiovascular disease and other disorders this direct toxic effect is not easy to envisage. It is thus more likely that plasma homocysteine serves to increase the intracellular levels of homocysteine, and that it either has direct toxic effects inside the cells or the toxicity results from a rise in SAH.

Another possibility is that plasma homocysteine is really only a marker for cardiovascular or other disease risks. For example, it may be that it reflects the status of nutrients such as folate or vitamin B_{12}, which are the factors involved in risk. A further possibility is that it is a risk *marker,* not causally involved but related to some other risk factor. For example, homocysteine might be increased in diets high in methionine, but if such diets are also high in cholesterol, the latter could be the real risk factor. Such confounding data can be taken into account from known risk factors, and this has been done in the studies discussed below, investigating homocysteine as a risk factor for cardiovascular disease. However, it leaves open the possibility that it is a marker that is causally related to some unknown risk factor for cardiovascular disease.

In summary, it seems possible that homocysteine is itself a potential risk, either intracellularly or extracellularly. Alternatively, it may be found in biologic association with SAH, which is an actual risk factor owing to its inhibition of methyltransferases. Furthermore, as already mentioned, homocysteine may not be a risk factor in itself but rather a biologic marker for some other determinant such as folate or nutritional status that is involved with the risk of disease.

In this latter context it is interesting to note that increased plasma homocysteine is associated with reduced status of any one of four nutrients. Reduced folate or vitamin B_{12} status prevents the proper functioning of the methylation cycle, the former by failing to keep it supplied with methyl groups, and the latter by directly reducing the activity of methionine synthase, one of the enzymes needed for the methylation cycle to turn. Reduced vitamin B_6 is also known to cause an accumulation of plasma homocysteine because the first two enzymes in the transsulfuration pathway—namely cystathionine synthase and cystathionase—are vitamin B_6–dependent. The vitamin riboflavin (vitamin B_2) is an essential precursor for the biosynthesis of flavin adenine mononucleotide (FMN) and flavin adenine dinucleotide (FAD). The latter is an essential cofactor for the folate dependent enzyme MTHFR and for an enzyme that activates the vitamin B_6 precursor, pyridoxal, to the biologically active form, pyridoxal 5-phosphate. Thus, in theory at least, reduction in the riboflavin status might give rise to increased plasma homocysteine. Finally, plasma homocysteine may be merely a risk marker of cardiovascular disease, its elevation occurring in parallel with some other unrelated cause.

The next section deals with the evidence for and against plasma homocysteine being a risk factor for cardiovascular disease. I will examine the circumstances in which homocysteine levels are raised, usually through inadequate provision of any one of the four nutrients involved in its metabolism, the most important being folate. Finally, if raised plasma homocysteine were truly a risk for cardiovascular disease, one would anticipate that genetic conditions where its level is known to be increased because of impairment of enzyme function would carry an increased risk of cardiovascular disease.

It will become clear that there is overwhelming evidence that homocysteine is either a risk factor (causes cardiovascular disease) or a risk marker (does not itself cause cardiovascular disease but faithfully reflects something that does). Which of these it is can only be resolved by properly conducted placebo-controlled intervention trials. Such trials must have sufficient numbers to allow conclusions to be drawn, bearing in mind the likely risk reduction in cardiovascular disease that reduction in plasma homocysteine would be expected to have.

RETROSPECTIVE AND PROSPECTIVE CASE—CONTROL STUDIES

The strongest evidence for a role of plasma homocysteine in cardiovascular disease comes from a range of retrospective case–control studies. The design and setting of these studies vary widely and have been the subject of several meta-analyses. A significant advance in this regard was the meta-analysis of Boushey *et al.* (8), who used 23 retrospective and four prospective case–control studies to conclude that homocysteine levels at the 90th or 95th centile of controls were associated with increased risks for the following conditions. The odds ratios and 95% confidence intervals (CIs) (in parentheses) are shown below:

- Cardiovascular disease, 1.7 (1.5–1.9)
- Cerebral vascular disease, 2.5 (2.0–3.0)
- Other vascular disease, 6.8 (2.9–15.8)

They also concluded that an increase in plasma homocysteine of 5 μmol/l would increase the risk of cardiovascular disease by about one third.

Wald *et al.* (9) performed a subsequent meta-analysis and again found a significant risk with retrospective studies. However, in that analysis it was becoming increasingly clear that the results from prospective (nested case–control) studies showed a lower magnitude of risk, or in some instances no apparent risk. Thus, of five prospective studies of ischemic heart disease, two showed no risk while three gave significant odds ratios (9).

Further meta-analyses followed as the results of new retrospective and prospective studies were published. Harkey and Eikelboom (10) reviewed the retrospective studies, adding 40 more and a further 150 cases since the review by Boushey *et al.* (8), again concluding that there was a strong risk association. Of particular value was a further large European study, which concluded that an increase of 5 μmol/l in plasma homocysteine was associated with a relative risk of 1.35 (CI, 1.1–1.6) in men and 1.42 (CI, 0.99–2.55) in women (11). However, in the same analysis, the results of 11 prospective case–control studies were not consistent. Five were clearly positive, whereas the others showed less significant risk associations or were negative.

More recently, Verhoef and Stampfer (12) have performed a meta-analysis involving some of the old studies but adding new ones for various forms of vascular disease. All of these were prospective. They carried out a risk estimate for the highest quartile of total plasma homocysteine *versus* the lowest quartile. Of 20 studies, three showed no association and four showed weak associations, which were not statistically significant. The remaining 13 studies all showed direct significant associations. It remains unclear why some studies are negative. However, vast quantities of data have been accumulated and, in most cases, the results have been positive.

GENETIC STUDIES

Rare and Drastic Genetic Mutations

Genetic disease or genetic variations of the enzymes involved in homocysteine have been the subject of many studies (13). In 1962, two patients with raised urinary homocysteine were reported by Carson and her colleagues in Belfast (14), followed shortly afterwards by further cases (15). Mudd and his colleagues (16) subsequently found that such cases were deficient in the enzyme cystathionine-β-synthase. These and other cases of deficiency of this enzyme had four major clinical features: dislocation of the lens, skeletal abnormalities, mental retardation, and thromboembolic disease, which usually caused premature death.

In 1969, Mudd *et al.* (17) described a different type of case in which there was methylmalonic aciduria as well as homocysteinuria. Unlike the earlier cases in which plasma methionine was abnormally low, in this case it was elevated. This disorder is now known to be caused by impaired vitamin B_{12} activation leading to reduced methionine synthase activity, with a rise in urinary and plasma homocysteine.

Reduction in the activity of methylmalonyl-CoA mutase—the other vitamin B_{12}–dependent enzyme—causes accumulation of methylmalonic acid. McCully (18) carried out a necropsy examination on a 7-week-old child and was struck by the presence of widespread vascular disease, very similar to that seen at postmortem in children with homocysteinuria. However, while such children have raised plasma methionine, McCully knew from Mudd's studies (17) that this particular child had remarkably reduced methionine levels. Both conditions had a common biochemical feature, namely an increase in plasma homocysteine. McCully then suggested that homocysteine was a cause of arteriosclerosis (18).

It subsequently transpired that there are two forms of classical homocysteinuria due to cystathionine-β-synthase deficiency. About half are responsive to the enzyme cofactor, vitamin B_6, if given in large doses. Mudd *et al* (19) showed that the reduction in homocystine in patients who have this treatment produced a highly significant lowering of the risk of cardiovascular disease. In the other form, betaine—a degradation product of choline—converts homocysteine to methionine using the enzyme betaine methyltransferase, independently of folate or vitamin B_{12}. Betaine was also found to lower plasma homocysteine in vitamin B_6–resistant patients. Again, such treatment was found to reduce the risk of cardiovascular disease (20).

Several other inborn errors of vitamin B_{12} absorption, activation, or metabolism have been discovered. In addition, nearly 50 patients with various severe deficiencies of the MTHFR enzyme have been identified (21). A common clinical feature of these diseases is thrombosis of the arteries and cerebral veins, which is the major cause of death. Raised plasma homocysteine is the usual feature of cases of MTHFR deficiency in which there is a genetic variant that causes a reduction in enzyme activity.

Such variations in the three enzymes that control plasma homocysteine levels are very rare. Historically, they would have been called inborn errors of metabolism, and all have in common a significantly raised plasma homocysteine level and an increased risk of cardiovascular disease. Reducing the level of plasma homocysteine in these cases reduces the risk. Thus, there is compelling evidence that elevations of plasma homocysteine of the order found in these inborn errors do cause cardiovascular disease.

Common and Mild Genetic Variations

MTHFR has recently been found to have a common genetic variant. This was first identified biochemically by Kang *et al* (22). These workers found that the isolated MTHFR enzyme from the lymphocytes of certain individuals was very heat sensitive. This so-called thermolabile variant of the enzyme was associated with increased plasma homocysteine. Some years later Gayette and colleagues in Canada found there was a change at base 677 from a cytosine (C) to thymine (T), 677C→T, causing an amino acid change at 227 from alanine (A) to valine (V), A227V (23). Being homozygous for this so-called thermolabile variant (TT) was very common in certain ethnic groups, its frequency ranging from 1.0% in African Americans to 25% in Hispanics (24). It is associated with reduced enzyme activity, and a mild but significant rise in plasma homocysteine.

While there is universal acceptance that this genotype does result in a phenotype with a mild elevation of plasma homocysteine, a meta-analysis of some 15 studies failed to show a significant association with cardiovascular disease (25). Brattstrom and Wilcken (26) have taken the view that the phenotype of increased plasma homocysteine found in individuals homozygous for the 677C→T variant provides evidence that moderate elevations of plasma homocysteine are possibly risk markers, but certainly not risk factors, for cardiovascular disease. A completely different stance is taken on this issue by Ueland *et al.* (27). These investigators argue that the combined studies on the involvement of 677C→T MTHFR lack statistical power, and that the numbers analyzed would not be expected to give a positive result for the anticipated level of increased cardiovascular risk.

A further consideration is that McNulty and his colleagues have shown that for this variant of MTHFR to malfunction and thus cause an increase in homocysteine, the homozygous or even heterozygous individual must also have diminished riboflavin (vitamin B_2) status (28). Molecular biologic studies on the variant of the enzyme have shown that its labile nature reflects the fact that it is 10 times more likely to lose the FAD cofactor from its active site—leading to irreversible inactivation—than the wild-type enzyme (29). These investigators also found that only in people with inadequate riboflavin status, by conventional criteria, is there an increase in plasma homocysteine. In other words, in subjects or populations where there is a good riboflavin status, there is no effect of having the variant enzyme. The many studies done on the risk of cardiovascular disease associated with the enzyme variant preceded this recent development and so failed to consider it. Thus, many of the cases examined for risk of cardiovascular disease—particularly in the United States—where the diet has traditionally been fortified with riboflavin—will probably be uninformative.

A further common variant of MTHFR has also been identified. This is a base change of adenine to cytosine at base 1298 (1298A→C), causing an amino acid change in the enzyme: glutamic acid (E) to alanine (A) at amino acid number 429, E429A (30). As in the case of the homozygous state for 677C→T, 1298A→C is also associated with some reduction in enzymatic activity but there appears to be no increase in plasma homocysteine (31,32). Although homozygosity for 1298A→C on its own does not increase plasma homocysteine, there is a suggestion that being a combined heterozygote with 677C→T may cause an increase (31). Currently, there is no evidence that 1298A→C, either on its own or in conjunction with 677C→T, increases the risk for cardiovascular disease. However, it is possible that there could be an increased risk with 1298A→C even in the absence of elevated homocysteine. It is possible that this or any other polymorphism of any folate-dependent enzyme could result in increased risk for cardiovascular diseases in the absence of any elevation in plasma homocysteine. This could happen if a polymorphism affected endothelial cells or cardiac cells but did not have an equal effect on liver cells. The former could lead to cardiovascular disease, whereas the liver, which mainly controls plasma homocysteine, would be unaffected and plasma homocysteine would not be raised. Therefore, one might not always expect that a polymorphism that causes disease will always be accompanied by an elevation in plasma homocysteine.

Methionine synthase is an enzyme in the methylation cycle. One common variant of the gene that codes for this enzyme has been reported (Table 1). It is a 2756A→G mutation involving a base change of adenosine (A) to guanosine (G), causing an amino acid change of D919G (aspartate (D) to glycine (G)) (6). It has been reported that being homozygous for this variant causes a modest but significant reduction in plasma homocysteine (33), although it has not been found to be related to cardiovascular disease risk (34,35).

Methionine synthase reductase (MTRR) is involved indirectly in the methylation cycle. During its normal functioning, a small percentage of the vitamin B_{12} cofactor that is attached to the active site of methionine synthase becomes oxidized (36). The MTRR enzyme is necessary to reduce this back to the active form. Reduction in MTRR activity would be expected to lead to reduced methionine synthase activity. A common variant of MTRR has been reported, 66A→G, causing an amino acid change I22M in the enzyme of isoleucine to methionine at amino acid 22. Being homozygous for the polymorphism did not appear to increase plasma homocysteine levels in the one small study done to date (37). More recently, in a large study of over 600 subjects, higher homocysteine levels were found in those who were AA for this polymorphism (38).

Cystathionine-β-synthase (CBS), as well as having severe mutations associated with the disease homocysteinuria, also has a variant that is present in about 5% to

TABLE 1. *Common polymorphisms in folate-related enzymes*

Enzyme	Change in gene	Amino acid change	Phenotype of variant	Reference
5,10-Methylene-tetrahydrofolate reductase (MTHFR)	677C to T	Alanine to valine (A227V)	• Lower enzyme activity • Lower folate status • Higher plasma homocysteine	24
MTHFR	1298A to C	Glutamic acid to alanine (E429A)	No changes	31,32
Methionine synthase (MS)	2756A to G	Aspartic acid to glycine (D919G)	• Higher plasma folate • Higher red cell folate • Lower plasma homocysteine	33,34
Methionine synthase reductase (MTRR)	66A to G	Isoleucine to methionine (I22M)	• No change in plasma homocysteine • Higher plasma homocysteine	37,38
Cystathionine-β-synthase (CBS)	844ins68	—	• No change in basal plasma homocysteine • Less accumulation of plasma homocysteine after a methionine load	39,40
Reduced folate carrier	80G to A	Arginine to histidine (R27H)	• No change in plasma homocysteine or red cell folate	41
Glutamate carboxypeptidase (FGCP) (polyglutamate hydrolase)	1561C to T	Histidine to tyrosine (H475Y)	• Lower enzyme activity • Lower plasma folate • Higher plasma homocysteine	42

13% of whites and 40% of sub-Saharan Africans, where there is a 68–base pair insertion at the intron 7 and exon 8 boundary at base 855. This so-called 844 INS 68 is associated with a reduction in the increase in plasma homocysteine that occurs after a methionine load (39,40).

There is also a polymorphism in the reduced folate carrier, 80G→A, changing an amino acid from arginine to histidine at amino acid number 27. This causes no change in plasma homocysteine (41). There is a polymorphism in the folate polyglutamate hydrolase 1561C→T, changing a histidine to a tyrosine (42).

RENAL DISEASE

As mentioned above, in renal disease there is a marked increase in plasma homocysteine for reasons that are not yet understood (5). Patients with renal disease also have a marked increase in the risk of cardiovascular disease. It is not clear what contribution the former makes to the latter, if any (6).

INTERVENTION TRIALS

From the above, it is clear that a raised level of plasma homocysteine is associated with cardiovascular disease in most studies. However, from the existing evidence—even if the effect of homocysteine is assessed in very similar types of study—it is impossible to be certain whether plasma homocysteine is merely a risk marker for cardiovascular disease or whether it is a risk factor that is involved—either directly or indirectly—in the etiology of some or all forms of cardiovascular disease. This can only be resolved by interventions to lower plasma homocysteine and by analysis of these interventions to determine the risk of cardiovascular disease. Several such trials are either ongoing (43) or at the planning stage and are summarized in Table 2.

COMMENT

There is virtually unanimous agreement that total plasma homocysteine is elevated in individuals with cardiovascular disease. This comes from retrospective studies, where the evidence is uniformly in favor of an association, and where the odds ratios of the risk are quite high—perhaps involving a benefit equivalent to that of lowering mean plasma cholesterol in the population (43). Such retrospective studies suffer from many design problems, the most important being various types of confounding. In addition, there is always concern about how appropriately controls are matched to the cases.

Prospective case–control studies do not suffer from this problem, in that the samples are taken from the cases and controls at the same time, when there is no evidence of cardiovascular disease. In addition, in studies of this type there is no possibility that the level of plasma homocysteine is in some way a result of cardiovascular disease rather than a cause. While some well-conducted case–control studies are negative and in a few the results were not significant, most favor a relationship between homocysteine and cardiovascular disease (12). There are several possible explanations for this: (a) it is a chance observation, which seems unlikely given the large number of studies; (b) homocysteine is a marker for some unrelated risk factor; (c) homocysteine is a marker for some causally related risk factor, such as the status of one or more of the vitamins involved; (d) homocysteine is either itself a cause or a metabolite of a cause (such as SAH) of cardiovascular disease. Only the outcome of the ongoing or planned controlled intervention trials using folic acid and other vitamins to lower plasma homocysteine can determine which of these four possibilities is the correct one.

Eight trials are so far involved (44). Two of these trials—the Vitamin Intervention for Stroke Prevention (VISP) and Women's Antioxidant and Cardiovascular Disease

TABLE 2. *Trials with folic acid and related vitamins to investigate the effect of lowering total plasma homocysteine on cardiovascular risk*

Name of study	Abbreviation	Size projection	Intervention	Comments
Norwegian study of homocysteine lowering with B vitamins in myocardial infarction, University of Tromso, Norway	NORVIT	3,000	Folic acid (5 mg × 2 weeks + 0.8 mg) + B_{12} (0.4 mg) *vs.* placebo in a 2×2 factorial design with B_6 (40 mg) *vs.* placebo	Small study, not likely to be conclusive on its own.
Western Norway B vitamin trial, University of Bergen, Norway	WENBIT	2,000	Folic acid (5 mg × 2 weeks + 0.8 mg) + B_{12} (0.4 mg) *vs.* placebo in a 2×2 factorial design trial with B_6 (40 mg) *vs.* placebo	Small study, not likely to be conclusive on its own.
Study of the effectiveness of additional reductions in cholesterol and homocysteine, University of Oxford, England	SEARCH	12,000	Folic acid (2 mg) + B_{12} (1 mg) *vs.* placebo in a 2×2 factorial design with Simvastatin (80 mg *vs.* 20 mg)	Large study, likely to yield result in 2004.
Cambridge heart antioxidant study, University of Cambridge, England	CHAOS-2	4,000	Folic acid (5 mg) *vs.* placebo	Small study, not likely to be conclusive on its own.
Women's antioxidant and cardiovascular disease study, Harvard Medical School, USA	WACS	8,000	Folic acid (2.5 mg) + B_6 (50 mg) + B_{12} (1 mg) *vs.* placebo	Compromised by fortification of the US diet with folic acid.
Vitamin intervention for stroke prevention, Wake Forest University, USA	VISP	3,600	Folic acid (2.5 mg) + B_6 (25 mg) + B_{12} (0.4 mg) *vs.* folic acid (0.02 mg) + B_6 (0.2 mg) + B_{12} (0.006 mg)	Compromised by fortification of the US diet with folic acid.
Heart outcomes prevention evaluation-2 study, McMaster University, Canada	HOPE-2	5,000	Folic acid (2.5 mg) + B_{12} (1 mg) + B_6 (50 mg) *vs.* placebo	Compromised because partly in Canada where diet is fortified with folic acid.
Prevention with a combined inhibitor and folate in coronary heart disease, University of Sydney, Australia	PACIFIC	10,000	Folic acid (0.2 mg or 2 mg) *vs.* placebo in a 2×2 factorial trial with omapatrilat. (Doses to be chosen after results of the pilot study.)	Pilot study done. Main study not started.

Study (WACS) trials, using, respectively, 3,600 and 8,000 subjects—are well under way (Table 2). Unfortunately, they are both taking place in the United States, where the fortification of flour with folic acid, introduced a few years ago, means that the folic acid intervention is not being compared with a true placebo group, or even with a non–folic acid control group. Thus, unless there is a significant dose effect between the high dose of folic acid used and that already achieved with fortification, a folic acid result will be missed. In this context, it was originally intended that fortification of 140 μg folic acid per 100 g of flour would increase the intake of folic acid by 100 μg per day (45). However, as mentioned above, it appears that the flour manufacturers have added significantly more folic acid to the flour. It is now clear that a low folate status or even a moderate elevation of plasma homocysteine is virtually nonexistent in the United States (46).

A further trial in Canada [Heart Outcomes Prevention Evaluation-2 (HOPE-2)] has recruited some 5,000 cases. About half are from Canada and half from Eastern Europe. The former will be exposed to the same level of fortification usage as the United States, significantly reducing the power of the trial.

A large trial is planned for Australia [Prevention with a Combined Inhibitor and Folate in Coronary Heart Disease (PACIFIC)]. A pilot phase involving a number of cases is complete, and the trial is ultimately planned to involve 10,000 cases. The main trial has not been started or funded.

There are two small trials ongoing in Norway (NORVIT and WENBIT) with 3,000 and 2,000 cases, respectively. A further small trial is ongoing in Cambridge [Cambridge Heart Antioxidant Study (CHAOS)]. The numbers in these three trials are rather small to reveal anything but a major effect.

A major trial [Study of the Effectiveness of Additional Reductions in Cholesterol and Homocysteine (SEARCH)] is in the final stages of recruitment in Oxford, which will include 12,000 cases and will compare 2.0 mg of folic acid plus 1.0 mg of vitamin B_{12} with a placebo. This trial is expected to yield results by 2004. This trial, with its large numbers, has the best chance of success.

If elevated homocysteine is a risk factor for cardiovascular disease, it would be anticipated that genetic alteration of the enzymes involved in its metabolism, where such elevation occurred, would be an increased risk. This clearly is the case in the severe mutations that cause a dramatic increase in plasma homocysteine (13–20). It does not, however, appear to be true for the common variants that cause only a small or moderate increase in homocysteine. As described above, it is not clear why the common variant 677C→T of MTHFR (25) does not appear to cause cardiovascular disease (26). Perhaps it is a question of the numbers needed to show such an effect (27), or of differences in risk owing to differences in riboflavin status (28). It is possible that the moderate increase in plasma homocysteine seen in this common variant only causes a small increase in risk.

Several other common variants of the genes involved in homocysteine metabolism have been described, namely methionine synthase, MTRR, and CBS. These may be associated with an increase in plasma homocysteine to a varying degree. As with 677C→T MTHFR, there are two ways to suggest whether they have a role in cardiovascular disease. The direct way is to see whether the prevalence is increased in affected individuals compared with matched controls. The most studied variant with a known phenotype causing a significant rise in plasma homocysteine is MTHFR 677C→T (24). Using the direct approach, there is no general agreement that this variant increases cardiovascular risk (25), either because no such increase in risk exists (26), or because insufficient numbers have been studied to show the association (27). An alternative but far less rigorous approach to this type of risk analysis is to take retrospective or prospective case–control studies, or a combination of the two, and calculate the increase in risk associated with a certain increase in plasma homocysteine. Thus, using the meta-analysis of Danesh and Lewington (47) it can be shown that an increase of 5 μmol/l in plasma homocysteine confers a 30% increased risk of cardiovascular disease. One could then look at the effects of the various polymorphisms and their ability to increase plasma homocysteine in various normal populations, and calculate the increased risk associated with the presence of these polymorphisms in the homozygous state. Such an approach was used recently by Gaughan *et al.* (38) to evaluate the risk of cardiovascular disease associated with the homozygous state for several polymorphisms of the genes involved in homocysteine metabolism (Table 1). Three common variants of genes that code for MTRR, methionine synthase, and MTHFR conferred the following percentage increase in risk of cardiovascular disease: MTRR (66A→G), AA, 4.2%; methionine synthase (2756A→G), AA, 2.4%; MTHFR (677C→T) TT, 16.2%.

One can go further, superimpose the homozygous prevalence in the population, and come up with the following analysis. Whereas MTRR 66AA enhances the risk by only 4%, because it has a high frequency it confers a risk of 29% for the population; and although MTRR enhances individual risk by a high 16%, it will confer a risk of only 10% for the population because of its lower frequency. Using surrogate plasma homocysteine values to obtain an idea of the involvement of a particular variant is attractive. However, this approach clearly makes several assumptions.

The classical mutations of cystathionine-β-synthase and the vitamin B_{12} metabolizing enzymes that affect methionine synthase and MTHFR all have in common a marked elevation of plasma homocysteine and an increased risk of cardiovascular disease. It is also clear that when such elevated levels are reduced, there is a reduction in cardiovascular disease (19,20). Thus. high levels of plasma homocysteine do appear to show a cause–effect relation with cardiovascular disease. That said, it is apparent that in many cases, plasma homocysteine levels remain raised for long periods without the development of cardiovascular disease (48). Consequently, the effect of elevated plasma homocysteine may not in itself be sufficient to cause cardiovascular disease, but may require the presence of another risk factor. Indeed, Graham *et al.* (11), in a major European study, suggested that, instead of being an independent risk factor *per se*, the role of elevated plasma homocysteine was to accelerate certain existing risk factors, such as cholesterol and hypertension. In either instance, a reduction in homocysteine might have considerable benefit.

Some investigators doubt whether a moderate elevation of plasma homocysteine causes cardiovascular disease. Such doubts are based on the lower risks found in prospective studies compared with those of the earlier retrospective studies. However, it has been suggested that the former may underestimate the risk because of the phenomenon of "regression dilution bias" (49). Other doubts are based on the lack of a clear increase in risk associated with the moderate increase in plasma homocysteine that occurs with the variant 677C→T of MTHFR (26). This may be explained by inadequate sample sizes (27) or by not taking into account the role of riboflavin status in the expression of the phenotype for this genotype (28). It appears this issue cannot be conclusively resolved until the intervention trials currently under way are completed (42), which will not be until the end of 2004 at the earliest.

ACKNOWLEDGMENTS

I wish to acknowledge the financial support of EU Demonstration Project BMH4-98-3549 and Abbott GMBH Wiesbaden–Delkenheim, Germany.

REFERENCES

1. Scott JM, Weir DG. Folic acid, homocysteine and one carbon metabolism: a review of the essential biochemistry. *J Cardiovasc Risk* 1998; 5: 223–7.
2. Scott JM, Weir DG. Folate/vitamin B_{12} interrelationships. In: Tipton KF, ed.*Essays in biochemistry*.Vol 28. London: Portland Press, 1994: 63–72.
3. Weir DG, Scott JM. Vitamin B_{12}: cobalamin. In: Shils ME, Olson JA, Shike M, Ross AC, eds. *Modern nutrition in health and disease*. Baltimore, MD: Williams & Wilkins.
4. Clarke S, Banfield K. S-adenosylmethionine-dependent methyltransferases. In: Carmel R, Jacobsen DW, eds. *Homocysteine in health and disease*. Cambridge: Cambridge University Press, 2001.
5. Van Guldener C, Kulik W, Berger R, *et al*. Homocysteine and methionine metabolism in ESRD: a staple isotope study. *Kidney Int* 1999; 56: 1064–71.
6. Guttormsen AB, Heland PM, Svarstad E, Refsum H. Kinetic basis of hyperhomocysteinemia in patients with chronic renal failure. *Kidney Int* 1997; 52: 495–502.
7. Mansoor MA, Svardal AM, Ueland PM. Determination of the in vivo redox status of cysteine, cysteinylglycine, homocysteine and glutathione in human plasma. *Anal Biochem* 1992; 200: 218–19.
8. Boushey a CJ, Beresford SAA, Omenn GS, Motulsky AG. A quantitative assessment of plasma homocysteine as a risk factor for vascular disease: probable benefits of increasing folic acid intakes. *JAMA* 1995; 274: 1049–57.
9. Wald NJ, Watt HC, Law MR, Weir DG, McPartlin J, Scott JM. Homocysteine and ischaemic heart diseases: results of a prospective study with implications regarding prevention. *Arch Intern Med* 1998; 158: 862–67.
10. Harkey GJ, Eikelboom JW. Homocysteine and vascular disease. *Lancet* 1999; 354: 407–13.
11. Graham IM, Daly LE, Refsum HMK, *et al*. Plasma homocysteine as a risk factor for vascular disease: the European Concerted Action Project. *JAMA* 1997; 277: 1775–81.
12. Verhoef P, Stampfer M. Epidemiology of vascular and thrombotic associations. In: Carmel R, Jacobsen DW, eds. *Homocysteine in health and disease*. Cambridge: Cambridge University Press, 2001.
13. Wilcken DEL, Wilcken B. Historical overview and recent perspectives. In: Carmel R, Jacobsen DW, eds. *Homocysteine in health and disease*. Cambridge: Cambridge University Press, 2001.
14. Carson NAJ, Cusworth DC, Dent CE, Field CMB, Neill DW, Westall RG. Homcystinuria: a new unborn error of metabolism associated with mental deficiency. *Arch Dis Child* 1963; 38: 425–36.
15. Mudd SH, Levy HL, Skouby F. Disorders of transsulfuration. In: Scriver CR, Beaudet AL, Sly WS, Valle D, eds. *The metabolic and molecular basis of inherited disease*. 7th ed. New York: McGraw-Hill, 1995.
16. Mudd SH, Finkelstein JD, Irreverre F, Laster L. Homocystinuria: an enzymatic defect. *Science* 1964; 143: 1443–5.
17. Mudd SH, Levy HL, Abels RH. A derangement in B12 metabolism leading to homocystinemia: cystathioninemia and methymalonic aciduria. *Biochem Biopages Res Commun* 1969; 35: 121–6.

18. McCully KS. Vascular pathology of homocysteinemia: implications for the pathogenesis of arterioscleréosis. *Am J Pathol* 1969; 56: 111–28.
19. Mudd SH, Srovky F, Levy HL, *et al.* The natural history of homocystinuria due to cystathionine-β-synthase deficiency. *Am J Hum Genet* 1985; 37: 1–31.
20. Wilcken DEL, Wilcken B. The natural history of vascular disease in homocystinuria and the effects of treatment. *J Inherit Metab Dis* 1997; 20: 295–300.
21. Rosenblatt DS. Inborn errors of folic acid and cobalamin metabolism. In: Carmel R, Jacobsen DW, eds. *Homocysteine in health and disease.* Cambridge: Cambridge University Press, 2001.
22. Kang SS, Wong PWK, Susmano A, Sora J, Norusis M, Ruggie N. Thermolabile methylenetetrahydrofolate reductase: an inherited risk factor for coronary disease. *Am J Hum Genet* 1991; 48: 536–45.
23. Goyette P, Sumner JS, Milos R, *et al.* Human methylenetetrahydrofolate reductase: isolation of cDNA, mapping and mutation identification. *Nat Genet* 1994; 7: 195–200.
24. Rosen R. Polymorphism of folate and cobalamin metabolism. In: Carmel R, Jacobsen DW, eds. *Homocysteine in health and disease.* Cambridge: Cambridge University Press, 2001.
25. Fletcher O, Kessling AM. MTHFR association with arteriosclerotic vascular disease? *Hum Genet* 1998; 103: 11–21.
26. Brattstrom L, Wilcken DEL. Homocysteine and cardiovascular disease: cause or effect? *Am J Clin Nutr* 2000; 72: 315–23.
27. Ueland PM, Refsum H, Beresford AA, Vollset SE. The controversy over homocysteine and cardiovascular risk. *Am J Clin Nutr* 2000; 72: 324–32.
28. McNulty H, McKinley MC, Wilson B, *et al.* Impaired functioning of thermolabile methylenetetrafolate reductase is dependent on riboflavin status: implications for riboflavin requirements. *Am J Clin Nutr* 2002; 76: 436–41.
29. Guenther BD, Sheppard CA, Tran P, Rozen R, Matthews RG, Ludwig ML. The structure and prospectives of methylenetetrahydrofolate reductase from *Escherichia coli* which suggests how folate ameliorates human hyperhomocysteinuria. *Nat Struct Biol* 1999; 6: 359–65.
30. Viel A, Dall'Agnese L, Simone F, *et al.* Loss of heterozygosity at the 5,10-methylenetetrahydrofolate-reductase locus in human ovarian carcinomas. *Br J Cancer* 1997; 75: 1105–10.
31. Van der Put NMJ, Gabreels F, Stevens EMB, Smeitink JAM. A second common mutation in the methylenetrahydrofolate reductase gene: an additional risk factor for neural-tube defects? *Am J Hum Genet* 1998; 62: 1044–51.
32. Weisberg I, Tran P, Christensen B, Sibani S, Rozen R. A second genetic polymorphism in methylenetetrahydrofolate reductase (MTHFR) associated with decreased enzyme activity. *Mol Genet Metab* 1998; 64: 169–72.
33. Harmon DL, Shields DC, Woodside JV, *et al.* Methionine synthase D919G polymorphism is a significant but modest determinant of circulating homocysteine concentration. *Genet Epidemiol* 1999; 17: 298–309.
34. Morita H, Kurihara H, Sugiyama, *et al.* Polymorphism of the methionine synthase gene: Association with homocysteine metabolism and late-onset vascular diseases in the Japanese population. *Arterioscler Thromb Vasc Biol* 1999; 19: 298–302.
35. Tsai MY, Welge BG, Hanson NQ, *et al.* Genetic causes of mild hyperhomocysteinemia in patients with premature occlusive coronary artery diseases. *Arteriosclerosis* 1999; 143: 163–70.
36. Ludwig ML, Matthews RG. Structure-based perspectives on B_{12} dependent enzymes. *Annu Rev Biochem* 1997, 66: 269–313.
37. Wilson A, Platt R, Wu Q, *et al.* A common variant in methionine synthase reductase combined with low cobalamin (vitamin B12) increases risk for spina bifida. *Mol Genet Metab* 1999; 67: 317–23.
38. Gaughan DJ, Kluijtmans LAJ, Barbaux S, *et al.* The methionine synthase reduction (MTRR) A66G polymorphism is a novel genetic determinant of plasma homocysteine concentrations. *Atherosclerosis* 2001; 157: 451–6.
39. Tsai MY, Yang F, Bignell M, Aras O, Hanson NQ. Relation between plasma homocysteine concentration, the 844ins68 variant of the cystathionine beta-synthase gene, and pyridoxal-5-phosphate concentration. *Mol Genet Metab* 1999; 67: 352–6.
40. De Stefano V, Dekou V, Nicaud V, *et al.* Linkage disequilibrium at the cystathionine beta synthase (CBS) locus and the association between genetic variation at the CBS locus and plasma levels of homocysteine: the Ears II Group: European Atherosclerosis Research Study. *Ann Hum Genet* 1998; 62: 481–90.
41. Chango A, Emery-Fillon, de Courcy GP, *et al.* A polymorphism (80G→A) in the reduced folate carrier gene and its associations with folate status and homocysteinemia. *Mol Genet Metab* 2000; 70: 310–15.

42. Devlin AM, Ling EH, Peerson JM, *et al.* Glutamate carboxypeptidase II: a polymorphism associated with lower levels of serum folate and hyperhomocysteinemia. *Hum Mol Genet* 2000; 9: 2837–44.
43. Clarke R, Daly L, Robinson K, *et al.* Hyperhomocysteinemia: an independent risk factor for vascular disease. *N Engl J Med* 1991; 342: 1149–55.
44. Clarke R. Design of clinical trials to test the homocysteine hypothesis of vascular disease. In: Carmel R, Jacobsen DW, eds. *Homocysteine in health and disease.* Cambridge: Cambridge University Press, 2001.
45. Rader JL, Weaver CM, Angyal G. Total folate in enriched cereal-grain products in the United States following fortification. *Food Chain* 2000; 70: 275–89.
46. Jacques PF, Selhub J, Boston AG, Wilson PWF, Rosenberg IH. The effect of folic acid fortification on plasma folate on total homocysteine concentrations. *N Engl J Med* 1999; 340: 1449–54.
47. Danesh J, Lewington S. Plasma homocysteine and coronary heart disease: a systematic review of published epidemiological studies. *J Cardiovasc Risk* 1998; 5: 229–32.
48. Yap S, Boers GHJ, Wilcken B, *et al.* Vascular outcome in patients with homocystinuria due to cystathionine-B-synthase deficiency treated chronically: a multicenter observational study. *Arterioscler Thromb Vasc Biol* 2001; 210: 2080–5.
49. Clarke R, Shipley M, Lewington S, Youngman L, Collins R, Marmot M, Peto R. Underestimation of risk associations due to regression dilution in long-term follow-up of prospective studies. *Am J Epidemiol* 1999; 150: 341–53.

DISCUSSION

Dr. Wanders: Homocysteine is very reactive. You mentioned that it might react with glutathione and with proteins. What is the effect on proteins if you have a raised homocysteine level? That's obviously a target to look for.

Dr. Scott: There are two possibilities, assuming that elevated homocysteine turns out to be a risk factor rather than a risk marker. The first is that circulating homocysteine with its free thiol group attaches itself to endothelial cells and onto proteins, glutathione, and so on. If that is the case, the evidence is rather confounded by the fact that in the circulation only about 4% of circulating homocysteine has a free −SH group. Most of it is so reactive that it's already bound to serum albumin, glutathione, cysteine, and so on. So while it's possible that the free thiol group does attack endothelial cells, I think it's far more likely that a raised plasma homocysteine is really just a reflection of a raised intracellular homocysteine level.

If you look at the biochemistry of an elevation of intracellular homocysteine, there are two types of effect. First, it could react with proteins, because most proteins in the cell are in a reduced state and most enzymes like the reduced state. If homocysteine were to attach itself to these proteins, enzymes, channels, or whatever, it could be detrimental to the cell. Second, there is a possibility that homocysteine, if it does accumulate, backs up to produce S-adenosyl homocysteine, which is a very potent inhibitor of many methyltransferases. Methyltransferases methylate DNA, lipids, proteins, and a whole range of other chemicals. Undermethylation of DNA is known to be a catastrophic occurrence because that's the way DNA is silenced.

It is pure speculation at this stage what the effects of an elevation of homocysteine may be. You can make up any theory you like, because it could be true. There's a wide variation in the possibilities, from activating the coagulation cascade to interfering with metabolism. I personally think its effects are mainly intracellular.

Dr. Wanders: It has been claimed that there is increased oxidative stress as a result of elevated homocysteine. What's your opinion?

Dr. Scott: When in doubt, there is a tendency to invoke oxidative stress. It's a very common explanation for many phenomena. I don't think the evidence for oxidative stress is particularly sound.

Dr. Fowler: I want to support the idea that the mechanism is unclear and that there are many theories. One problem has been the lack of suitable animal models—they have used the wrong kinds of homocysteine and in very high concentrations. However, there are some more promising models coming out, based on genetic knockouts that reflect the situation in mild hyperhomocysteinemia, and they do show evidence of damage to the endothelium. There is still a lot of work to be done in this area.

Dr. Pencharz: You've talked a lot about genetics, so now let's discuss nutrition. You said the diet contains a 60% excess of methionine, so that raises the issue of methionine/cysteine ratios. That in turn leads to the function of the sulfur amino acids in the methionine cycle, which is probably transmethylation, which you just mentioned. We've done some work that will be appearing in the *American Journal of Clinical Nutrition* looking at maximum methionine requirement and then minimum methionine requirement (in other words, an excess of cysteine). The maximum is around 19 mg/kg and the minimum is around 5 or 6 mg/kg. Could you say something about the issue of the methionine/cysteine ratio, and about creatine. There are people who are considering giving creatine, which is one of the end products of the transmethylation pathway, to try to slow down the formation of homocysteine.

Dr. Scott: We've done experiments where we have given healthy human volunteers extra methionine in the diet to see whether there is a resulting elevation of plasma homocysteine. It turns out that in this experimental situation, the amount of methionine you would have to give is far higher than you would ever find in the diet. So it looks as though the three existing enzymes are sufficient, even if they're not working optimally, to reduce or get rid of methionine, and in practice you don't find elevation of homocysteine after a methionine load—although you do if the person is vitamin B_6–deficient, and there's a diagnostic methionine loading test for that situation. Thus, the body appears capable of tolerating very large intakes of methionine, but I don't know what effect that may have on cysteine metabolism. I'm sure that that's a critical issue, because cysteine is involved in glutathione synthesis and so on. Obviously the more methionine you take, the more homocysteine you generate, and the assumption is that because it's not elevated it must be catabolized. One would predict that a high methionine would produce transiently high homocysteine, which may not be obvious in the plasma, and that in turn would produce elevation of cysteine, which would undoubtedly change the ratio of the two amino acids.

Dr. Pencharz: I was thinking about it a different way. There are alternative dietary approaches providing a lower methionine intake, for example the use of soy protein; and there are rabbit studies from the University of Western Ontario showing that the atherogenesis produced by casein is reduced by feeding a lower sulfur amino acid diet. Vernon Young introduced the double methionine model, which we've also been using to look at the rates of transmethylation, remethylation, and transulfuration. We have data in humans showing how the different ratios of methionine to cysteine affect the plasma homocysteine level.

Dr. Scott: In our experience, plasma homocysteine doesn't seem to be raised with a high methionine intake, but of course homocysteine may be increased in the cells, which would be a completely different matter.

Dr. Böhles: I would like you to say a bit more about the high reactivity of homocysteine. Obviously it's related to the −SH group, but that same −SH group is present in glutathione and cysteine. So, what causes the difference? Is it true that homocysteine *per se* is toxic, or is it some other compound? I think we are too free with the term "high reactivity." What is beneficial in glutathione we call a bad thing in homocysteine!

Dr. Scott: I think it is to do with the redox state of the cell whether or not thiols are formed. I can only tell you what is known experimentally, which is that there is virtually no free homocysteine in the circulation (only 1%–4%). Therefore, it's a fact that it dimerizes with other things. I don't know enough about glutathione or cysteine metabolism to say whether they do the same thing, and I'm sure there are people here who could answer that question. Whether homocysteine dimerizes in the cell is a moot point because it is difficult to show whether it is free or not in the intracellular location because of the need to disrupt the cell. However, I think the assumption is that in normal cells, homocysteine would be quite reactive, and I think that is probably to do with the redox state.

Dr. El-Baroudy: I have a comment rather than a question. In the case of homocystinuria, we are talking about a condition—that is, methionine synthase deficiency—that is inherited in a strictly mendelian way. However, in the case of neural tube defects associated with folate deficiency, the inheritance is nonmendelian and polygenic, involving many factors that combine to cause vulnerability to folate deficiency.

My second comment is that I agree with you that there aren't any normal people any more! We now know that the genome consists of between 33,000 and 34,000 genes, and each of these genes produces more than one protein by splicing, as you explained, so we are talking about a *proteome* rather than a genome, with 33,000 genes yielding about a million proteins. During the splicing journey, many abnormal proteins are produced, but they are discarded on the way, and most people turn out to be normal.

Dr. Scott: In relation to neural tube defects, my main research interest is in that area. What we are trying to do is to track down the genes that are responsible for these defects. Because 80% of neural tube defects are preventable by giving extra folic acid, we obviously believe that they are going to turn out to be folate-related genes. So far, we've found only one, which is 5,10-methylene-THF-reductase. This accounts for 13% of the population-attributable risk. We hope to find more-because there is no doubt that neural tube defects represent an interaction between genetic predisposition on the one hand and dietary intake on the other. That means that many women who have infants with neural tube defects have polymorphisms that don't show themselves clinically. Women who have had a child with a neural tube defect are clinically perfectly healthy, with normal folate levels. In our study in Ireland (1), we took 50,000 samples and identified 100 pregnancies affected by neural tube defects. Only 5% of those women were folate-deficient, but so mild that we would not have treated them if they had come to our clinic. Thus, in the face of what appears to be a normal folate status, women have babies with neural tube defects. Why is that? It is probably because they have genes that code for suboptimal folate-related activity. If you give extra folic acid, the protein does well enough to prevent neural tube defects from occurring.

Dr. Bachmann: As a follow up to what you just said, the expression of genetic polymorphism may be regulated, but in the end we need to know how the pathway is challenged. So my question is, what is the impact of causal factors such as adenosyl methionine, adenosyl homocysteine, and so on during fetal development? We should be looking for a polymorphism that causes a rate-limiting effect on something in fetal life but not later on.

Dr. Scott: If one just looks at the facts to start with, 80% of neural tube defects are preventable if the woman gets extra folic acid periconceptionally. That tells us that some event in the closure of the neural plate, as it forms the neural tube that will eventually become the spinal column, is folate-sensitive and folate-responsive—because if you give extra folate you can prevent the defect. What that event is, and whether it's related to homocysteine or methionine metabolism, is pure speculation at this stage. There are various models, but I don't have much confidence in them because there is no evidence that the causes of neural tube defects in mice are the same as in humans. I know that some of these murine models are folate-responsive, but the folate responsive element may not be the same as in the human. I suspect that there is some event in the closure of the neural plate that is very folate-sensitive, and my best guess as to what that might be is cell division, because folates are necessary for purine and pyrimidine biosynthesis. It might also be something to do with the methylation cycle—there are dozens of enzymes that are methyltransferases, which switch on receptors and switch off DNA, and so on. There are many possibilities.

Dr. Endres: Thirty years ago we did a lot of loading studies with methionine, up to 100 mg/kg body weight with single-bolus loading (2), and at that time we were not able to detect homocysteine or homocystine in the plasma, possibly because the column chromatography we used was not adequate; nowadays we use gas chromatography mass spectrometry (GCMS). But what we did detect was that many healthy people had an increased renal excretion of cystathionine when given methionine, even premature babies, and when we gave vitamin B_6 at the same time as the methionine loading, urinary cystathionine excretion decreased (3). In the treatment of homocystinuria, we know that up to 6 g of betaine a day is very helpful for remethylation (4). Was this used in any of the interventional studies you showed, or did they only use B vitamins and folate?

Dr. Scott: Right. The background to the question is that in inborn errors where those enzymes are deficient, betaine has been found to be a successful treatment (5). Betaine converts homocysteine back to methionine by a method that does not depend on methionine synthase, folate, or B_{12}. In the trials you refer to, it is possible that betaine would have been effective, but what those trials were trying to do was to look for a public health response. The public health response will be either folic acid added to the diet—perhaps to everyone's diet—or treatment with supplements, or perhaps both. The challenge of trying to prove the involvement of folic acid is the big one in the first instance, because that would be the most amenable to public health measures. If you take people in this room and give them folate, most will have a fall in their homocysteine level. If that's a risk factor for cardiovascular disease, we'd be nuts not to take folic acid. I take it myself, but I'm not suggesting everyone do that—it's obviously a matter of choice.

The difficulty in these trials is that if you're going to put someone into a placebo-controlled trial and your standard method of treatment is to reduce plasma homocysteine when it is elevated, then ethically you couldn't do such a trial. In practice, it is not standard treatment to reduce plasma homocysteine, so it remains ethical to do these trials in many places. However, if any of these trials comes out positive, then—as with folate and neural tube defects—everyone will have to be treated if they have a raised homocysteine.

Dr. Fowler: Are you convinced that if these trials do reduce the incidence of vascular disease that proves that homocysteine is the risk factor, or does it not show that folate does the job and, at the same time, incidentally brings down homocysteine?

Dr. Scott: I agree it leaves open that question. Perhaps the actual event in cardiovascular disease is folate-related—that's entirely possible. I think the result would be the same. If folate reduces cardiovascular disease, I don't think the public will be too concerned about whether homocysteine is the etiologic factor or folate. However, I agree that homocysteine could be just a biochemical marker.

Dr. Marquardt: You make it a strong point that we don't have enough folate in our diet unless we live in the United States and eat supplemented flour. I'd like to know the frequency of increased levels of homocysteine in the German population. In addition, what is your personal recommendation? You just stated that you take folic acid. Should we all take folic acid? Should we take vitamin B_6 and vitamin B_{12} as well? What would be your recommendation for the kind of dosage that we would need to reduce our homocysteine levels?

Dr. Scott: In the studies that have been done—and they've been in a variety of different communities and settings—perhaps one third of young healthy adults have homocysteine levels that are not lowered by folic acid, one third have moderate elevation, and one third rather higher. The degree of lowering that you find depends on your starting level. Homocysteine increases markedly with age, but the reason for that is not entirely clear. The kidney is certainly very important in homocysteine metabolism, and it is possible that renal metabolism decreases with age. If you took a group of elderly people—say the Framingham group in the United States before treatment—70% to 80% of them would have homocysteine levels in the range likely to be reduced by folic acid. The problem about not knowing whether this is a risk factor or a risk marker is the key. It may be that elevation of homocysteine to, say, around 12 μmol/l doesn't really matter. Perhaps a level of around 20 doesn't matter. However, a level of 100, as occurs in homocystinuria, does matter. The problem is in the intermediate level, and this is where the debate about risk factor *versus* risk marker becomes important. My own view is that if there are other risk factors, one should probably take folate. There is evidence that the presence of other risk factors can increase the risk-associated effect of a raised homocysteine level from about three-fold to more than eight-fold. A large European study showed that the concomitant presence of high blood pressure and high homocysteine may give a 30-fold increase in risk (6).

As to dose, taking a folate supplement of 400 μg a day would certainly get the homocysteine concentration down as low as it will go, and that dose represents the recommended daily allowance (RDA). I don't think there is anyone who says that taking 400 μg of folic acid a day carries any risk. From a public health point of view, one hopes that these trials will go ahead and be completed, because there's no doubt that the only way we will ever get a consensus about this is from a successful trial. Successful trials in the United States cannot be done any more because of folate fortification of flour. In the United Kingdom, it looks as though folate fortification will not occur because there have been objections by people who claim, quite rightly, that their choice is being removed. It's like the fluoride story. If it turns out that homocysteine is a risk factor for cardiovascular disease, then I think public opinion will change quite rapidly.

Dr. Sies: I have a minor comment about the structure. It is an α amino acid, and there's only one sulfur in the compound.

There's another player of potential of interest for cardiovascular disease and that's *S*-nitroso-homocysteine. Do you think that might also be of interest in explaining potential cardiovascular effects?

Dr. Scott: There's always a chemist in the audience! Well, the answer as to what the causative agent might be is wide open. It might be an analog of homocysteine or it might be SAH. Until a lot more is known—perhaps from animal models, perhaps through tissue culture studies—it will remain speculation.

Dr. Yamashiro: I think most people agree that a high consumption of fat leads to a high prevalence of cardiovascular disease. Is there any epidemiologic evidence that a high consumption of protein also leads to cardiovascular disease?

Dr. Scott: There could be a connection between high protein intake and high cholesterol intake. Also, a high protein intake may equal a high methionine intake. Our evidence is that the body is good enough at getting rid of a high methionine intake, and it doesn't allow the plasma homocysteine to go up. Cysteine may go up and the pyruvate may go up, but it looks as though most people can handle any elevation of methionine in the diet. So I think that will probably not be the connection, but it does bring up another interesting point: even if homocysteine is a risk factor for cardiovascular disease, it's only one risk factor, and it probably accounts for maybe 20% of the risk at most. In public health terms, that is a huge proportion, but in terms of sorting out an individual case study, it is very small. In the analyses that have been done in a large study sponsored by the European Union (7), homocysteine elevation and cholesterol elevation seemed to be independent risk factors for cardiovascular disease, with no overlap, while by contrast the presence of raised homocysteine and high blood pressure was a multiplicative risk factor. Therefore, it looks as though cholesterol and raised homocysteine act by a different route. Maybe there is a dietary association, I don't know.

Dr. Steinmann: As I remember from the old literature, the parents of patients with classical homocystinuria were reported to have increased homocysteine levels but to have a normal lifespan. Does this still hold true?

Dr. Scott: I think the problem is that the numbers are so small that it's impossible to be sure about the lifespan of heterozygotes.

Dr. Superti-Furga: I wonder why you did not to comment on the effect of homocysteine on the coagulation system. You started out by telling us that patients with inborn errors show this hypercoagulation state, which may be responsible for the cardiovascular effects.

Dr. Scott: I believe that homocysteine causes cardiovascular disease in homocysteinuria and this could be through activating the coagulation cascade or through endothelial damage; there are several different possible mechanisms. There is evidence experimentally that elevation of plasma homocysteine activates factor VIIa and switches on the tissue factor coagulation pathway (8). So there's some *in vitro* and indeed some *in vivo* evidence that the coagulation cascade may be switched on. The reason I don't warm to that explanation for the cardiovascular effects is that I think it would have to take place in the plasma, and in the plasma I can't imagine that homocysteine bound to albumin would activate factor VIIa.

REFERENCES

1. Daly LE, Kirke PN, Molloy A, Weir DG, Scott JM. Folate levels and neural tube defects. Implications for prevention. *JAMA* 1995; 274: 1698–1702.
2. Bremer HJ, Endres W. Primary cystathioninuria, methionine load tests and response to pyridoxine. *Helv Paediat Acta* 1972; 27: 525–36.
3. Endres W, Vogt R, Riegel KP, Bremer HJ. The influence of vitamin B6 on cystathioninuria in premature infants. *Clin Chim Acta* 1978; 86: 89–93.
4. Smolin LA, Benevenga NJ, Berlow S. The use of betaine for the treatment of homocysteinuria. *J Pediat* 1981; 99: 467–72.
5. Yap S, Boers GHJ, Wilcken B, *et al.* Vascular outcome in patients with homocysteinuria due to cystathionine–B-synthor deficiency treated chronically: a multicenter observational study. *Arterioscler Thromb Vasc Biol* 2001; 21: 2080–5.
6. Graham IM, Daly LE, Refsum HM, *et al.* Plasma homocysteine as risk factor for vascular disease: the European concerted action project. *JAMA* 1997; 277: 1775–81.
7. Graham IM, Daly LE, Refsum HM, *et al.* Plasma homocysteine as risk factor for vascular disease: the European concerted action project. *JAMA* 1997; 277: 1775–81.
8. Carmel R and Jacobsen DW. Homocysteine on hemostasis in homocysteine in health and disease. *Cambridge University Press UK* 2001; 415–24.

Genetic Expression and Nutrition, edited by Claude Bachmann and Berthold Koletzko. Nestlé Nutrition Workshop Series, Pediatric Program, Vol. 50. Nestec Ltd., Vevey/Lippincott Williams & Wilkins, Philadelphia, © 2003.

Minimum Protein Requirements in Infancy and Childhood: Insights from Patients with Protein-Restricted Diets

Helene Ogier de Baulny, L. François, M. O. Grenèche, M. Giraud, and *E. Depondt

*Service de Neurologie, Maladies Métaboliques, Hôpital Robert Debré, Paris; *Service de Génétique Médicale, Hôpital Necker Enfants Malades, Paris, France*

Knowledge about the normal child's requirements for protein and essential amino acids is still limited. Two approaches have been used to estimate protein requirements. The first involves estimating the protein intake of exclusively breastfed infants, which is supposed to represent the true requirement on an evolutionary basis. The second involves factorial analyses, based on the calculated amounts of protein required for growth, stool and skin losses, and obligatory catabolism. As infants get older, the requirement for protein and essential amino acids for growth decreases as growth rate declines, while protein loss from the body and obligatory catabolism maintenance account for an increasingly large proportion of the total requirement.

Genetic errors of enzyme function resulting in disturbed metabolism of proteins or essential amino acids provide opportunities for estimating the amounts of protein needed for normal growth and development. In theory, during such disorders the amount of specific protein or essential amino acid required for obligatory catabolism is nil. To meet this condition, the enzymatic block must be complete, without any residual enzymatic activity. In addition, the enzymatic block has to lack the ability to stimulate alternate pathways, and cannot be amenable to treatments that involve such pathways. In practice, the degree of residual enzyme activity is unknown and is evaluated *a posteriori* on the basis of the individual tolerance for the controlled amino acid given in the diet. The lowest tolerances indicate complete block while higher tolerances suggest residual enzymatic activity. However, such diseases are useful tools only if the following conditions pertain: (a) the ability to make reliable metabolic assessments by accurate measurements of metabolites in plasma or urine, with a known range of safe values; (b) the availability of a satisfactory dietary treatment allowing normal nutritional status, growth, and development.

Only a few disorders meet these conditions and so can be studied in this way. The best examples are phenylketonuria (PKU), due to phenylalanine hydroxylase deficiency, and maple syrup urine disease (MSUD), due to branched chain decarboxylase deficiency. Most of the other amino acid disorders have characteristics that make them inappropriate. For example, tyrosinemia type I, caused by fumarylacetoacetase deficiency, is treated with NTBC 2-(2-nitro-4-trifluoro-methylbenzoyl)-1,3-cyclohexanediol, a molecule that blocks the second step of tyrosine catabolism. This model would have been ideal to compute the requirement for phenylalanine plus tyrosine. The therapeutic diet in these patients, however, results in high plasma tyrosine levels ($\approx$400 μM), which are obviously well outside the normal range (43 $\pm$ 16 μM). Methylmalonicaciduria and propionic aciduria are two other disorders that are treated with a protein-restricted or a valine-restricted diet. However, the site of the catabolic block in these conditions allows adjacent metabolic disturbances with accumulation of various additional organic acids. Except for general well-being, we have no clear criteria to assess metabolic control in these patients, and despite treatment, severely affected patients have poor growth and development. Urea cycle defects caused by various enzyme deficiencies in the urea cycle are responsible for hyperammonemia. Most patients with these conditions present in the neonatal period and die within a few days. Because of the presence of residual enzymatic activity, some patients have late-onset forms. These individuals can have normal growth and normal blood ammonia levels when on a protein-restricted diet and when given drugs such as sodium benzoate and sodium phenylbutyrate that permit nitrogen to be excreted in compounds other than urea.

Our first step in the study described in this chapter was to analyze the protein, phenylalanine, and leucine intakes in patients affected with PKU (n = 34) and MSUD (n = 12), using the dietary records at 3, 6, 9, 12, 24, and 36 months of age. These patients were obtained from cohorts diagnosed and treated during the last decade in the two main metabolic units in Paris. We compared our results with previously published data and with the results obtained by the two classic methods.

CLASSIC METHODS FOR COMPUTING PROTEIN AND ESSENTIAL AMINO ACID REQUIREMENTS

The Human Milk Model

Breastfeeding is the agreed reference of all international committees formed to assess nitrogen and amino acid needs during the first 6 months of life (1). Few data are available from 6 to 12 months of age and we use these for comparison (2). Requirements are derived from the average volume of milk absorbed by healthy breastfed infants and the mean bioavailable protein and amino acid content of mature human milk. These studies report a protein intake of 1.5 to 1.6 g/kg/d during the second month of life, decreasing gradually to about 1g/kg/d between the third and 12th months (2–5).

Assuming a birth weight of 3.5 kg and a normally growing infant, protein intakes average 6 to 7 g/d. Based upon estimated energy requirements (5) and the average body weight according to references for the French population (6), the daily protein requirement expressed in terms of energy intake averages 1 g per100 kcal from the third to 12th months of life, while being higher in the first month (1.6 g/100 kcal) and in the second month (1.4 g/100kcal) (Table 1).

TABLE 1. *Estimates of the daily bioavailable protein and amino acid requirements, based upon the human milk model and factorial analysis*

Age (months)	0–1	1–2	2–3	3–4	4–5	5–6	6–9	9–12	12–24	24–36
Mean body weight (kg)	3.55	4.35	5.125	5.875	6.625	7.1	8.1	9.05	10.75	13.75
Mean energy intake (kcal/d)	430	480	520	560	600	640	730	820	1000	1250
Protein requirement (HM/FA)										
g/day	6.9/7.6	6.8/7.8	6/8	6/8	6.3/8	6.6/8.5	7.3/8.6	7.2/10.5	–/13.4	–/14.6
g/100 Kcal	1.6/1.8	1.4/1.6	1.1/1.5	1/1.4	1/1.3	1/1.4	1/1.2	0.9/1.3	–/1.3	–/1.2
PA requirement (HM/FA)										
mg/day	180/300	180/310	224/320	228/320	239/320	277/340	277/340	275/420	–/540	–/585
mg/100 Kcal	42/70	70/38	43/60	41/57	40/53	39/53	38/46	33/51	–/54	–/48
Leu requirement (HM/FA)										
mg/day	620/540	570/550	550/570	560/547	590/570	610/600	675/745	675/745	–/950	–/1040
mg/100 Kcal	144/125	119/114	106/109	100/98	98/97	95/94	82/91	82/91	–/95	–/83

Leu, leucine; PA, phenylalanine; HM, human milk model; FA, factorial analysis.

The estimated minimum daily requirement for individual amino acids is obtained by multiplying the total daily protein consumption by a composite estimate of the lowest amino acid content of human milk reported by the Food and Agriculture Organization (FAO) (7). The study by Janas and Picciano (3) evaluated human milk consumption for the first 2 months of life in parallel with measurement of its amino acid content at 2, 4, and 8 weeks of age. This study gives an accurate estimate of the daily intakes of amino acids. The data obtained from these studies show that the daily intakes of phenylalanine could be as low as 180 mg/d in the first 2 months of life, increasing to 225 and 275 mg/d over the following 10 months. Intakes of leucine average 600 to 675 mg/d during the first and after the fourth months of life. Intakes are lower (550–570 mg/d) between the ages of 1 and 4 months. Expressed in terms of energy intake, the phenylalanine intake averages 40 mg/100 kcal and the leucine intake averages 100 mg/100 kcal, except for the first 2 months of life, during which it is estimated to be 145 and 120 mg/kcal, respectively.

Factorial Analysis

Estimation of the protein requirements by the factorial method is highly controversial. This method implies that maintenance needs are known precisely and that there is agreement on the way to assess the needs for growth. However, based upon Fomon's calculations for infants aged 0 to 12 months and by applying the method for 24- and 36-month-old children, protein requirements amount to 1.98 g/kg/d for the first month, and decrease to 1.14 g/kg/d from the second to the 12th month, with a further decrease to 1.1 g/kg/d during the second and third years of life (5). An estimate of minimum amino acid requirements can be derived from this factorial approach using the amino acid content in the whole body protein (8) and assuming that there is no difference between the amino acid profiles for maintenance and growth.

The results expressed as gram per 100 kcal for protein and milligram per 100 kcal for phenylalanine and leucine are summarized in Table 1. Except for the first 3 months of life, phenylalanine and leucine requirements are constant and average 50 and 90 mg/100 kcal, respectively. These results are in good agreement with the data previously obtained by the human milk method. During the first 3 months, the estimated leucine requirement is higher, averaging 100 to 125 mg/100 kcal. In comparison with the human milk model, the factorial analysis could underestimate the leucine requirement.

INBORN ERRORS OF METABOLISM AS A MODEL

Owing to the defective phenylalanine and branched chain amino acid catabolism, the treatment of patients with PKU and MSUD relies on diets restricted in phenylalanine and leucine, respectively. To prevent acute or chronic neurologic deterioration, the aim of the diet is to keep blood phenylalanine and leucine levels below 5 mg/dl. The desired amounts of phenylalanine and leucine are given, using quantified amounts of milk in infancy and, progressively, quantified amounts of milk, cereals, vegetable, and fruits in childhood. In parallel, to allow normal growth and development, the diet is supplemented with specific mixtures that contain all the amino acids except for the harmful phenylalanine in PKU and the branched chain amino acids in MSUD. The proportions of the amino acids given in these mixtures are close to those in human milk, except that the PKU mixtures are enriched with tyrosine. In addition, the mixtures contain all the minerals, vitamins, and trace elements in concentrations proportional to the protein content. The prescription of daily amounts is calculated to cover the recommended protein requirements for age (1.5–2.0 g/kg/d), while appropriate daily energy is supplied by additional carbohydrate and lipid. In summary, these patients receive a normal diet for age except that they have a minimal intake of phenylalanine or leucine and thus minimal intakes of natural protein.

The PKU Model

During the first year, the distinction between typical PKU (without residual phenylalanine hydroxylase activity) and atypical PKU (with some residual enzyme activity) is imperfect. The ability to define phenylalanine tolerance (that is, the maximum quantity of phenylalanine that a child can ingest during a period of normal growth without an increase in blood phenylalanine level beyond what is considered ideal) is most often attained between the second and third years of treatment. During this period, some children (four of the 34 in our cohort) have phenylalanine intakes close to, or even greater than, the lowest estimated requirement. These patients obviously have atypical PKU, and their protein and phenylalanine intakes are not informative for estimating the minimum requirement. Conversely, patients whose intakes are lower than the estimated requirements have typical PKU (unless no attempt was made to increase their phenylalanine intake above their minimum needs).

From our results, sufficient phenylalanine intakes could be estimated to be around 25 to 35 mg/100 kcal (250 mg/d) on average during the first 3 years of life. This is substantially less than the estimated requirements for normal children, but very close to the lower range of phenylalanine intakes of breastfed infants in the age range of 3 to 12 months (Table 2). These results are not in conflict with those obtained by others (7,9–12). In particular, our lower intakes are in good agreement with the American collaborative study (9), which showed that phenylalanine needs vary between 55 ± 16 mg/kg/d (46 ± 13 mg/100 kcal) during the first trimester and 27 ± 8 mg/kg/d during the fourth trimester in a group of children for whom the objective of treatment was to maintain blood phenylalanine levels between 1.0 and 5.4 mg/dl. Similarly, by computing the amount of natural protein intakes in our patients affected by typical PKU, the average intakes decrease progressively from 0.76 to 0.54 g/100 kcal, which means a progressive increase in the daily intake from 5 ± 1 to 7 ± 1 g/d. These amounts are about 45% to 50% of the recommended requirement and the minimum requirement, deduced from the factorial analysis and from the human milk model.

TABLE 2. *Natural protein, phenylalanine and leucine intakes in patients affected with phenylketonuria and maple syrup urine disease*

Age (months)	3	6	9	12	24	36
PKU patients (n = 34)						
Mean body weight (kg)	5.920	7.470	8.680	9.980	12.430	14.330
Energy intakes (Kcal)	655 ± 70	785 ± 85	890 ± 105	965 ± 100	1120 ± 90	1300 ± 165
Natural protein intakes						
g/day	4.7 ± 0.8	5.6 ± 1.1	6.3 ± 1.1	6.5 ± 1.3	6.9 ± 1.1	7 ± 1.2
g/100 Kcal	0.72	0.71	0.71	0.67	0.62	0.54
PA intakes						
mg/day	235 ± 40	245 ± 47	246 ± 45	250 ± 51	264 ± 43	266 ± 44
mg/100 Kcal	36	31	27	26	24	20
MSUD patients (n = 12)						
Mean body weight (kg)	5.2	7	7.960	8.710	11.100	13.820
Energy intakes (Kcal)	600 ± 90	800 ± 85	870 ± 66	970 ± 66	1200 ± 117	1350 ± 81
Natural protein intakes						
g/day	3.7 ± 0.88	4.3 ± 1	5.3 ± 1	5.9 ± 1.4	6.5 ± 1.4	7.6 ± 1.5
g/100 Kcal	0.62	0.5	0.6	0.6	0.55	0.55
Leu intakes						
mg/day	340 ± 80	340 ± 80	370 ± 85	370 ± 85	400 ± 90	470 ± 95
mg/100 Kcal	57	42	42	38	33	35

Values are mean ± SD. Leu, leucine; MSUD, maple syrup urine disease; PA, phenylalanine; PKU, phenylketonuria.

The MSUD Model

The dietary tolerance of leucine in 12 children affected with the neonatal onset form of MSUD decreased from 57 to 35 mg/100 kcal during the first 3 years of treatment (340–470 mg/d) (Table 2). Whatever form of expression is used, these results are in the lower range of the intakes reported in previous studies (7,10,12). In previous studies, the aim of treatment was to maintain blood leucine levels between 5 and 10 mg/dl, whereas in the present study, blood leucine levels were maintained below 5 mg/dl. Similarly, by computing the amount of the natural protein intake in our patients with MSUD, intakes averaged 0.6 g/100 kcal, with daily intakes from 4.0 ± 0.9 to 7.5 ± 1.5 g/d. These amounts are somewhat lower than those recorded for children with PKU during the first year of life. The reason for this could be the risk of acute metabolic imbalance that makes clinicians more cautious about prescribing leucine in MSUD than they are with the treatment of PKU. As with the PKU model, the protein intakes in patients with MSUD are between 35% and 50% of both the recommended requirements and the minimum requirement, calculated by factorial analysis and from the human milk model.

COMMENT

The results of factorial analysis and the PKU or MSUD models are in good agreement over the daily amounts of protein and essential amino acids needed to meet the requirements for growth and non-urinary nitrogen losses (Fig. 1). In absence of catabolism of phenylalanine and leucine, patients with PKU and MSUD require 4.0 to 7.5 g/d (0.55–0.75 g/100 kcal) of protein to achieve normal growth and development. These intakes are between 35% and 50% lower than the minimum estimated requirement. This difference could represent the fraction of obligatory catabolism present in healthy infants fed on a normal diet (7). Whether these results can be applied

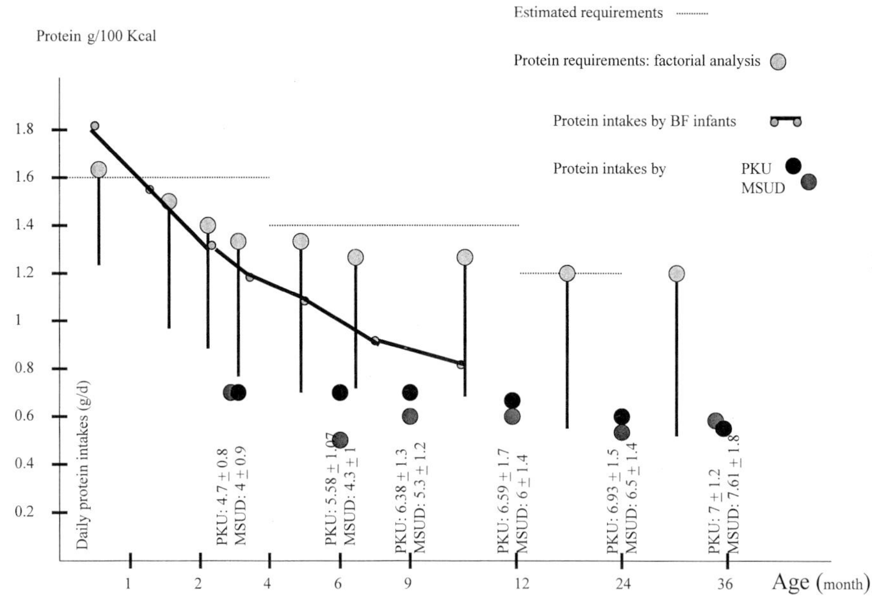

FIG. 1. Natural protein intakes in 34 patients with PKU and 12 patients with MSUD compared with: **(A)** protein intakes in exclusively breastfed infants; **(B)** protein requirements estimated by factorial analysis; and **(C)** joint committee estimates. Exclusively breastfed infants drastically decrease their protein intakes from 1.8 to 0.8 g/100 kcal/d. Results of factorial analysis are in good agreement with the breastfed model during the first 2 or 3 months of life. Thereafter, the estimated requirements are 18% to 24% greater. These results have led Fomon *et al* to revise their recommended dietary intakes (5). If we consider the theoretical minimum protein requirements (needs for growth plus nonurinary nitrogen losses) represented by the lower part of the vertical bars, these are about 0.75 to 0.57 g/100 kcal from the third month of age up to the third year. Our PKU and MSUD patients have similar natural protein intakes.

to normal infants is unknown. Patients with PKU and MSUD have diets that provide large amounts of nitrogen and nonessential and essential amino acids, except for those that are controlled. Under such conditions, we do not know whether normal infants are able to cease their "obligatory" catabolism. However, this estimate of the fraction of obligatory catabolism suggests that a sufficient protein intake is around 1 g/100 kcal in normal children. This is well below the recommended dietary protein allowances. It is not unlikely that this estimate may also apply to normal adolescents and adults.

REFERENCES

1. World Health Organisation. WHO technical report series No 724. *Energy and protein requirements.* Geneva: FAO/WHO/UNU, 1985: 64–112.
2. Salmenperä L, Perheentupa J, Siimes MA. Exclusively breastfed healthy infants grow slower than reference infants. *Pediatr Res* 1985; 19: 307–12.
3. Janas LM, Picciano MF. Quantities of amino acids ingested by human milk-fed infants. *J Pediatr* 1986; 109: 802–7.
4. Räihä N. Milk protein quantity and quality and protein requirements during development. *Adv Pediatr* 1989; 36: 347–68.

5. Fomon SJ. Requirements and recommended dietary intakes of protein during infancy. *Pediatr Res* 1991; 30: 391–5.
6. Sempé M, Pédron G, Roy-Pernot MP. Auxologie : méthodes et séquences. *Eds Théraplix*, Paris, 1979.
7. Ruch T, Kerr D. Decreased essential amino acid requirements without catabolism in phenylketonuria and maple syrup urine disease. *Am J Clin Nutr* 1982; 35: 217–28.
8. Young VR, El-Khoury AE. Can amino acid requirements for nutritional maintenance in adult humans be approximated from the amino acid composition of body mixed protein? *Proc Natl Acad Sci USA* 1995; 92: 300–4.
9. Acosta PB, Wenz E, Williamson M. Nutrients intakes in treated infants with phenylketonuria. *Am J Clin Nutr* 1977; 30: 198–208.
10. Kindt E, Halvorsen S. The need of essential amino acids in children: an evaluation based on the intake of phenylalanine, tyrosine, leucine, isoleucine, and valine in children with phenylketonuria, tyrosine aminotransferase defect, and maple syrup urine disease. *Am J Clin Nutr* 1980; 33: 279–86.
11. Wendel U, Ullrich K, Schmidt H, Batzler U. Six-year follow-up of phenylalanine intakes and plasma phenylalanine concentrations. *Eur J Pediatr* 1990; 149[Suppl 1]: S13–6.
12. Bresson JL, Rey F, Poggi F, *et al.* Inborn errors of metabolism: a model for evaluation of essential amino acid requirements. In: Räihä NCR, ed. *Protein metabolism during infancy.* Nestlé Nutrition Workshop Series, Vol 33. New York: Raven Press, 1994: 211–27.

DISCUSSION

Dr. Pencharz: I was a little confused by your presentation. At first you talked about intakes of protein in the younger age group of I think around 2 g/kg, and then it went down to about 0.5 g/kg, but in one of your slides towards the end you were talking about intakes of between 0.4 and 0.6 g/100 kcal. I'm not quite sure I understand the idea you were trying to put across.

Dr. Ogier de Baulny: I switched from g/kg/day to g/100 kcal/day because it's obviously not very satisfactory to go on expressing intake as g/kg/day with increasing age. In a 6 year old child, for example, you wouldn't express intake as g/kg/day. Also, when you transform the requirement to g/100 kcal/day, the protein requirement becomes quite constant.

Dr. Pencharz: What you are talking about is protein intake as percentage of energy intake, which you say is constant through the life cycle of the child, but it isn't—at least not in the FAO studies or in the North American process, and we've looked at this very closely in normal children. There is not a fixed relation between protein intake and energy intake, so I wouldn't agree with your point of view.

Dr. Ogier de Baulny: I think that when you are treating patients on a protein restricted diet, it is easier to find the correct level of protein intake when it is expressed in terms of energy intake.

Dr. Pencharz: Well, this is something we need to talk about later in more detail. Is it your practice to give as low an amount of protein as you can and then you look at growth as your outcome variable, or have you been looking at nitrogen balance? What outcome variables have you been using?

Dr. Ogier de Baulny: We don't carry out balance studies. We use growth parameters and nutritional labels like albumin, prealbumin, and plasma amino acids. The patients we have studied with PKU and MSUD have all had normal growth and nutritional status throughout their lives. This is in contrast to patients with methymalonicaciduria and propionicaciduria who have decreased linear growth velocity in spite of having the same protein and energy intake. I don't know the reason for this unless it is related to the harmful effects of propionic acid or other organic acids.

Dr. Superti-Furga: Although our patient numbers are smaller, we have the same experience as Dr. Ogier. I would like to respond to what Dr. Pencharz just said. It's not that we try to give as small an amount of protein as possible; in fact we try to give as much protein as possible, but we have to draw the line when patients decompensate or in the case of PKU show high phenylalanine levels. In general, though, our experience is exactly the same, and if you look at recent trends it is clear that protein requirements and RDAs have been constantly corrected downwards. I think the "truth" may still be a bit lower.

Dr. Böhles: Wouldn't we learn more if we could relate the protein intake to the desirable growth rate—"desirable" meaning the optimal growth rate that can be achieved? I think it is rather too static a concept to relate protein intake only to body weight or to energy intake later on. I think we need to change our focus to what is desirable and what can be achieved. Would the recommended protein intakes be altered if we related them to growth rate rather than to attained weight or energy intake?

Dr. Ogier de Baulny: I don't know. I haven't looked at it in that way. But our patients with PKU and MSUD grew strictly normally until the age of 10 or more. We don't have complete data for adolescence, however. During that period growth velocity is similar to that between 9 months and 12 months of age, so it is very likely that an increased protein intake will be required.

Dr. Bachmann: Your approach stresses that with a total enzyme block in PKU and MSUD, you won't have any obligate oxidation. This is true, but you probably still need protein or nitrogen just for gluconeogenesis. Did you do similar calculations for patients with urea cycle disorders put on protein restriction plus essential amino acids?

Dr. Ogier de Baulny: Yes, we have some patients with urea cycle defects. However, most of these are late onset forms so they have some residual activity. Secondly, there is a great deal of subclassification among these disorders, so few of them are exactly the same. This means that for a particular condition we rarely have more than a few patients. Also, we treat our patients with urea cycle disorders with a protein-restricted diet and with sodium phenylbutyrate and benzoate supplementation but we don't give essential amino acids as a supplement. On this regimen we get a protein intake of around 1–1.3 g/100 kcal/day at 3 years of age. They receive about 150 mg/kg/day of phenylbutyrate.

Dr. Bachmann: What the organism needs is a sum of optimally balanced essential amino acids plus some non-essential nitrogen. In this discussion I think there has been some confusion about this issue. The main thing that is relevant for optimal metabolism of protein is its quality, and a lack of imbalance is what is important for achieving the lowest protein intake compatible with normal growth. Dr. Ogier's message mainly concerns two amino acids where we get a feel for the minimum quantities necessary, compared with the non-specific data available on most nutrient mixtures where we only have a sum of all the amino acids or the total protein content. I think in this discussion we should clearly distinguish between specific amino acids and total amino acids or total protein composition. Would you agree?

Dr. Ogier de Baulny: Yes, I do.

Dr. Fowler: Isn't the quality of protein equal in all these comparison studies?

Dr. Ogier de Baulny: No, it's not. Early on the patients receive high quality protein through human milk or infant formula, but then they are progressively switched to less efficient protein by 4 to 6 months of age. Even so they still grow well and their total protein intake remains much the same.

Dr. Fowler: We talk about minimum requirements, but what about *optimal* requirements? What is optimal and does it matter if we have too much protein or above the minimum? What do nutritionists say about this? Can anyone answer that?

Dr. Bier: I'll try to answer the question if you can define optimum for me!

Dr. Fowler: Okay, what is maximal then? Too high is bad, is it not?

Dr. Ogier de Baulny: When you give too much protein to young infants, you get hyperaminoacidemia and a high blood urea. This is not good for the kidney or the liver.

Dr. Fowler: We're all aware, of course, that in the old days we used to see many cases of methioninemia and tyrosinemia caused by high protein intakes. I'm sure there is a maximum level that should be aimed for, and this might be close to the minimum level—which would be the optimal level too.

Dr. Endres: There are many studies showing that infants and toddlers receive too much protein. I will cite three. First, 10 years ago a study from Milan (1) showed that the protein intake was more than three times higher than that recommended by European and American nutrition societies. Second, 4 years ago a study from Denmark (2) showed that the intake was about two times higher than recommended. Third, a study from Germany carried out 3 years ago then found much the same (3). Thus I think the message Dr. Ogier gave us—that is, of values of about 0.6–0.8 g protein/100 kcal compared with the current range of 1.8–3 g/100 kcal for starter formulas and up to 4.5 g/100 kcal for follow on formula—shows there is a great discrepancy between what may be needed and what is actually given. To answer the question about possible harm caused by this high protein intake, it is evident that plasma urea nitrogen is much higher in infants receiving a formula containing 2.25 g of protein/100 kcal or more compared with breast-fed infants. I think this renal burden should be considered.

To return to Dr. Böhles's comment, it is difficult to determine standards for protein intake after the breast-feeding age, though up to, say, 6 months, it is possible to use breast milk protein content as the standard. However, I think it is very helpful to look at plasma amino acids and urea nitrogen. My aim is to decrease the protein content in formulas—and we just launched one containing only 1.8 g protein/100 kcal—because we are convinced that in most formulas the protein content is too high. The practical point of Dr. Ogier's presentation is that she showed that the requirement for protein is much lower than even that recommended by national nutritional bodies.

Dr. Pencharz: I can help answer the question about minimum and maximum protein because. This is something we've looked at in terms of dietary recommendations. The lower end is the RDA, and then there is also an upper limit. We regard the bottom end of the range as a safe intake. The reason I was having problems with the protein intake per 100 kcal is that it goes up through childhood. Thus even if you take the absolute obligatory minimum intake in healthy populations, which is around 0.7 g/100kcal by the time the child is 6 or 8, it goes up to 1.4 g/100 kcal. In North America at the moment we are saying there is a minimum for an individual, which is the average or the EAR, and then there is the minimum for a population, which is the RDA. We think in children the maximum is lower than it is in adults. For adults, we're talking about 30% of energy intake; for children, we're talking about 15% to 20% of energy intake.

Dr. Böhles: I would like to bring another point into the discussion that's always forgotten when talking about protein requirements, and that is that there are no protein stores in the body. This means you only have the protein you need. If a child is sick in bed and doesn't move about much, naturally that child has a smaller protein requirement than when normally active. If the entire population always ran instead of walking calmly, it would have a totally different protein requirement. This point is never emphasized in discussions on the subject. We should always define what activities are being done, which is especially important in childhood—children never walk about calmly! We should take these factors into consideration when discussing protein requirements.

Dr. Ogier de Baulny: Of course, I agree about that, but that is always going to be a problem with standards.

Dr. Borum: We need also to remember the old concept of "conditionally essential" nutrients, which was based on the idea that something occurring during development, or in old age, or when there is altered organ function, such as renal disease, that may alter the nutrient requirements. We've spoken about using dietary components to respond to altered nucleotide sequences in DNA, but perhaps we should also think about how dietary components, or ratios of dietary components, might actually affect transcription and translation, and might thereby influence or control gene expression directly. Such influences may be acting in addition to what you describe as part of the inborn error of metabolism.

Dr. Ogier de Baulny: I agree, and probably this is one of the reasons why human milk has optimal nutrient quality, because besides amino acids there are many other proteins and non-protein nitrogen containing components that may have a role in growth. These are absent from infant formulas.

REFERENCES

1. Bellu R, Ortisti P, Mazzoleni V, *et al.* Nutritional survey on a sample of one-year-old infants in Milan: Intake of macronutrients. *Nutr Res* 1991; 11: 1221–9.
2. Michaelsen KF. Nutrition and growth during infancy. The Copenhagen Cohort Study. *Acta Paedriatr Suppl* 1997; 420: 1–36.
3. Kersting M, Sichert-Hellert W, Alexy U, Manz F, Schöch G. Macronutrient intake of 1 of 18 year old German children and adolescents. *Z Ernährungswiss* 1998; 37: 252–9.

Genetic Expression and Nutrition, edited by Claude Bachmann and Berthold Koletzko. Nestlé Nutrition Workshop Series, Pediatric Program, Vol. 50. Nestec Ltd., Vevey/Lippincott Williams & Wilkins, Philadelphia, © 2003.

How to Assess Amino Acid Requirements (With Particular Reference to Inborn Errors of Amino Acid Metabolism)

Paul B. Pencharz and Ronald O. Ball

Department of Pediatrics, The Hospital for Sick Children; and Department of Nutritional Sciences, University of Toronto, Toronto, Ontario, Canada

The 20 amino acids for which transfer RNA exists are the components from which all proteins are synthesized. If they cannot be synthesized endogenously in sufficient amounts then they are needed in the diet (1). Classically in healthy adults, nine amino acids are regarded as essential (or indispensable) (Table 1). During growth or in disease, several others are regarded as conditionally indispensable; in neonates only five of the 20 can be regarded as fully indispensable (Table 1). If amino acids are not present in the right proportions, then protein synthesis will be reduced (2,3). Furthermore, it has been shown that whole body protein breakdown is increased when less-balanced mixtures of amino acids are fed (4). The endogenous amino acids then provide the missing amino acids for protein synthesis, but net protein accretion and nitrogen balance (retention) is lower when a less than ideal mixture of amino acids is given, either as a protein or as a mixture of free amino acids (4,5).

In inborn errors of amino acid metabolism the requirement for certain amino acids may be altered. For example, in phenylketonuria (PKU), owing to the block in the hydroxylase pathway in the classic form, tyrosine formation from phenylalanine is reduced to less than 2% of normal (6). As a result, tyrosine becomes an indispensable amino acid (7). Similarly, phenylalanine needs are reduced (8).

Until recently, the requirements for the amino acids involved in inborn errors of metabolism have been estimated by plasma amino acid concentrations (8)—phenylalanine concentrations in the case of patients with PKU and branched chain amino acids (BCAA) in patients with maple syrup urine disease. However, plasma amino acid concentrations are generally recognized as insensitive indicators of amino acid requirements (2).

METHODS TO DETERMINE AMINO ACID REQUIREMENTS

Determination of amino acid requirements involves feeding graded levels of the test amino acid to the subject and looking for a clearly definable change in a relevant biologic variable. Classically, outcome variables such as nitrogen balance and growth

TABLE 1. *Categorizing dietary amino acids by their degree of indispensability*

Dispensable	Conditionally indispensable	Indispensable
Alanine	Arginine	Histidine
Aspartate	Cysteine	Isoleucine
Asparagine	Glutamine	Leucine
Glutamate	Glycine	Lysine
Serine	Proline	Methionine
—	Tyrosine	Phenylalanine
—	—	Threonine
—	—	Tryptophan
—	—	Valine

Adapted from Pencharz et al. (1) While this table was first developed for neonates, the amino acids listed as conditionally indispensable may become so in older age groups during times of disease.

have been used in infants and children, and nitrogen balance in adults (2,3,9). Clearly both of these variables are indirect and hence may be insensitive. Furthermore, the body urea pool may need 7 to 10 days to adapt to changes in the dietary intake of protein or amino acids (10). This becomes a problem in infants and children, who cannot be maintained on either deficient or excessive intakes for an extended time (11). Thus, alternative methods are needed.

Although carbon oxidation methods had previously been developed in animals (12–14), it was the pioneering work of Meguid and coworkers (15) which first applied carbon oxidation methods to the determination of essential amino acids in humans. They used the direct oxidation approach in which the test amino acid was also used as the tracer. Their initial studies were in the fed state only, and questions were raised as to whether this was an accurate representation of amino acid requirements on a 24-hour basis. This led them to conduct 24-hour direct oxidation amino acid balance studies (16), which provided similar estimates of leucine requirements compared with the earlier direct oxidation studies (15).

Before the development of the 24-hour balance method, a different carbon oxidation model was introduced (17) to study adult amino acid requirements—namely, indicator amino acid oxidation (IAAO). This model had first been developed and validated in growing pigs, and was shown to provide an estimate of requirement very similar to that determined by the classical techniques of nitrogen balance and growth (13,14). IAAO uses another essential amino acid such as lysine or phenylalanine (in the presence of an excess of tyrosine) as a tracer, which is independent of the changes in the levels of intake of the test amino acid (2,13,14,17–20). As the intake of the limiting amino acid increases from deficient to adequate, the indicator amino acid is partitioned between incorporation into protein and oxidation (2,3,13,14). In the early animal studies, it was shown that the IAAO method did not require adaptation to the level of the test amino acid (13,14), which is a basic requirement for using nitrogen balance to determine amino acid requirements (2,9).

When we decided to adapt the IAAO to humans, one of the first steps was to study the effects of adaptation on the requirement estimate. This was done with subjects fed either

4.2 mg/kg/d or 14 mg/kg/d of phenylalanine for two 9-day periods, during which the test levels of phenylalanine were fed on days 3, 6, and 9. There was no effect of previous adaptation to the two levels of phenylalanine on the rate of oxidation within test intakes of amino acid, and therefore no effect on the estimate of requirement (21).

Based on these studies and the earlier animal work, we concluded that previous adaptation to the level of the test amino acid was not needed. It appears that the response of protein synthesis and oxidation of essential amino acids occurs very rapidly following meals providing varying intakes of one amino acid, compared with the very slow response of whole-body nitrogen pools to a similar change. We have also shown in free living subjects, over a wide range of total protein intake, that it may be necessary to standardize the level of protein intake for 2 days before conducting an indicator study (22). All of these steps are important because they enabled us to develop a minimally invasive indicator model to study amino acid requirements under clinical conditions (11).

Most recently, the concept of a 24-hour indicator amino acid balance (IAAB) has been applied to the indicator oxidation model (23). While this approach is theoretically the most satisfactory method yet developed for determining amino acid requirements, it has the drawback of requiring the subject to be detained for 24 to 29 hours for each study, which compares with the 8 to 10 hours needed for the standard fed-state IAAO study. Furthermore, careful inspection of the 24-hour indicator oxidation data shows that oxidation of the indicator amino acid in the fasted state is constant across the four levels of lysine fed (23), and hence does not alter the requirement determined from the fed oxidation data alone. In addition, the users of the 24-hour IAAB method still believe that adaptation to the level of the test amino acid is needed (23), despite the fact that estimates based on 24-hour amino acid balances, using both direct oxidation and indicator oxidation approaches, are not different from the nonadapted, short-term studies.

Users of the IAAB method have chosen leucine as their indicator amino acid. Up to now we have been reluctant to do this in our IAAO studies, because of the common degradative pathway for the three BCAAs and hence the possibility of interactions between the three. This would introduce an uncontrolled variable that might affect the linearity of the oxidation response to decreasing intakes of the limiting (test) amino acid, which in turn could alter the estimate of requirement for the test amino acid. Studies have been conducted in piglets to show that lysine and phenylalanine give comparable values (2,13,14). Similar studies have not been conducted to confirm that leucine is a satisfactory indicator amino acid.

Our group also introduced a two-phase linear crossover regression analysis for determining amino acid requirements (17,21). This approach provides an objective statistical estimate of the mean requirement and an estimate of the population variance that can be used to calculate a safe dietary intake for the particular test amino acid. Before the introduction of this approach, and indeed for many subsequent experiments, the number of levels of the test amino acid studied had been inadequate for proper statistical derivation of the requirement. For the IAAO and IAAB studies, the subjects are fed amino acid–based diets, which are unpalatable. For the IAAO studies using the minimally invasive model, the subjects are on the test diet only on the day of the oxidation study (7,11). Conversely, for the IAAB study, depending on the length of adaptation, the subjects are on the amino acid based test diet for 8 days for each test level (23). The two-phase method needs each subject to be studied at six or seven levels of the test amino acid. The reason for this large number of levels is to ensure that at least three data points are used for determination of the slope of the line. In practical terms, this means that subjects need to take the study diet for 6 or 7 days using the IAAO model, and for between 54 and 63 days for the IAAB model, as presently used. Clearly the IAAB method is much more demanding and probably totally impractical in children.

The change in the plasma level of the test amino acid has also been used to define essential amino acid requirements, most successfully for tryptophan (24); however, on balance it is an insensitive method and has been supplanted by the carbon oxidation method (9,25–27). Clinically, the blood levels of essential amino acids are used to monitor the control of the diet in patients with PKU or maple syrup urine disease (8). We have found that in children with PKU, the blood concentration of phenylalanine does not change in a quantitative way in response to graded levels of dietary phenylalanine (28). Conversely, the change in blood phenylalanine (fed minus fasting) responds in a linear fashion to phenylalanine intake (28). The possible implication of these data is that clinical monitoring of patients with PKU could be improved by comparing the fed to fasted levels.

Over the past 15 years, since the original direct oxidation studies, there has been controversy over how best to determine essential amino acid requirements and what the correct values are for adults (2,3,9,25–27). In general, the older values, based on nitrogen balance, were two to three times lower than some of the new values derived by carbon oxidation techniques. To aid the reader, a comparison of the methods is shown in Table 2, with a listing of strengths and weaknesses. Strong support for the estimates obtained by direct oxidation of the test amino acid was provided by the introduction of indicator amino acid oxidation (2,17), which is based on different assumptions. Finally, it was realized that part of the problem with the older nitrogen balance estimates was that most of the studies did not examine each individual subject at a series of graded levels. This is important because we have shown in every one of our requirement studies, using repeated measurements on individuals, that the subject is always a highly significant variable. This source of variability can only be accounted for if the subjects are studied at several levels. Reanalysis of one study—in which nitrogen balances were performed in women at different levels of lysine intake, using regression analysis and taking into account miscellaneous nitrogen losses—resulted in a lysine requirement estimate of about 30 mg/kg/d (29). This value is comparable with the estimates obtained using direct oxidation (30), indicator oxidation (17), 24-hour lysine balance (31), and, most recently, 24-hour IAAB (23). Given the convergence of these different methods the carbon oxidation methods are now regarded as having replaced the traditional nitrogen balance technique, at least for studies in adults.

We have thus moved to make the IAAO model as noninvasive as possible (11) so that it can be used in a wide variety of age groups, physiologic states, and clinical conditions. We have shown that the tracer can be given orally (rather than intra-

TABLE 2. *Comparison of methods used to determine amino acid requirements*

Method	Strengths and limitations
Growth	• Useful only in rapidly growing young animals.
Nitrogen balance	• Classical method, but it needs great care to make sure the balances are as precise as possible and miscellaneous losses must also be measured or estimated from literature values. • Requires 5 to 7 days adaptation to each level of amino acid for equilibration of the body nitrogen pools.
Direct oxidation	• Unclear as to whether, or how much, previous adaptation is needed—indeed Zello (21) suggests that at most only 2 days of adaptation are needed. • Can only be used to determine the requirements of amino acids whose carboxyl group is directly released to the bicarbonate pool and so will appear in breath (lysine, phenylalanine, leucine, isoleucine, valine, and possibly threonine). • Fed state studies appear to provide similar requirement estimates to (24 h) balances.
24-h Direct oxdiation	• Unclear as to whether previous adaptation to the level of the test amino acid is needed (see direct oxidation above), but adaptation to the experimental diet is needed for at least 4 before starting the study. • Can only be used to determine the requirements of amino acids whose carboxyl group is directly released to the bicarbonate pool and so will appear in breath (see above).
Indicator oxidation	• There is still some debate as to the best indicator to use and to the present phenylalanine (in the presence of an excess of tyrosine), lysine and leucine (for 24-h indicator balance studies see below) have been used; the only other possible choices are valine and isoleucine. • The evidence (discussed in the text) suggests that previous adaptation to the level of the test amino acid is not needed. • The requirements of any essential amino acid or conditionally indispensable amino acids can also be studied with indicator oxidation.
24-h Indicator balance	• This is a combination of the features of 24-h direct balance with the indicator oxidation approach and so has many theoretical advantages. So far its users have made its use more limited by insisting on adaptation to the level of the test amino acid for 8 days. Leucine was used as the indicator, which has some potential problems that are discussed in the text. • As discussed in the text, it is more invasive and time-consuming for the subjects and costly for the investigative team. As such, unless it is modified its application will be limited to studies in healthy adult volunteers and exceptionally well-funded scientists. • The requirements of any essential amino acid or conditionally indispensable amino acids can also be studied with 24-h indicator oxidation and balance.

venously); that there is no need for previous adaptation to the level of the test amino acid [although a 2-day adaptation to the level of the protein fed may be necessary (22)]; and finally, that the subjects do not have to be adapted to the experimental diet, although if they are not, this extends the length of the study from about 5 hours to about 9 hours. The additional 4 hours are needed to allow for equilibration of the background $^{13}CO_2$ enrichment in the breath, in addition to the 5 hours needed for the oxidation study itself (11). Children with inborn errors of amino acid metabolism are already on standardized dietary protein intakes; therefore, in these subjects even adaptation to the level of protein intake is not needed.

Although we developed the minimally invasive model to permit the study of children with PKU, we have more recently been able to use the model to determine total BCAAs in 6- to 10-year-old healthy children (32). The way is now open to a more general application of this modified IAAO model to the determination of amino acid requirements in vulnerable groups such as infants, children, adolescents, and pregnant and lactating women (3,7,11).

DETERMINATION OF TYROSINE AND PHENYLALANINE REQUIREMENTS IN CHILDREN WITH PKU

PKU is a disorder of aromatic amino acid metabolism in which phenylalanine cannot be converted to tyrosine, or only to a very limited extent (6). Thus, tyrosine is an indispensable amino acid in PKU because it is not supplied endogenously through phenylalanine hydroxylation (33), or only to a very limited degree (6).

The aim of dietary treatment of PKU is to maintain plasma concentrations of phenylalanine, tyrosine, and other amino acids within the normal range, thereby allowing optimal growth and brain development. Treatment consists of supplying adequate energy, other amino acids, and nutrients, while phenylalanine intake is restricted and tyrosine intake is supplemented (8). At present, the dietary management of patients with PKU is empirical because it is based on monitoring plasma amino acid concentrations, blood urea nitrogen, and growth indices, and not on direct measures of tyrosine and phenylalanine requirements.

The recommended daily aromatic amino acid intakes (phenylalanine plus tyrosine) in healthy infants and children (34), and the currently recommended levels for phenylalanine and tyrosine for children with PKU (8), are shown in Table 3 for three age groups. In PKU, the supplemental tyrosine intake when added to the phenylalanine intake far exceeds the recommendations for aromatic amino acid intake in the general, healthy population (8,34). In fact, the median recommended tyrosine intake across the different age groups represents five to seven times the corresponding phenylalanine intake. This suggests that, of the total of aromatic amino acid requirement, phenylalanine contributes around 20% and tyrosine around 80%. These proportions are significantly different from the relative balance of phenylalanine and tyrosine—and their respective contributions to meeting aromatic amino acid requirements—in animals (35–39) and healthy humans (40), in whom dietary tyrosine was shown to spare 40% to 50% of the phenylalanine requirement. This almost-equivalent contribution of phenylalanine and tyrosine to total aromatic amino acid intake is consistent with the plasma (33,41) and mixed body protein (42) ratio of phenylalanine to tyrosine. Therefore, we hypothesized that the recommended phenylalanine and tyrosine intakes for children with PKU were incorrect.

We determined tyrosine requirements using lysine as the indicator amino acid (7). Based on the ratios of phenylalanine to tyrosine described above, we hypothesized

TABLE 3. *Recommended daily aromatic acid (phenylalanine plus tyrosine) requirements in healthy children and those with phenylketonuria*

	Infants	Preschool	Older children
Healthy[a]			
Total aromatics ($mg \cdot kg^{-1} \cdot d^{-1}$)	25–90	69	22
PKU[b]			
Phenylalanine ($mg \cdot kg^{-1} \cdot d^{-1}$)	15–70	15–40	15–35
Tyrosine ($mg \cdot kg^{-1} \cdot d^{-1}$)	250–350	115–200	90–160
Total aromatics ($mg \cdot kg^{-1} \cdot d^{-1}$)	265–420	130–240	105–195

PKU, phenylketonuria.
[a] From FAO/WHO/UNU, 1985 (34).
[b] From Elsas and Acosta, 1994 (8).

that the tyrosine requirement in PKU would account for approximately 45% of the total aromatic amino acid requirement. When tyrosine intake was below the requirement level, lysine oxidation was high. As the tyrosine intake was increased, lysine oxidation fell until a constant low level of oxidation was reached. These data were analyzed by two-phase linear crossover analysis and a breakpoint was defined as the point where lysine oxidation was first minimized. The mean dietary requirement was determined to be between 16.3 and 19.2 mg/kg/d, depending upon whether lysine oxidation or label appearance ($F^{13}CO_2$) was used as the endpoint. These values would represent 40.4% and 44.4% of the total aromatic intake in children with PKU, respectively. The current recommendations for tyrosine intake in patients with PKU are thus overestimated by a factor of around 5. The findings of this study have significant implications for the dietary treatment of individuals with PKU.

Dietary phenylalanine restriction has been the mainstay of the treatment of PKU for more than 40 years (8). Its main aim is to maintain a phenylalanine intake that will allow optimum growth and brain development, by supplying adequate energy, protein, and other nutrients while restricting phenylalanine. Implementation of such a diet soon after birth usually prevents most of the overt clinical manifestations of PKU. Nevertheless, there is evidence that neuropsychologic and cognitive function is not entirely normalized in individuals with PKU receiving current treatment regimens (43–46). Present estimates of phenylalanine requirements are based on plasma phenylalanine levels and growth rate in relation to dietary intake, and not on direct and sensitive measurements of amino acid metabolism.

Because patients with classical PKU have a negligible or minimal capacity to oxidize phenylalanine (6), we reasoned that their dietary phenylalanine requirements would be lower than those in healthy children by an amount equal to the obligatory losses of phenylalanine. Dietary requirements for phenylalanine in healthy children or in those with PKU have yet to be defined by direct measurements of amino acid metabolism. We therefore referred to our previous study using indicator amino oxidation, which showed that the mean tyrosine requirement of children with PKU was 19.2 mg/kg/d (7). Next we used a ratio of 55:45 between phenylalanine and tyrosine in the tissues of humans and animals (18,21,35,36,47), which, when factored to the mean tyrosine requirements, predicts a mean phenylalanine requirement for healthy children of approximately 23.5 mg/kg/d. The requirement for phenylalanine or any other indispensable amino acid is the sum total of that needed for protein synthesis plus irreversible losses (10). Because there are no data in children, we had to turn to a study of phenylalanine requirements in adult male subjects (21), in whom obligatory oxidation was estimated to be approximately 26%. Using the predicted phenylalanine requirement for children, and the estimated obligatory oxidation for phenylalanine, the obligatory loss was calculated to be approximately 6.1 mg/kg/d. Subtracting the estimate for obligatory loss from the predicted mean requirement of 23.5 mg/kg/d, the resulting value of 17.4 mg/kg/d became the mean phenylalanine requirement predicted for children with PKU.

The objectives of this study were to determine the phenylalanine requirement of children with classical PKU using the technique of indicator amino acid oxidation, and to compare the results with our previous estimate of the tyrosine requirement obtained using the same technique. This study has only been presented in abstract form (28), but the mean estimate of phenylalanine needs is 14 mg/kg/d. This compares with the tyrosine estimate obtained using the same technique of label oxidation ($F^{13}CO_2$) of 19.2 mg/kg/d. Therefore in children with PKU we estimate that the total mean aromatic amino acid requirement is about 33 mg/kg/d, and that the proportions of phenylalanine and tyrosine are 42% and 58%, respectively. This is the reverse of the relative proportions in the tissues of normal individuals (47).

DETERMINATION OF AMINO ACID REQUIREMENTS IN OTHER INBORN ERRORS OF AMINO ACID METABOLISM

Clearly from the success of the minimally invasive indicator oxidation method (11) in defining tyrosine (11) and more recently phenylalanine needs (28) in children, the way is open to using this technique in studying the next most prevalent inborn error of amino acid metabolism, maple syrup urine disease (8). Before attempting to study patients with this condition, we felt it necessary to determine total branched chain amino acid needs in healthy children (32) and adults (48). Previous work by others had only determined requirements for leucine and valine (26), and has merely estimated isoleucine requirements. We reasoned that, because of the common degradative pathway for the three BCAAs and the known interactions between them, it is better to study them collectively, in a balance approximating that in egg protein, using the indicator amino acid oxidation model.

Now that we have control data, we are poised to study patients with maple syrup urine disease. Again our reasoning is similar to that already outlined for the study of phenylalanine requirements in children with PKU—namely, to try and estimate obligatory oxidation of the three BCAAs and subtract that value from the requirement estimates made in healthy children and adults. This theoretical exercise is important because it allows the estimation of levels of intake above and below the predicted requirement breakpoint. The statistical definition of the breakpoint is optimized when the levels are evenly distributed so that three are above it and three below.

SUMMARY AND CONCLUSIONS

The advent of the minimally invasive indicator amino acid oxidation model has opened the way to objective and sensitive determination of amino acid requirements in patients with inborn errors of amino acid metabolism. Requirements for tyrosine and phenylalanine have been determined for children with PKU. From these estimates, it is possible to state that published estimates of tyrosine requirements in PKU are a five-fold overestimate. Using our minimally invasive IAAO method, it is now possible to extend quantitative studies of amino acid requirements into vulnerable groups such as infants, children, pregnant women, and patients with diseases.

ACKNOWLEDGMENTS

This research was supported by a grant from the Canadian Institutes of Health Research (formerly the Medical Research Council of Canada) (grant MT 10321). Mead Johnson Canada generously provided the protein-free powder.

REFERENCES

1. Pencharz PB, House JD, Wykes LJ, Ball RO. What are the essential amino acids for the preterm and term infant? In: *10th Nutricia Symposium*. Dordrecht: Kluwer Academic Publishers, 1996; 21: 278–96.
2. Zello GA, Wykes LJ, Ball RO, Pencharz PB. Recent advances in method of assessing amino acid requirements for adult humans. *J Nutr* 1995; 125: 2907–15.
3. Brunton JA, Ball RO, Pencharz PB. Determination of amino acid requirements by indicator amino acid oxidation: applications in health and disease. *Curr Opin Clin Nutr Metab Care* 1998; 1: 449–53.
4. Duffy B, Gunn T, Collinge J, Pencharz PB. The effect of varying protein quality and energy intake on the nitrogen metabolism of parenterally fed very low birth weight. *Pediatr Res* 1981; 15: 1040–44.
5. Fuller MF, Reeds PJ, Cadenhead A, Seve B, Preston T. Effects of the amount and quality of dietary protein on nitrogen metabolism and protein turnover in pigs. *Br J Nutr* 1987; 58: 287–300.
6. Van Spronsen FJ, Reijngoud DJ, Smit GP, *et al*. Phenylketonuria: the in vivo hydroxylation rate of phenylalanine into tyrosine is decreased. *J Clin Invest* 1998; 101: 2875–80.
7. Bross R, Ball RO, Clarke JTR, Pencharz PB. Tyrosine requirements in children with classical PKU determined by indicator amino acid oxidation. *Am J Physiol* 2000; 278: E195–201.
8. Elsas LJ, Acosta PB. Nutritional support of inherited metabolic diseases. In: Shils ME, Olson JA, Shike M, eds. *Modern nutrition in health and disease*. Malvern: Lea & Febiger, 1994: 1147–206.
9. Fuller MF, Garlick PJ. Human amino acid requirements: can the controversy be resolved? *Annu Rev Nutr* 1994; 14: 217–41.
10. Rand WM, Young VR, Scrimshaw NS. Change in urinary nitrogen excretion in response to low-protein diets in adults. *Am J Clin Nutr* 1976; 29: 639–44.
11. Bross R, Ball RO, Pencharz PB. Development of a minimally invasive protocol for determination of phenylalanine and lysine kinetics in humans during the fed state. *J Nutr* 1998; 128: 1913–19.
12. Waterlow JC, Stephen JM. Adaptation of the rat to a low-protein diet: the effect of a reduced protein intake on the pattern of incorporation of L-(14C) lysine. *Br J Nutr* 1966; 20: 461–84.
13. Kim KI, McMillan I, Bayley HS. Determination of young pigs using and indicator amino acid. *Br J Nutr* 1983; 50: 369–82.
14. Ball RO, Bayley HS. Tryptophan requirement of the 2.5-kg piglet determined by the oxidation of an indicator amino acid. *J Nutr* 1984; 114: 1741–6.
15. Meguid MM, Matthews DE, Bier DM, Meridith CN, Soeldner JS, Young VR. Leucine kinetics at graded leucine intakes in young men. *Am J Clin Nutr* 1986; 43: 770–80.
16. El-Khoury AE, Fukagawa NK, Sanchez M, *et al*. The 24-h pattern and rate of leucine oxidation, with particular reference to tracer estimates of leucine requirements in healthy adults. *Am J Clin Nutr* 1994; 59: 1012–20.
17. Zello GA, Pencharz PB, Ball RO. The dietary lysine requirement in humans determined by the oxidation of an indicator amino acid, L-(1-13C) phenylalanine. *Am J Physiol* 1993; 264: E677–85.

18. Roberts SA, Thorpe JM, Ball RO, Pencharz PB. Tyrosine requirement of healthy adult males receiving a fixed phenylalanine intake determined using amino acid oxidation. *Am J Clin Nutr* 2001; 73: 276–82.
19. Lazaris-Brunner G, Rafii M, Ball RO, Pencharz P: Tryptophan requirement in young adult women determined by indicator amino acid oxidation, using L-(13C)-phenylalanine. *Am J Clin Nutr* 1998; 68: 303–10.
20. Wilson D, Rafii M, Ball RO, Pencharz PB. Threonine requirement in young adult men determined by indicator amino acid oxidation, using L-(1-13C)-phenylalanine. *Am J Clin Nutr* 2000; 71: 757–64.
21. Zello GA, Pencharz PB, Ball RO. Phenylalanine flux, oxidation and conversion to tyrosine in humans studied with L-(1-13C)-phenylalanine. *Am J Physiol* 1990; 259: E835–43.
22. Thorpe JM, Roberts SA, Ball RO, Pencharz PB. Effect of prior protein intake on phenylalanine kinetics. *J Nutr* 1999; 129: 343–8.
23. Kurpad AV, Raj T, El-Khoury AE, *et al.* Lysine requirements of healthy adult Indian subjects, measured by indicator amino acid balance technique. *Am J Clin Nutr* 2001; 73: 900–7.
24. Tontisirin K, Young VR, Scrimshaw NS. Plasma tryptophan response curve and tryptophan requirements of children with Down's syndrome. *Am J Clin Nutr* 1972; 25: 976–9.
25. Millward DJ. Metabolic demands for amino acids and the human dietary requirement: Millward and Rivers (1988) revisited. *J Nutr* 1998; 128: 2563–76S.
26. Young VR, Borgonha S. Nitrogen and amino acid requirements: the Massachusetts Institute of Technology amino acid requirement pattern. *J Nutr* 2000; 130: 1841–9.
27. Metges CC, Petzke KJ, Young VR. Dietary requirements for indispensable amino acids in adult humans: new concepts, methods of estimation, uncertainties and challenges. *Ann Nutr Metab* 1999; 43: 267–76.
28. Courtney-Martin G, Rafii M, Ball RO, Clarke JTR, Bross R, Pencharz PB. Phenylalanine requirement in children with clasical phenylketonuria [abstract]. *FASEB J* 1999; 13: A678.
29. Rand WM, Young VR. Statistical analysis of nitrogen balance data with reference to the lysine requirements in adults. *J Nutr* 1999; 129: 1920–6.
30. Meridith CN, Wen ZM, Bier DM, Matthews DE, Young VR. Lysine kinetics at graded lysine intakes in young men. *Am J Clin Nutr* 1986; 43: 787–94.
31. El-Khoury AE, Pereira PC, Borgonha S, *et al.* Twenty-four-hour oral tracer studies with L-(1-^{13}C)lysine at a low (15mg.kg^{-}1.d^{-}1) and intermediate (29 mg.kg^{-}1.d^{-}1) lysine intake in healthy adults. *Am J Clin Nutr* 2000; 72: 122–30.
32. Mager R, Ball RO, Pencharz PB. Branched chain amino acid requirements in school aged children determined by indicator amino acid oxidation [abstract]. *FASEB J* 2001; 15: A266.
33. Scriver CR, Kaufman S, Woo SL. The hyperphenylalanemias. In: Scriver CR, Beaudet AL, Sly WS, Valle D, eds. *The metabolic basis of inherited disease.* New York: McGraw-Hill, 1989: 495–546.
34. FAO/WHO/UNU. *Energy and protein requirements.* Geneva: World Health Organization, 1985.
35. House JD, Pencharz PB, Ball RO. Tyrosine kinetics and requirements during total parenteral nutrition in the neonatal piglet: the effect of glycyl-L-tyrosine supplementation. *Pediatr Res* 1997; 41: 575–83.
36. House JD, Pencharz PB, Ball RO. Phenylalanine requirements determined by using L-(1-^{14}C) phenylalanine in neonatal piglets receiving total parenteral nutrition supplemented with tyrosine. *Am J Clin Nutr* 1997; 65: 984–93.
37. Milner JA, Garton RL, Burns RA. Phenylalanine and tyrosine requirements of immature beagle dogs. *J Nutr* 1984; 114: 2212–16.
38. Stockland WL, Lai YF, Meade RJ, Sowers JE, Oestemer G. L-phenylalanine and L-tyrosine requirements of the growing rat. *J Nutr* 1971; 101: 177–84.
39. Williams JM, Morris JG, Rogers QR. Phenylalanine requirement of kittens and the sparing effect of tyrosine. *J Nutr* 1987; 117: 1102–7.
40. Burrill LM, Schuck C. Phenylalanine requirements with different levels of tyrosine. *J Nutr* 1964; 83: 202–8.
41. Mitchell GA, Lambert G, Tanguay RM. Hypertyrosinemia. In: Scriver CR, Beaudet AL, Sly WS, Valle D, eds. *The metabolic basis of inherited disease.* New York: McGraw-Hill, 1995: 1077–106.
42. Munro HN, Fleck A. Analysis of tissues and body fluids for nitrogenous constituents. In: Munro HN, ed. *Mammalian protein metabolism.* Vol III. New York: Academic Press, 1969: 423–525.
43. Smith I, Beasley MG, Ades AE. Intelligence and quality of dietary treatment in phenylketonuria. *Arch Dis Child* 1990; 65: 472–8.
44. Thompson AI, Smith I, Brenton D, *et al.* Neurological deteriorations in young adults with phenylketonuria. *Lancet* 1990; 336: 602–5.
45. Thompson AJ, Smith I, Kendall BE, Youl BD, Brenton D. Magnetic resonance imagine changes in early treated patients with phenylketonuria. *Lancet* 1991; 337: 1224.

46. Thompson AJ, Trittotson S, Smith I, Kendall B, Moore SG, Brenton DP. Brain MRI changes in phenylketonuria: association with dietary status. *Brain* 1993; 116: 811–82.
47. Widdowson EM. Changes in body composition during growth. In: David J, Dobbing J, eds. *Scientific foundations of paediatrics*. London: Heinemann, 1981: 330–42.
48. Riazi R, Ball RO, Pencharz PB. Total branched chain amino acid requirements in young healthy adult men determined using indicator amino acid oxidation using L(1-13C) phenylalanine [abstract]. *FASEB J* 2001; 15: A266.

DISCUSSION

Dr. Koletzko: You said that adaptation occurs within a few hours. We know from many other examples that the organism has several ways to adapt to variations in substrate intake. I wonder how good the arguments are that this adaptation really occurs within a few hours, and there is no effect of interpretation on the measurements.

Dr. Pencharz: Well the estimates obtained for lysine—which has been studied the most—show that no adaptation *versus* adaptations for as long as 40 days gives you the same estimate, so in practical terms it looks as though the minimally invasive model with 6 hours of adaptation is sufficient. For phenylalanine, we've shown that it makes no difference whether you adapt for no time, 2 days, 6 days, or 9 days. So whereas with nitrogen balance it makes a very big difference, the adaptation of amino acid metabolism is very rapid, a matter of hours.

Dr. Endres: I think the motivation behind ranges of tyrosine intake in infants or children wit PKU is quite variable, and some people give additional tyrosine (1). When you determined the breakpoint, as you call it, does that really answer the question of how much tyrosine is needed by a PKU patient? It's not the tyrosine that matters, it's the dopamine that's formed. Could you comment on that? I think your answer is perhaps a little too easy for this disease.

Dr. Pencharz: Your comment is well taken and I apologize if I wasn't sufficiently clear in my caveats. I said that the indicator would suggest how much is needed for protein synthesis; it won't necessarily measure the other pathways. I did mention, however, that Joe Clarke, who is a coinvestigator in our PKU studies, had been involved in a study where the endpoint was neurologic behavior, and was unable to show any benefit in a double-blind, placebo-controlled trial with 100 mg of tyrosine (2). I am aware that there are other people who don't agree, and we could argue at length about the validity of various data. I think there are different endpoints, and I remain open to using others. But in terms of whether you need additional tyrosine to deal with competition from large neutral amino acids affecting uptake into cells and then the synthesis of dopamine—or other brain endpoints—your point is well taken. I accept that's not the end of the story.

Dr. Bachmann: Could you also clarify what protein synthesis we are talking about? In an adult, there is about 15 kg of muscle mass, 1.5 kg of liver, a similar amount of brain, and the kidneys. Are there not some functionally important protein synthesizing organs that might be completely ignored if we look only at the total sum? I'm not criticizing your approach, but perhaps it won't show up the weakest points in the process of protein synthesis in tissues that have a rapid turnover. We have only averages, which are dominated by muscle.

Dr. Pencharz: I agree that although we talk about, let's say, lysine being 7% or 8% of mixed body protein, the amount of lysine in different proteins in the body varies quite markedly, and this applies to other amino acids. All available methods ultimately reflect what's happening at a whole body level. None of the studies that I've referred to looks at something occurring across skeletal muscle or across the kidney. These techniques are analogous to resting metabolic rate for energy metabolism. We are not trying to determine the tyrosine requirement of the kidney; rather, we are attempting to determine the tyrosine requirement for the whole body. From a dietary point of view, we are really saying how much the diet should contain. In energy expenditure studies, we can only say how much energy the whole body needs, not what is the energy expenditure of the heart or liver. These protein studies are very analogous. Whether they involve nitrogen balance, direct oxidation, indicator techniques, or whatever, they are still reflections of whole body metabolism in an integrated fashion. However, the minimally invasive technique can be applied, for example, to patients with chronic renal disease. Dr. Scott talked about the sulfur amino acids earlier, and I would say that all methionine requirements are reduced if you have chronic renal disease, because the enzymes and pathways for sulfur amino acid metabolism exist in the kidney as well as in the liver. Therefore, at a dietary level you can use this technique to define whether methionine requirements are indeed reduced if you have chronic renal insufficiency.

Dr. Micheli: Eighty percent of very preterm babies receive antenatal corticosteroids. By giving these, we modify genetic programming not only of the lungs but also of the circulation in the heart and brain. Can you tell us how this interferes with amino acid metabolism and requirements?

Dr. Pencharz: You raise a good point. Stephanie Atkinson, who works down the road from me in Hamilton, has shown very clearly that antenatal steroids have adverse affects on skeletal muscle. This is likely to be detrimental in such infants, in whom skeletal muscle is already rather a small component of total body protein. I don't know of any good studies that show what this does to the leucine turnover *versus* phenylalanine, and so on. It is a problem in examining such effects that we have very few data on requirements for essential amino acids in children—only three studies in fact (3), which are the sum total of worldwide studies in this area.

Dr. Bier: I'd like to comment on Dr. Micheli's question. I don't believe anybody has systematically studied amino acid kinetics in newborn infants as a function of maternal steroid administration. We've studied amino acid kinetics in a lot of neonates and preterm infants for other reasons, and without systematic analysis, we've been unable to distinguish between the mothers who had received antenatal steroids and those who had not on the basis of kinetic events measured in those neonates.

Dr. Scott: Dr. Pencharz, your model predicts that the rate of protein synthesis is relatively unchanged over the periods of measurement. That surprises me a bit because I would have thought the intake of other nutrients—carbohydrates and so on—would change the levels of hormones such as insulin and glucagon, which would have catabolic or anabolic effects. Do you control for other macronutrients during the study periods? I think you would expect some fluctuation in protein synthesis.

Dr. Pencharz: You are quite right. Experimentally what one does is to feed the subjects at hourly intervals, which minimizes fluctuation. This subject was explored in a recent article in the *American Journal Clinical Nutrition* (4). There are differences in the rates of oxidation of the indicator amino acid (the investigators used leucine in that paper) in the fasted state *versus* the fed state. However, they tried to keep as many variables constant as possible to come up with the kinds of answers they did.

Dr. Baerlocher: Did you mention the phenylalanine concentration in your patients with PKU when you did your study with tyrosine? High levels of phenylalanine could influence tyrosine metabolism.

Dr. Pencharz: The subjects were all maintained in what, in Toronto, is regarded as good control—that is, plasma phenylalanine in the range of 400 to 600 μmol/l. We did a *post hoc* correlation analysis and couldn't find any effect of phenylalanine concentration.

Dr. Bier: You went rather quickly through the fact that lysine oxidation went up in the phenylalanine study once you got above the requirement level. Would you like to discuss that? It shouldn't happen theoretically.

Dr. Pencharz: It was a big surprise to us. The way we are presently interpreting this is as follows. The indicator study is dependent on the concept that the indicator amino acid is partitioned between on one hand, incorporation into protein and, on the other hand, oxidation. Because oxidation increased statistically, that implies that something is happening within the whole body of those children as you give phenylalanine intakes above 14 mg/kg/d, and that the utilization of the indicator—that is lysine—for protein synthesis is less and hence there is more oxidation. This is really saying that whole body protein synthesis is being inhibited. Why should that be? Well, we are saying in effect that the availability of all 20 amino acids at the point of polyribosomes for protein synthesis is somehow being affected. Being more specific is impossible based on our data. From other published reports, you can hypothesize that as you increase plasma phenylalanine, you may affect the uptake of tryptophan, for example. Tryptophan may then become a limiting factor intracellularly in terms of protein synthesis. This is all speculation.

Dr. Kern: You gave very precise recommendations for intake of amino acids based on hourly feeding. In practice, patients with PKU take their powder supplements two or three times a day. Does it matter if they take their amino acid mixture in this way and not every hour?

Dr. Pencharz: This has been looked at by various people, including VernonYoung. The answer is that it probably doesn't matter. The model that we're using, where we try to keep everything else as constant as possible by feeding hourly, still seems to reflect the requirements even when you switch to meal feeding.

Dr. Superti-Furga: I think your observation that protein synthesis goes down when you increased the phenylalanine intake is very interesting and may be a clue to the pathogenesis of the disorder. Did you cause these controlled subjects to lose their metabolic control by giving the higher dosage?

Dr. Pencharz: No, they didn't go out of clinical control. We figured out the highest quantity we could safely use. We didn't see any metabolic consequences, any change in acid–base status, electrolytes, and so on.

Dr. Superti-Furga: I think this is very important because here in Switzerland we're still struggling to find the optimal target for phenylalanine blood values. Knowing that increasing the phenylalanine intake, even if this is not reflected at blood level, may have adverse effects is very significant. Thank you.

Dr. Pencharz: Clearly there will be the adverse effects that we are all aware of in terms of the central nervous system, but we were just trying to see whether we could define at a biochemical or physiologic level what the requirements are. They appear to be lower than previously reported, which I think is worth knowing, but now we have to decide how to apply the knowledge. There's work still to be done before we can apply it to a PKU population.

Dr. Borum: I understand that the experimental design tightly controls carbohydrate and lipid intake and involves hourly feeding, but what is the impact of carbohydrate and lipid intake on protein metabolism and protein requirements in particular situations such as childhood obesity? If a child has an excessive energy intake, is there an altered amino acid requirement? And, what about other situations such as ketogenic diets for seizures or diets for various inborn errors of metabolism?

Dr. Pencharz: I can answer that in various ways. We directly studied this in parenterally fed neonates where we held amino acids constant and varied the nonprotein energy, either giving glucose only or replacing 33% of the glucose energy by fat. In that particular model, we were unable to show any differences in protein turnover. We weren't defining requirements, merely glycine and leucine turnover. Therefore, we weren't able to show any differences. When we looked at the literature on dietary requirements, the main question seemed to be what is the minimum carbohydrate intake, because we know that the carbon skeletons of protein are necessary for gluconeogenesis. Providing you give enough carbohydrate to meet the minimum necessary for glucose production, that seems to be all that is required. The balance between fat and carbohydrate doesn't appear to matter.

Dr. Wahba: In a pregnant woman with PKU who is on a phenylalanine-restricted diet, what is the prognosis for the baby with regard to intellectual development?

Dr. Pencharz: It has been well established that if you don't treat the mother the baby will be brain damaged. These babies do not grow as well, and their brains are abnormal. If the woman is on a properly controlled phenylalanine intake during pregnancy, the outcome appears to be normal.

Unidentified participant: You show that a safe level of phenylalanine in patients with PKU is around 17 to 20 mg/kg/d. Have you studied this "safe" level in terms of neurologic outcome?

Dr. Pencharz: You are asking me whether, having defined this breakpoint, we have then gone on and done long-term studies while feeding children at that level. The answer is no, and that's what I meant when I said that more work needs to be done. The phenylalanine data have not yet been published in a peer-reviewed journal, but we are presently thinking about doing studies with tighter control.

Dr. Fowler: I'm sure we still need to measure blood phenylalanine levels to control our patients and not just rely on this worked out level of minimum requirements. I have a question: can you extrapolate this new level of phenylalanine that you've elegantly determined to a minimum level of protein requirement, as we discussed in the previous talk?

Dr. Pencharz: Not really. Protein requirements come down to the issue of how much you need of the other 19 amino acids.

Dr. Fowler: Perhaps my approach is naïve, but if I take your 20 mg of phenylalanine as a minimum requirement, and if we assume that 4% of protein is phenylalanine, then you come up to level of 500 mg/kg as the minimum protein requirement, which is not far off Hélène Ogier's value of 600 mg.

Dr. Pencharz: Well, that assumes 100% utilization. Peter Reeds and I reviewed the world literature on this fairly thoroughly, and *maintenance*—that is, no growth at all—is about 680 mg/kg. Whether you actually need 1.5 g isn't clear, and there is also the issue of activity. Exercise probably doesn't make a big difference, but it may increase your value to around 800 mg/kg.

Dr. Koletzko: We need to be cautious about implying that the proportions of amino acids in proteins determine the metabolic requirements in the organism. Also, in the real world, at least in infants, we not only have the true protein nitrogen intake, but we also have quite a lot of non-protein nitrogen, and we don't really know to what extent that is utilized by the organism.

Dr. Bier: I want to advise caution about the use of the term "requirement." We use it to mean a variety of things. For example, there's the maintenance requirement; there's the minimum requirement for something or other, which might include growth; there's an estimated average requirement; and there's a recommended dietary allowance. They're all remarkably different, and we have used the word "requirement" very casually throughout the discussions.

Dr. Pencharz: We're talking about studies involving means for a population. When you talk about going from a mean to a *requirement* for a population, you usually imply the mean plus two standard deviations. The value of 0.68 g/kg which I just gave, which rounds up to 0.7, becomes 1.1 g when you add two standard deviations, so this is the population requirement rather than the mean estimate for an individual.

REFERENCES

1. Lou HC, Likkelund C, Gerdes AM, Udesen H, Bruhn P. Increased vigilance and dopamine synthesis by large doses of tyrosine or phenylalanine restriction in phenylketonuria. *Acta Paediatr Scand* 1987; 76: 560–5.
2. Smith ML, Hanley WB, Clarke JT.. Randomised controlled trial of tyrosine supplementation on neuropsychological performance in phenylketonuria. *Arch Dis Child* 1998; 18: 116–21.
3. FAO/WHO/UNU. *Energy and protein requirements*. Geneva: World Health Organization, 1985.
4. Kurpad AV, Raj T, El-Khoury A, *et al.* Lysine requirements of healthy adult Indian subjects, measured by an indicator amino acid balance technique. *Am J Clin Nutr* 2001; 73: 900–7.

Genetic Expression and Nutrition, edited by Claude Bachmann and Berthold Koletzko. Nestlé Nutrition Workshop Series, Pediatric Program, Vol. 50. Nestec Ltd., Vevey/Lippincott Williams & Wilkins, Philadelphia, © 2003.

Long-Chain Omega-3 Fatty Acids are Conditionally Essential Substrates in Children: Effects on Visual Function in Children With Phenylketonuria

Berthold Koletzko, Skadi Beblo, Ania Muntau-Heger, *Wolfgang Mueller-Felber, and Hannes Reinhardt

*Division of Metabolic Diseases and Nutritional Medicine, Dr. von Hauner Children's Hospital, Kinderklinik and Kinderpoliklinik, University of Muenchen, Muenchen, Bavaria; and *Friedrich-Baur-Institut, Ludwig-Maximilians-University of Munich, Germany*

Classical phenylketonuria (PKU) is one of the most common inborn errors of metabolism and affects about 1 in 7,000 white neonates. PKU is caused by deficient activity of hepatic phenylalanine hydroxylase. Untreated PKU causes severe psychomotor handicap from early childhood onward (1), but if the diagnosis is made by neonatal screening and dietary treatment is initiated early within the neonatal period, largely normal physical and cognitive development can be achieved. The therapeutic diet is based on a strictly limited intake of natural proteins, and is adapted to the patient's individual phenylalanine tolerance. It is supplemented with phenylalanine-free synthetic amino acid mixtures to meet the age-adapted nitrogen requirements.

Although this approach is of great benefit, several studies have indicated that even when treated, patients with PKU continue to have subtle but detectable neurologic deficits. In particular, they tend to have slightly lower intelligence quotients than their healthy siblings (2,3), with inferior school achievement (4). Many patients suffer from a reduced ability to concentrate and show prolonged reaction times (5,6).

The reason for these subtle dysfunctions is not clear. While fluctuations in plasma phenylalanine concentrations influence cognitive performance (6), additional variables seem to contribute as well. The mainstay of treatment—strict dietary management—may induce metabolic imbalances. Depletion of various vitamins and trace elements has often been reported in PKU patients (7–12).

The long-chain polyunsaturated fatty acids (LC-PUFAs)—which include the long-chain ω-3 fatty acid docosahexaenoic acid (C22:6n-3; DHA)—form an important group of micronutrients that is usually in poor supply in PKU patients. The natural food sources of LC-PUFA are meat, liver, fish, and eggs, which are rich in protein and hence are avoided in the PKU diet. Plasma or serum concentrations of LC-PUFA are low in patients with classical PKU (13–15). LC-PUFAs are essential components of all cell membranes and modulate membrane functions (16–18). Animal studies have clearly established that the availability of LC-PUFAs during early life affects visual and behavioral development (19,20).

Long-chain ω-3 fatty acids influence gene expression by interacting with several nuclear transcription factors, such as the peroxisome proliferator activator receptor α (PPARα), the nuclear factor κB, the sterol regulatory binding protein 1, and the polyunsaturated fatty acid regulatory element. Recent experimental studies indicate that DHA is the most potent ligand and activator of the retinoid X receptor in brain tissue and hence regulates its gene expression, a signaling pathway that is thought to influence neural function (21).

In both premature and full-term infants, several controlled clinical trials have shown benefits for visual and psychomotor development from dietary supplementation with preformed LC-PUFA (22–29). However, little is known about the potential effects of LC-PUFA status on neurologic function beyond infancy. We hypothesized that children with PKU who have a poor LC-PUFA supply may benefit from supplementation with fish oil, which provides n-3 LC-PUFA. We chose as the primary outcome variable the latency of visual evoked potentials (VEPs), because VEPs can be reliably and repeatedly measured, are largely independent of the subject's cooperation and motivation, and reflect a representative aspect of the information processing capacity of the central nervous system.

TRIAL DESIGN

Our clinical trial was performed in patients with classical PKU aged between 1 and 11 years (30). Inclusion was restricted to patients with a documented average phenylalanine level, determined at least monthly over the previous 6 months, of less than 360 μmol/l (<6 mg/dl). This inclusion criterion was chosen because the mere participation in a clinical trial tends to motivate patients with chronic diseases to adhere more strictly to their recommended treatment (31). In this trial, such an effect might have influenced the results because of the impact of acute or chronic increases in phenylalanine concentrations in plasma and tissues on brain function (1,6). To avoid this potential confounding effect, we chose to include only patients with optimal metabolic control. We excluded patients with additional diseases other than PKU, or with overt abnormal physical or neurologic signs.

At baseline, patients underwent a detailed physical and neurologic examination, blood was taken for routine blood tests and estimation of phenylalanine concentration, and VEPs were measured as described below.

The protocol was reviewed by the ethics committee of the Bayerische Landesaerztekammer (Bavarian Board of Physicians), Munich, Germany. The aims and nature of the study were explained in detail to all parents or guardians, as well as to the children who were school-age or older, and written informed consent was obtained from the parents.

VISUAL EVOKED POTENTIALS

VEPs were measured using a Toennies NeuroScreen 1.63 (Jaeger-Toennies, Hoechberg, Germany). Subjects were exposed to a 3.2-Hz alternating black-and-white checkerboard pattern displayed on a video screen. Pattern size was modified to achieve stimulation of the fovea (5′), the entire retina (30′), and two intermediate sizes (10′ and 15′). All recordings were performed in duplicate with 50 stimulations each. Small children were sitting on their parent's lap to achieve optimal position and relaxation. Latencies of the P100 potential were determined independently by two experienced physicians according to standard criteria.

SUPPLEMENTATION

After the baseline examination, patients were supplied with fish oil capsules (Ameu; kindly provided by Omega-Pharma, Berlin, Germany) in a dose providing 15-mg DHA per kilogram of body weight daily. Each capsule contained 500 mg of purified salmon oil, which yielded 35% of the fatty acids as ω-3, including 18% as eicosapentaenoic acid and 12% as DHA. The gelatin coating of the capsules contained 3 mg of phenylalanine. All other aspects of dietary treatment remained unchanged. After 90 days of fish oil supplementation, a follow-up examination was performed in the same way as the baseline examination.

HEALTHY CONTROLS

Healthy age-matched children were examined as controls, provided they were on a mixed diet; children on a vegetarian or other special diet were excluded. Controls were examined clinically, and VEPs were measured at baseline and after 90 days in a subgroup who volunteered to participate in a second examination. For ethical reasons, no blood sample was taken in control children, and they did not receive fish oil supplements.

SUBJECTS

Of the 65 patients in the respective age range treated at this center and screened for participation in the study, 38 patients were included, of whom 36 completed the protocol (age 6.3 ± 0.55 years; 19 girls and 17 boys; Fig. 1) (30). The diagnosis of PKU had been established by neonatal screening and was confirmed by further clinical testing and molecular genetics. Dietary treatment was initiated within the first 3 weeks of life in each patient. All patients were regularly followed up in our outpatient clinic for inborn metabolic diseases and were on a strictly protein-restricted diet. The protein substitution by a synthetic amino acid mixture followed the recommendations of the German Society of Nutrition for protein intake in children, with a 20% surplus to account for possible differences in the biologic value of synthetic amino acid mixtures relative to natural proteins. Thirty children aged 6.6 ± 0.5 years (15 girls and 15 boys) volunteered as control subjects.

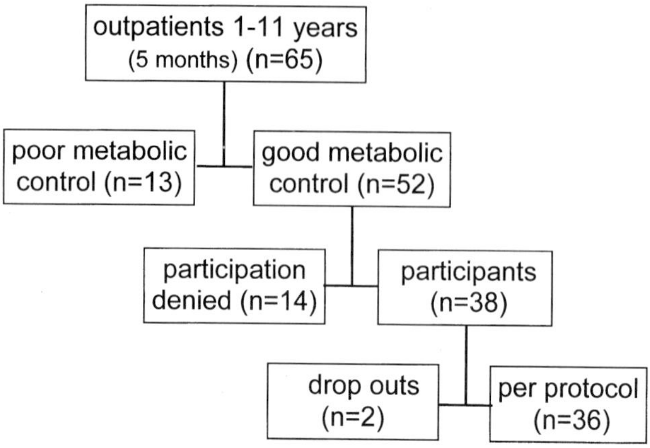

FIG. 1. Flow chart of patient selection and treatment.

DATA ANALYSIS

Data were analyzed for comparisons between patients and controls using Student *t* test; *t* tests for paired samples were used to evaluate the treatment effects in the patients. The level of statistical significance was set at $p < 0.05$. Data are given as mean ± SEM unless otherwise stated.

RESULTS

VEPs were recorded in 36 patients with PKU and in 30 healthy controls. The quality of the VEP recordings was very satisfactory, with no appreciable difference between duplicate recordings of each checkerboard pattern. However, some recordings could not be evaluated with sufficient reliability, usually due to inattentiveness, which led to patterns with small, indistinguishable amplitudes. Thus, the number of VEPs evaluated differs from the number of subjects investigated. The evaluations of the two examiners agreed in 65% to 85% of the tracings that could be analyzed reliably. Only these recordings were accepted for further analysis.

The mean latency of VEP P100 was significantly longer in patients with PKU than in the controls in all stimulation modes (Fig. 2). The largest and most significant difference of the P100 latency was shown for isolated foveal stimulation with a very small checkerboard pattern (5′), with a difference between the mean latencies of 9.8 ms. Although the difference between the mean latencies of the other stimulation patterns used was smaller, the results show that the abnormality is not confined to the fovea and its fiber tracts.

In patients with PKU, supplementation with fish oil capsules for 90 days led to a shortening of P100 latencies. Mean latencies were shorter than at baseline in all stimulation modes. Statistically significant differences were noted for the isolated foveal stimulation (5′) and for one of the intermediate sizes (15′) (Fig. 3). Plasma phenylalanine concentrations did not change significantly during the intervention period,

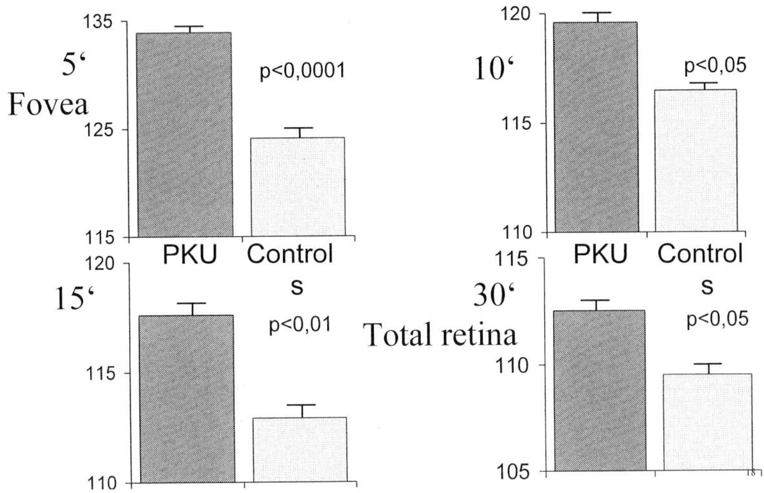

FIG. 2. Latencies of visual evoked potentials (VEP) in 36 children with phenylketonuria (hatched bars, left) are significantly longer than in 30 age-matched healthy controls (stippled bars, right).

even though the fish oil capsules contained small amounts of phenylalanine (3 mg/capsule). Plasma phenylalanine values (mean ± SEM) were 0.27 ± 0.01 mmol/l at baseline and 0.27 ± 0.02 mmol/l after 3 months of supplementation [not significant (NS)].

To exclude an effect of time on P100 latency, the healthy controls were asked to undergo a second VEP recording 90 days after the baseline examination. In contrast to the PKU patients treated with fish oil, there was no change in P100 latencies in the

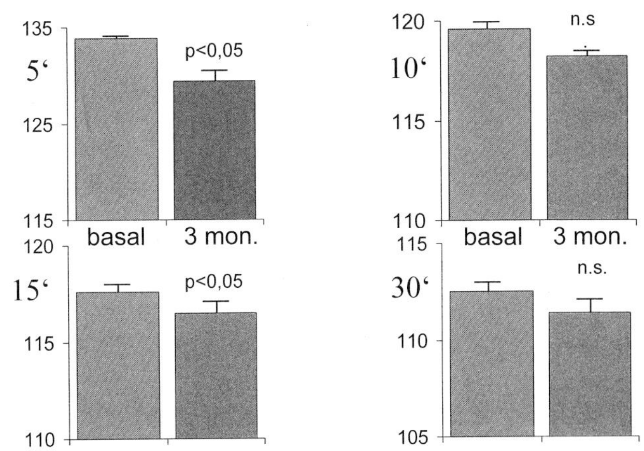

FIG. 3. Visual evoked potential latencies in patients with phenylketonuria at baseline and after 3 months of supplementation with fish oils. Latencies are significantly improved at stimulation with 5′ (foveal stimulation) and 15′ (intermediate), while there are nonsignificant trends for the other two pattern sizes.

TABLE 1. *Latencies of visual evoked potentials in healthy children at baseline and follow up after 3 months*

Checkerboard pattern	5′		10′		15′		30′	
	9		12		12		12	
n	Mean	SEM	Mean	SEM	Mean	SEM	Mean	SEM
Baseline	126.7	3.0	117.8	1.7	113.2	1.7	109.8	1.9
Follow up	127.5	2.2	118.2	1.6	114.0	1.8	109.4	1.7
p	n.s.		n.s.		n.s.		n.s.	

controls who agreed to be re-examined, on either foveal stimulation or entire retinal stimulation (Table 1) (30).

TOLERANCE AND SAFETY OF FISH OIL ADMINISTRATION

During the period of fish oil administration, mild manifestations of self-limiting acute diarrhea occurred in five children with PKU. In each of these five cases, other family members were affected by diarrhea at about the same time. Therefore, we consider that acute infectious gastroenteritis was the likely cause of the diarrhea in those cases. One case of slight epistaxis was observed at the beginning of fish oil administration, which resolved spontaneously without further intervention under continued fish oil supply. In none of the cases was it considered necessary to reduce the daily dose or to interrupt fish oil administration. No other adverse events were reported by the patients or their parents.

COMMENT

Our results show that children with early and continuously treated PKU have significantly longer P100 VEP latencies than a matched group of healthy controls. Administration of fish oil providing some 15 mg of DHA per kilogram of body weight daily for 3 months led to a marked shortening of VEP latencies, while plasma phenylalanine concentrations remained unchanged. Thus, our study strongly supports a causal effect of LC-PUFA supply on neurologic function and the speed of information processing in patients with PKU.

We chose to perform an open trial to enable us to apply strict inclusion criteria and still recruit a sizable number of patients, in view of the limited numbers of PKU patients available for study, even at this large treatment center. The electrophysiologic nature of the outcome measure, which is not expected to be altered by a placebo effect, and the remarkably stable results in the healthy controls make us confident that we have shown a real effect of the intervention.

Although patients with PKU showed a marked delay of VEP latencies compared with healthy, age-matched controls, the variance of the test results was remarkably similar in patients and controls for all stimulation patterns. Thus, the overall result cannot be attributed to a few patients with markedly pathologic recordings, but rather to an abnormality that characterizes the group of patients with PKU in general. The clinical significance of these results relates not only to visual function but extends to other neurologic functions, because delays in VEP latency appear to represent retarded information transmission in other parts of the central nervous system (CNS). Indeed, standardized test results on more complex functions also improved with supplementation of ω-3 LC-PUFA (unpublished data). Our finding may be due to a subtle subclinical defect or delay in the myelination of the visual pathways or a disturbance of synaptic information processing related to a reduction in the physiologically high LC-PUFA concentrations in synaptosomal membranes (32). VEP latencies are primarily related to conduction velocity in the optic nerve, and inflammatory demyelination is the prototype for their pathologic delay. Abnormalities in cerebral white matter, consistent with disturbed myelination, are found in untreated and even early-treated PKU (33–37) on magnetic resonance imaging. In this respect, our electrophysiologic findings provide a functional counterpart to those anatomic studies.

The observed alterations in patients with PKU appear to be related to the mode of dietary treatment. Depending on the individual phenylalanine tolerance, the PKU diet provides roughly one sixth of the usual intake of natural proteins found in healthy children, and consequently also of other components of the protein-rich foods that are avoided. In patients with PKU, an inadequate supply—with clinically relevant consequences—has been recognized for calcium, iron, zinc, and vitamin B_{12}, which are supplemented in current amino acid substitution diets for patients with PKU. More recently, LC-PUFA depletion of plasma lipids has been reported in PKU (13,14). DHA is considered functionally the most relevant LC-PUFA, because it has decisive functions in the assembly, maintenance, and proper function of cell membrane lipids (38), and is a major PUFA in the brain phospholipid pool (39). Although DHA is not an essential nutrient in the strict sense, its endogenous synthesis from its precursor α-linolenic acid, found in vegetable oil, is very inefficient in humans (40). Several studies in newborn infants have recognized that DHA supply is an important factor in the postnatal development of the visual system and cognition (22–29).

The causal role of the DHA supply is confirmed by the significant improvement in VEP latencies upon administration of fish oil, which provides relatively large amounts of DHA and some other n-3 LC-PUFAs. The change in VEP latencies over the study period of 3 months clearly does not constitute a normal developmental process, because an age-matched control group had unchanged VEP latencies after the same follow-up period. Thus, the iatrogenic induction of DHA depletion in patients with PKU provided with an apparently unbalanced diet does impair the development of proper signal transmission in the optic system. This underlines the importance of LC-PUFA for physiologic brain function in general.

After the completion of our trial, Agostoni and coworkers (41) reported similar results in a smaller group of patients participating in a longer-term controlled trial. Supplementation of 10 PKU patients with fish oil, providing some 10-mg DHA per kilogram of body weight daily for 12 months resulted in normalization of an initially delayed 15′ pattern-reversal VEP and a 2-Hz flash VEP; a 1-Hz flash VEP was also normalized. In a placebo-treated control group of patients with PKU, these recordings were unchanged.

The results of our study indicate that DHA supplementation has beneficial effects on information processing within weeks, which is in accordance with results from recent animal studies (42). However, DHA supplementation of our patients over 3 months only reduced the mean delay in VEP latencies relative to the healthy controls and failed to induce complete normalization (for example, for 5′: 10 ms at baseline *vs.* 5 ms at follow up). In contrast, Agostoni *et al* reported complete normalization after 12 months of supplementation with a 30% lower dose of n-3 LC-PUFA per kg body weight (41). Thus, a long-term supply of DHA may be necessary for complete normalization of the observed delay in signal transduction in PKU.

In view of the functional benefit and the absence of appreciable adverse effects, we consider it desirable to provide PKU patients with preformed n-3 LC-PUFA. Further evaluation of supplementation in larger groups of patients with classical PKU are needed to test its potential effects on other functional outcomes and to investigate the matter of the optimum dosage. The observations reported here are relevant beyond this specific group of patients with an inborn error of metabolism because the physiologic effects are obviously a result of the dietary substrate supply. Thus, the results obtained lead us to conclude that dietary preformed long-chain ω-3 polyunsaturated fatty acids, such as DHA, are conditionally essential substrates for humans not only in infancy, but also during the preschool and school-age period.

ACKNOWLEDGEMENTS

We are grateful for the invaluable support by the patients and their families, the staff of the Dr. von Hauner Children's Hospital, and the Friedrich-Baur-Institute. The study was financially supported in part by SHS Gesellschaft für Klinische Ernährung, Heilbronn, Germany and by the Child Health Foundation, Munich, Germany (http://www.kindergesundheit.de). S. B. was the recipient of a scholarship awarded by the University of Munich. Fish oil capsules were generously provided by Omega Pharma, Berlin, Germany.

REFERENCES

1. Scriver CR, Kaufman S. Hyperphenylalaninemia: phenylalanine hydroxylase deficiency. In: Scriver CR, Beaudet AL, Sly WS, Valle DL, eds. *The metabolic and molecular basis of inherited disease.* New York: McGraw Hill, 2001: 1667–724.
2. Burgard P, Schmidt E, Rupp A, Schneider W, Bremer HJ. Intellectual development of the patients of the German Collaborative Study of children treated for phenylketonuria. *Eur J Pediatr* 1996; 155 [Suppl 1]: 33–8.
3. Azen CG, Koch R, Friedman EG, *et al.* Intellectual development in 12-year-old children treated for phenylketonuria. *Am J Dis Child* 1991; 145: 35–9.
4. Stemerdink BA, Kalverboer AF, van der Meere JJ, *et al.* Behaviour and school achievement in patients with early and continuously treated phenylketonuria. *J Inherit Metab Dis* 2000; 23: 548–62.
5. Burgard P, Rey F, Rupp A, Abadie V, Rey J. Neuropsychologic functions of early treated patients with phenylketonuria, on and off diet: results of a cross-national and cross-sectional study. *Pediatr Res* 1997; 41: 368–74.
6. Schmidt E, Burgard P, Rupp A. Effects of concurrent phenylalanine levels on sustained attention and

calculation speed in patients treated early for phenylketonuria. *Eur J Pediatr* 1996; 155 [Suppl 1]: 82–6.

7. Acosta PB, Yannicelli S. Plasma micronutrient concentrations in infants undergoing therapy for phenylketonuria. *Biol Trace Elem Res* 1999; 67: 75–84.
8. Artuch R, Vilaseca MA, Moreno J, Lambruschini N, Cambra FJ, Campistol J. Decreased serum ubiquinone-10 concentrations in phenylketonuria. *Am J Clin Nutr* 1999; 70: 892–5.
9. Artuch R, Colome C, Vilaseca MA, *et al.* Plasma phenylalanine is associated with decreased serum ubiquinone-10 concentrations in phenylketonuria. *J Inherit Metab Dis* 2001; 24: 359–66.
10. Bodley JL, Austin VJ, Hanley WB, Clarke JT, Zlotkin S. Low iron stores in infants and children with treated phenylketonuria: a population at risk for iron-deficiency anaemia and associated cognitive deficits. *Eur J Pediatr* 1993; 152: 140–3.
11. Gropper SS, Acosta PB, Clarke-Sheehan N, Wenz E, Cheng M, Koch R. Trace element status of children with PKU and normal children. *J Am Diet Assoc* 1988; 88: 459–65.
12. Acosta PB, Stepnick-Gropper S, Clarke-Sheehan N, *et al.* Trace element status of PKU children ingesting an elemental diet. *J Parenter Enter Nutr* 1987; 11: 287–92.
13. Galli C, Agostoni C, Mosconi C, Riva E, Salari PC, Giovannini M. Reduced plasma C-20 and C-22 polyunsaturated fatty acids in children with phenylketonuria during dietary intervention. *J Pediatr* 1991; 119: 562–7.
14. Sanjurjo P, Perteagudo L, Rodriguez-Soriano J, Vilaseca A, Campistol J. Polyunsaturated fatty acid status in patients with phenylketonuria. *J Inherit Metab Dis* 1994; 17: 704–9.
15. Giovannini M, Agostoni C, Biasucci G, *et al.* Fatty acid metabolism in phenylketonuria. *Eur J Pediatr* 1996; 155 (suppl 1): 132–5.
16. Koletzko B, Decsi T. The role of long-chain polyunsaturated fatty acids for infant growth and development. In: Bendich A, Deckelbaum R, eds. *Preventive nutrition: primary and secondary prevention.* Totowa, NJ: Humana Press, 2001: 237–52.
17. Innis SM. The role of dietary n-6 and n-3 fatty acids in the developing brain. *Dev Neurosci* 2000; 22: 474–80.
18. Heird WC. The role of polyunsaturated fatty acids in term and preterm infants and breastfeeding mothers. *Pediatr Clin North Am* 2001; 48: 173–88.
19. Neuringer M. Infant vision and retinal function in studies of dietary long-chain polyunsaturated fatty acids: methods, results, and implications. *Am J Clin Nutr* 2000; 71[Suppl]: 256–67S.
20. Neuringer M, Reisbick S, Janowsky J. The role of n-3 fatty acids in visual and cognitive development: current evidence and methods of assessment. *J Pediatr* 1994; 125[Suppl]: S39–47.
21. de-Urquiza AM, Liu S, Sjoberg M, *et al.* Docosahexaenoic acid, a ligand for the retinoid X receptor in mouse brain. *Science* 2000; 290: 2140–4.
22. Birch EE, Garfield S, Hoffman DR, Uauy R, Birch DG. A randomized controlled trial of early dietary supply of long-chain polyunsaturated fatty acids and mental development in term infants. *Dev Med Child Neurol* 2000; 42: 174–81.
23. Birch EE, Hoffman DR, Uauy R, Birch DG, Prestidge C. Visual acuity and the essentiality of docosahexaenoic acid and arachidonic acid in the diet of term infants. *Pediatr Res* 1998; 44: 201–9.
24. Carlson SE, Werkman SH. A randomized trial of visual attention of preterm infants fed docosahexaenoic acid until two months. *Lipids* 1996; 31: 85–90.
25. Willatts P, Forsyth JS, DiModugno MK, Varma S, Colvin M. Effect of long-chain polyunsaturated fatty acids in infant formula on problem solving at 10 months of age. *Lancet* 1998; 352: 688–91.
26. SanGiovanni JP, Parra CS, Colditz GA, Berkey CS, Dwyer JT. Meta-analysis of dietary essential fatty acids and long-chain polyunsaturated fatty acids as they relate to visual resolution acuity in healthy preterm infants. *Pediatrics* 2000; 105: 1292–8.
27. SanGiovanni JP, Berkey CS, Dwyer JT, Colditz GA. Dietary essential fatty acids, long-chain polyunsaturated fatty acids, and visual resolution acuity in healthy full-term infants: a systematic review. *Early Hum Dev* 2000; 57: 165–88.
28. Uauy R, Hoffman DR. Essential fat requirements of preterm infants. *Am J Clin Nutr* 2000; 71[Suppl 1]: 245–50S.
29. Koletzko B, Agostoni C, Carlson SE, *et al.* Long chain polyunsaturated fatty acids (LC–PUFA) and perinatal development. *Acta Paediatr* 2001; 90: 460–4.
30. Beblo S, Reinhardt H, Muntau AC, Mueller-Felber W, Roscher AA, Koletzko B. Fish oil supplementation improves visual evoked potentials in children with phenylketonuria. *Neurology* 2001; 57: 1488–91.
31. Horwitz RI, Horwitz SM. Adherence to treatment and health outcomes. *Arch Intern Med* 1993; 153: 1863–8.

32. Koletzko B, Aggett PJ, Bindels JG, *et al.* Growth, development and differentiation: a functional food science approach. *Br J Nutr* 1998; 80[Suppl 1]: S5–45.
33. Weglage J, Bick U, Schuierer G, *et al.* Progression of cerebral white matter abnormalities in early treated patients with phenylketonuria during adolescence. *Neuropediatrics* 1997; 28: 239–40.
34. Ullrich K, Moller H, Weglage J, *et al.* White matter abnormalities in phenylketonuria: results of magnetic resonance measurements. *Acta Paediatr Suppl* 1994; 407: 78–82.
35. Bick U, Fahrendorf G, Ludolph AC, Vassallo P, Weglage J, Ullrich K. Disturbed myelination in patients with treated hyperphenylalaninaemia: evaluation with magnetic resonance imaging. *Eur J Pediatr* 1991; 150: 185–9.
36. Weglage J, Schuierer G, Kurlemann G, Bick R, Ullrich K. Different degrees of white matter abnormalities in untreated phenylketonurics: findings in magnetic resonance imaging. *J Inherit Metab Dis* 1993; 16: 1047–8.
37. Bick U, Ullrich K, Stober U, *et al.* White matter abnormalities in patients with treated hyperphenylalaninaemia: magnetic resonance relaxometry and proton spectroscopy findings. *Eur J Pediatr* 1993; 152: 1012–20.
38. Chen Y, Houghton LA, Brenna JT, Noy N. Docosahexaenoic acid modulates the interactions of the interphotoreceptor retinoid-binding protein with 11-cis-retinal. *J Biol Chem* 1996; 271: 20507–15.
39. Farquharson J, Cockburn F, Patrick WA, Jamieson EC, Logan RW. Effect of diet on infant subcutaneous tissue triglyceride fatty acids. *Arch Dis Child* 1993; 69: 589–93.
40. Pawlosky RJ, Hibbeln JR, Novotny JA, Salem NJ. Physiological compartmental analysis of alpha-linolenic acid metabolism in adult humans. *J Lipid Res* 2001; 42: 1257–65.
41. Agostoni C, Massetto N, Biasucci G, *et al.* Effects of long-chain polyunsaturated fatty acid supplementation on fatty acid status and visual function in treated children with hyperphenylalaninemia. *J Pediatr* 2000; 137: 504–9.
42. Tsukada H, Kakiuchi T, Fukumoto D, Nishiyama S, Koga K. Docosahexaenoic acid (DHA) improves the age-related impairment of the coupling mechanism between neuronal activation and functional cerebral blood flow response: a PET study in conscious monkeys. *Brain Res* 2000; 862: 180–6.

DISCUSSION

Dr. Matthieu: You measured the fatty acid in the plasma and not in the membrane. In the plasma, it mainly reflects the diet, and not really what has been synthesized. I would like to have seen values in blood cells, which would reflect the membrane values.

Dr. Koletzko: When using plasma fatty acid composition as a marker of tissue pools, we obviously have to live with some limitations. I do agree with you that we have to be extremely careful in not overinterpreting data obtained from the plasma compartment only. With respect your suggestion that it might have been advantageous to use blood cells, various people have investigated that, including ourselves. The overall conclusion is that concentrations in plasma and cells run in parallel following a dietary intervention, although the kinetics are different, and in young children both plasma and red blood cell contents of docosahexaenoic acid correlate with visual and cognitive functions (1–4). One might reasonably ask whether blood cell fatty acid contents are a better indicator of changes in nervous tissues, but there's no indication of that whatsoever from the available studies. However, the analysis of red blood cell fatty acids is prone to larger errors, partly due to a higher risk of oxidation of polyunsaturated fatty acids in the iron-containing erythrocytes. Also, one lipid transported in plasma lipoproteins, but not blood cell lipids, are the major source of brain lipid uptake. Another more feasible option in young children is to analyze oral mucosa cells, which we found to be a good marker of dietary polyunsaturated fatty acid supply, even beyond the end of an intervention, but again their composition seems to follow other kinetic rules than plasma and blood cell compartments (5). Nonetheless, Hoffman and coworkers found infant visual development to be correlated to DHA contents in oral mucosa cells (6). With regard to this study on children with PKU supplemented with marine oil for 3 months, all available evidence would lead us to predict that both plasma and blood cell composition will show a DHA increase. I do not see any evidence that analysis of blood cell composition would provide any advantage over plasma lipid analysis in testing the hypotheses we raised.

Dr. Matthieu: My second query is that for your cognitive task—if I got the message correctly—you did not have real controls. The controls were not patients with PKU, they were healthy children. This introduces bias because when psychologists examine children with PKU, they look at them in a different way from healthy children.

Dr. Koletzko: Obviously it would be preferable to have a fully double-blind study in children with PKU who are randomized to ω-3 or not. We chose not to do that because at the outset we wanted to do a pilot study using the limited patient population at our unit. Our goal was to determine whether there is any detectable effect at all with a precise electrophysiologic measurement. We were unaware of another study by Agostoni *et al* that was published after our trial was completed (7). Therefore, we certainly did not expect to find such striking and large effects, and when planning our study we felt we should not dilute the power of our study by separating our patient population into two groups. Also, it did not seem feasible to perform a cross-over study because there are marked carryover effects with fatty acids that would require long washout periods, and one might hypothesize that correction of a tissue DHA depletion possibly would not be reversed in weeks or even months. However, the results of the VEP study are certainly valid even under the conditions of an open trial because these are objective measurements, and the evaluation of the VEPs was performed by blinded investigators who just received the tracings.

Dr. Matthieu: My last point is about the way you showed your data. On your graphs, you started at a level of 100, so your differences look very large. In fact, they are not. I made a rough calculation, and the difference in evoked potential latencies is 7.5%, and the improvement is only 3.5%. I wonder if these differences are really of clinical significance.

Dr. Koletzko: One should consider that it would not make sense to start the scale at 0, there is no such thing as VEP latencies of 0 in human beings. The difference in 5′ P100 ms latencies between patients and controls is 9.8 ms or 7.9% of control values, and the improvement achieved in patients with DHA supplementation is 3.6 ms or 2.9% of baseline values in controls. This is indeed quite a large difference—7.5% represents quite a substantial delay in visual latency. Also, I am not aware which other intervention can achieve such a marked change of the speed of visual information processing in only 3 months. I find that quite remarkable and certainly worthwhile if it can be achieved with such a simple intervention that only corrects an iatrogenic deficiency syndrome.

Dr. Wanders: Is it your hypothesis now that what explains the lower C22:6 is a lower rate of synthesis in patients with PKU? If so, what could be the mechanism behind that?

Dr. Koletzko: I don't think we have any evidence that the rate of synthesis is lower in PKU patients than in healthy controls, but we also don't have much data in healthy children. I assume that healthy controls on a regular diet get so much preformed DHA from different foods that they do not depend entirely on an endogenous synthesis of DHA to meet their tissue needs.

Dr. Wanders: What about arachidonic acid supplementation in patients with PKU?

Dr. Koletzko: We haven't seen the same degree of depletion of plasma arachidonic acid in children and we don't yet know the functional relevance of arachidonic acid beyond infancy in the range that we have seen (8). During intervention, even with the relatively very high dose of fish oil that we gave, there was no effect on circulating arachidonic acid levels. Biochemically the synthesis of arachidonic is much shorter, with only three enzymatic steps, than for DHA (9). Thus, it is quite possible that there is a more active endogenous production of arachidonic acid.

Dr. Yamashiro: Did any of the children have liver dysfunction?

Dr. Koletzko: These were healthy children with a metabolic defect involving phenylalanine hydroxylase. They did not have any liver abnormality or other disease. We have no indication that the defect in phenylalanine hydroxylase has any direct impact on metabolic pathways in lipid or fatty acid metabolism. Our assumption is that these patients do not have any defect in turnover, but obviously, we have no direct measurements for comparison with healthy children. We suspect that the children with PKU, even with normal desaturase activity, are not able to match the DHA status of healthy children who receive preformed DHA, and this has functional consequences.

Dr. Micheli: How much PUFA crosses the placenta? In addition, is there any way to quantify the effect on membranes using the new brain imaging techniques? I was thinking particularly of nuclear magnetic resonance (NMR) spectroscopy.

Dr. Koletzko: We would very much like to know how much crosses the placenta. There are indications that there is active transfer from the mother to the fetus—for example, cord blood concentrations are higher than maternal blood concentrations; there are high concentrations in fetal tissue; and in the perfused placenta, there is preferential transfer from the maternal to the fetal side (10–13). Mechanisms have been proposed, including specific binding proteins and translocases that would facilitate transfer (14). We are trying to look at this more closely but we have no quantitative analyses as yet.

As for brain imaging, my understanding is that Professor Harry Lafeber and coworkers at Amsterdam have studied preterm infants, who were randomized to diets with and without long-chain PUFA, with sequential magnetic resonance imaging (MRI). As far as I have seen, no morphologic differences were found. Cunnane and coworkers in Toronto gave labeled fatty acids to small animals and attempted to follow their utilization with MRI spectroscopy (15), but the methodologic challenges in bringing this to useful *in vivo* studies in humans should not be underestimated.

Dr. Lütschg: I have a question about visual evoked potentials. The latencies depend on myelination. Did you look at age dependency? Also, you said that with the size 5 and 15 patterns there was a significant difference, but not with size 10 and 30. Why not?

Dr. Koletzko: We had age-matched controls, so there was an identical age distribution in the two populations. I don't think there was much of an age effect. However, there were limitations in evaluating the traces in the youngest preschool children, in some of whom VEP amplitudes were very small and hence we had to discard those tracings. With respect to the effects detected with different pattern sizes, a limitation may be our limited sample size. The trend was in the same direction for all pattern sizes, but it was only significant for two of them. I guess that if we had doubled the sample size, we might have found a significant difference for all of them.

Dr. Superti-Furga: You started by saying that children with PKU have some loss of IQ, so why did you then focus on the effect on VEPs? Why haven't you been looking for other, perhaps clinically more relevant, effects of these metabolites?

Dr. Koletzko: I agree that visual latency is not a measurement that translates directly into a clinically relevant outcome. However, the attraction of this measurement is that it is an objective measure of the speed of information processing in the central nervous system (CNS) that can be precisely determined. With my background in biochemical measurements, I always tend to be a bit careful about behavioral measurements because of my uncertainties about their precision and reproducibility. Here we started with the hypothesis that DHA would influence signal transduction because of such observations in animal studies and in young infants. It is conceivable that there might be also an effect on IQ but that remains to be tested. We have no real indication that IQ is influenced at all by fatty acids, other functions apparently are—for example, the speed with that one can solve a maze problem. The speed of information processing, however, could very well be of clinical relevance, particularly in a world where speed of response becomes increasingly important.

Dr. Superti-Furga: That is precisely where you probably have a learning affect. When you do the same test 3 months later, you remember how to do it. That's the first thing that comes to my mind.

Dr. Koletzko: It is a valid concern, but we did not see any learning effect at all in our controls. Nevertheless, I should emphasize again that this was a pilot study because we did not expect to find such a large response. If we had anticipated the marked effects, we certainly would have chosen a somewhat different study design.

Dr. Scott: Do you know anything about the effects of extreme diets—for example, children on a strict vegan diet? Could you anticipate an effect from their known intakes? I suppose you must have made a guestimate of that.

Dr Koletzko: Biochemical studies in vegetarian and vegan women showed that they tend to have low concentrations of DHA in plasma and also in their breast milk, and that is reflected in low plasma DHA in the breastfed babies (16,17). Potential functional consequences are certainly something one would want to look at.

Dr. Steinmann: From animal studies, do we have any idea about the turnover of fatty acids in the cerebral membranes, and are there any regional differences in the composition of the fatty acids?

Dr. Koletzko: There are a number of animal studies that investigated the fatty acid uptake and turnover in brain, and effects of dietary and other exogenous factors (18–22), but I am not sure how much is really known on the variance of fatty acid metabolism and turnover in different areas of the brain.

Dr. Kunz: If you give a certain amount of ^{13}C-labeled fatty acid, how much of the label will you find in the plasma fatty acids?

Dr. Koletzko: We did some studies in lactating women. When we gave 2 mg of labeled linoleic acid per kilogram of body weight, about 20% of it was oxidized to CO_2 and exhaled over a period of 5 days; 13% of the label appeared in the breast milk over 5 days; and two thirds disappeared into other body compartments with slow turnover, such as adipose tissue and possibly other compartments (23,24). The limitation of these types of *in vivo* studies in humans is that we cannot gain direct access to some of the pools of interest, such as the liver.

Dr. Böhles: If we argue that your findings are related to Δ6-desaturase activity, then every decrease in DHA should be balanced by an increase in the precursor fatty acid. Did you find increases in other fatty acids?

Dr. Koletzko: We found relatively high levels of linoleic and α-linolenic acids, but I think that was because children with PKU have a high intake of those fatty acids. They consume far more vegetable oil and less animal fat than healthy children. They also had higher concentrations of elongation products, but not of the long-chain metabolites.

Dr. Böhles: The Δ6-desaturases are very much influenced by other cofactors—insulin, vitamins, and so on. Could it be that a deficiency of coproducts influences Δ6-desaturase activity?

Dr. Koletzko: The activity of desaturases is indeed modulated by exogenous factors, and it is quite possible that, for example, trace elements or other factors that are altered in the PKU diet could have an influence. That is certainly something to look at.

Dr. Bachmann: Do you think you're observing a fish oil effect or a DHA effect? And, is the absorptive capacity and motility of the gut the same during fish oil supplementation as without supplementation? After all, arachidonic acid is a precursor of prostaglandins and so on.

Dr. Koletzko: Obviously, we've given fish oil and not DHA, so we can only say that we have seen an effect of fish oil supplementation. To give isolated DHA would be a nice experiment, but rather expensive. As for gastrointestinal function, surprisingly we did not find dyspepsia, other discomfort, or any untoward effects on bowel movements.

Dr. Bachmann: Was peak phenylalanine the same in both groups? You didn't get much steeper peaks in the nonsubstituted group than in the fish oil group?

Dr. Koletzko: We measured the phenylalanine concentrations on a regular basis and found no change from baseline (266 $\pm$ 14 μmol/l) to study end (271 $\pm$ 21 μmol/l), even though the gelatin capsules contained small amounts of 3 mg of phenylalanine per capsule.

Dr. Bachmann: So your DHA measurements were also made in the morning after an overnight fast, before the capsule?

Dr. Koletzko: Blood samples for fatty acid measurements were taken during outpatient visits, but we the plasma phospholipid composition is not much changed between pre- and postprandial states.

Dr. Bachmann: But you didn't look for changes in the red blood cell membranes. I think Dr. Matthieu's earlier question was a good one, because measurements in red cells would integrate the incorporation of DHA over time rather than being a transient view, as it is with plasma DHA levels.

Dr. Koletzko: I think that might be useful if one wanted to study a population with uncharacterized and rapidly changing dietary habits. However, we had a closely supervised population that took fish oil capsules on a daily basis. On theoretical grounds, your view is valid, but I don't think there is any evidence available that there would be a differential effect on the red blood cells under the conditions of our study. Indeed, I'm not aware of any conclusive evidence to show that red blood cells are a better measure of tissue composition or functional effects than plasma. In fact, the measurement of red blood cell membranes is far more prone to error, primarily because you have so much iron in them and hence rather unstable conditions.

Dr. Borum: When focusing on polyunsaturated fatty acids, we usually consider their synthesis or their role as a precursor of something else. However, is it possible that the degradation of the fatty acids—and by that I mean oxidation or the degradation of unsaturated bonds in other ways—is different in patients with PKU and healthy controls?

Dr. Koletzko: It is certainly possible. That could be relevant to these differences.

Dr. Endres: Where did you get your nonsupplemented PKU formula? The ones I am aware of all have a healthy provision of oils.

Dr. Koletzko: That is true for the infants. For older age groups, as far as I know none of the amino acid products marketed in Europe today are supplemented with DHA.

Dr. Endres: However, your results could promote the industry to add DHE to products for all children. What is your view on this?

Dr. Koletzko: I hope that will be the result, but I would want to add a word of caution: I think it would be a mistake to put DHA into all products without having obtained further data. We need far more information, for example on appropriate dosing.

REFERENCES

1. Birch EE, Hoffman DR, Castaneda YS, *et al.* A randomized controlled trial of long-chain polyunsaturated fatty acid supplementation of formula in term infants after weaning at 6 wks of age. *Am J Clin Nutr* 2002; 75: 570–80.
2. Makrides M, Neumann MA, Jeffrey B, Lien EL, Gibson RA. A randomized trial of different ratios of linoleic to alpha-linolenic acid in the diet of term infants: effects on visual function and growth. *Am J Clin Nutr* 2000; 71: 120–9.

3. Birch EE, Garfield S, Hoffman DR, Uauy R, Birch DG. A randomized controlled trial of early dietary supply of long-chain polyunsaturated fatty acids and mental development in term infants. *Dev Med Child Neurol* 2000; 42: 174–81.
4. Bakker EC, van Houwelingen AC, Hornstra G. Early nutrition, essential fatty acid status and visual acuity of term infants at 7 months of age. *Eur J Clin Nutr* 1999; 53: 872–9.
5. Koletzko B, Knoppke B, von Schenck U., Demmelmair H, Damli A. Non-invasive assessment of essential fatty acid status by cheek cell analysis. *J Pediatr Gastroenterol Nutr* 1999; 29: 467–74.
6. Hoffman DR, Birch EE, Birch DG, Uauy R. Fatty acid profile of buccal cheek cell phospholipids as an index for dietary intake of docosahexaenoic acid in preterm infants. *Lipids* 1999; 34: 337–42.
7. Agostoni C, Massetto N, Biasucci G, *et al.* Effects of long-chain polyunsaturated fatty acid supplementation on fatty acid status and visual function in treated children with hyperphenylalaninemia. *J Pediatr* 2000; 137: 504–9.
8. Koletzko B, Decsi T. The role of long-chain polyunsaturated fatty acids for infant growth and development. In: Bendich A, Deckelbaum R, eds. *Preventive Nutrition: primary and secondary prevention.* Totowa, NJ: Humana Press, 2000.
9. Demmelmair H, Iser B, Rauh PA, Koletzko B. Comparison of bolus versus fractionated oral applications of [13C]-linoleic acid in humans. *Eur J Clin Invest* 1999; 29: 603–9.
10. Berghaus TM, Demmelmair H, Koletzko B. Fatty acid composition of lipid classes in maternal and cord plasma at birth. *Eur J Pediatr* 1998; 157: 763–8.
11. Berghaus T, Demmelmair H, Koletzko B. Essential fatty acids and their long-chain polyunsaturated metabolites in maternal and cord plasma triglycerides during late gestation. *Biol Neonate* 2000; 77: 96–100.
12. Otto SJ, Houwelingen AC, Antal M, *et al.* Maternal and neonatal essential fatty acid status in phospholipids: an international comparative study. *Eur J Clin Nutr* 1997; 51: 232–42.
13. Kuhn DC, Crawford MA, Stevens P. Transport and metabolism of essential fatty acids by the human placenta. *Contr Gynec Obstetr* 1985; 13: 139–40.
14. Minda H, Larque E, Koletzko B, Decsi T. Systematic review of fatty acid composition of plasma phospholipids of venous cord blood in full-term infants. *Eur J Nutr* 2002; 41: 125–31.
15. Cunnane SC, Nadeau CR, Likhodii SS. NMR and isotope ratio mass spectrometry studies of in vivo uptake and metabolism of polyunsaturates by the developing rat brain. *J Mol Neurosci* 2001; 16: 173–80.
16. Sanders TA, Reddy S. The influence of a vegetarian diet on the fatty acid composition of human milk and the essential fatty acid status of the infant. *J Pediatr* 1992; 120: 71–7.
17. Specker BL, Wey HE, Miller D. Differences in fatty acid composition of human milk in vegetarian and nonvegetarian women: long-term effect of diet. *J Pediatr Gastroenterol Nutr* 1987; 6: 764–8.
18. Rapoport SI. In vivo fatty acid incorporation into brain phosholipids in relation to plasma availability, signal transduction and membrane remodeling. *J Mol Neurosci* 2001; 16: 243–61.
19. Contreras MA, Chang MC, Rosenberger TA, *et al.* Chronic nutritional deprivation of n-3 alpha-linolenic acid does not affect n-6 arachidonic acid recycling within brain phospholipids of awake rats. *J Neurochem* 2001; 79: 1090–9.
20. Su HM, Bernardo L, Mirmiran M, *et al.* Bioequivalence of dietary alpha-linolenic and docosahexaenoic acids as sources of docosahexaenoate accretion in brain and associated organs of neonatal baboons. *Pediatr Res* 1999; 45: 87–93.
21. Rapoport SI. In vivo fatty acid incorporation into brain phospholipids in relation to signal transduction and membrane remodeling. *Neurochem Res* 1999; 24:1403–15.
22. Green P, Kamensky B, Yavin E. Replenishment of docosahexaenoic acid in n-3 fatty acid-deficient fetal rats by intraamniotic ethyl-docosahexaenoate administration. *J Neurosci Res* 1997; 48: 264–72.
23. Demmelmair H, Baumheuer M, Koletzko B, Dokoupil K, Kratl G. Metabolism of U13C-labeled linoleic acid in lactating women. *J Lipid Res* 1998; 39: 1389–96.
24. Del Prado M, Villalpando S, Elizondo A, Rodriguez M, Demmelmair H, Koletzko B. Contribution of dietary and newly formed arachidonic acid to human milk lipids in women eating a low-fat diet. *Am J Clin Nutr* 2001; 74: 242–7.

Genetic Expression and Nutrition, edited by Claude Bachmann and Berthold Koletzko. Nestlé Nutrition Workshop Series, Pediatric Program, Vol. 50. Nestec Ltd., Vevey/Lippincott Williams & Wilkins, Philadelphia, © 2003.

Fatty Acid Regulation of Gene Expression: A Genomic Mechanism to Improve the Metabolic Syndrome

Steven D. Clarke, Carolanne Nelson, Daniela Gasperikova, *James DeMar, *Alexandre Lapillonne, and *William C. Heird

*Institute for Cellular and Molecular Biology and the Division of Nutritional Sciences, University of Texas, Austin, Texas, USA; *USDA/ARS Children's Nutrition Research Center, Baylor College of Medicine, Houston, Texas, USA*

The development of obesity and associated insulin resistance involves a multitude of gene products, including proteins involved in lipid synthesis and oxidation, thermogenesis, and cell differentiation. The genes encoding these proteins are in essence the blueprints that we have inherited from our parents. However, what determines the way in which blueprints are interpreted is largely dictated by a collection of environmental factors. The nutrients we consume are among the most influential of these environmental factors.

During the early stages of evolutionary development, nutrients functioned as primitive hormonal signals that allowed the early organisms to turn on pathways of synthesis or storage during periods of nutrient deprivation or excess. As single-cell organisms evolved into complex life forms, nutrients continued to be environmental factors that interacted with hormonal signals to govern the expression of genes encoding the proteins involved in energy metabolism, cell differentiation, and cell growth. Nutrients govern the tissue content and activity of different proteins by functioning as regulators of gene transcription, nuclear RNA processing, messenger RNA (mRNA) degradation, and mRNA translation, as well as functioning as posttranslational modifiers of proteins.

One dietary constituent that has a strong influence on cell differentiation, growth, and metabolism is fat. The fatty acid component of dietary lipid not only influences hormonal signaling events by modifying membrane lipid composition, but fatty acids also have a very strong direct influence on the molecular events that govern gene expression. In this chapter we discuss the influence of n-9, n-6, and n-3 fatty acids on gene expression in the liver and skeletal muscle, and the impact this has on intra- and interorgan partitioning of metabolic fuels (Fig. 1).

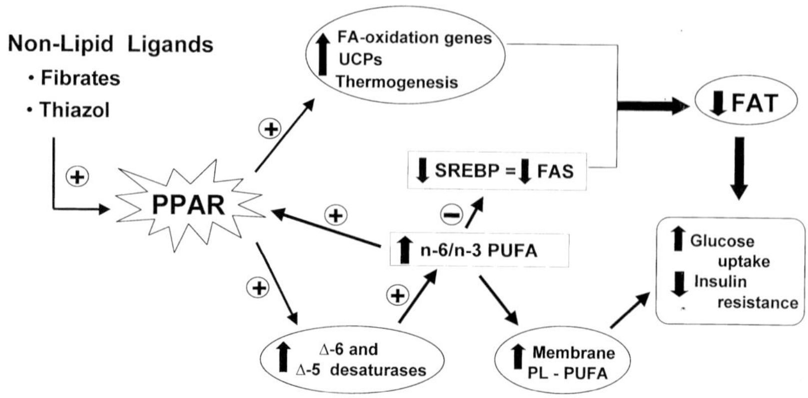

FIG. 1. Polyunsaturated fatty acids (PUFA) as hepatic fatty acid repartitioning agents. This schematic describes the mechanisms by which dietary polyunsaturated fatty acids redirect fatty acids away from triglyceride storage and toward fatty acid oxidation. The data supporting the schematic have been derived from studies with the liver. AOX, peroxisomal acyl-CoA oxidase; ACC, acetyl-CoA carboxylase gene; CPT, liver type carnitine palmitoyltransferase; ER, endoplasmic reticulum; FA, fatty acids; FAS, fatty acid synthase gene; PL, phospholipid; TG, triglycerides; VLDL, very low density lipoproteins.

POLYUNSATURATED FATTY ACIDS AS FUEL REPARTITIONING AGENTS

In 1967 Nichaman *et al.* (1) reported that raising the linoleate content of the diet from 4% of energy to 18% reduced plasma triglyceride concentrations in normal and hyperlipemic subjects by 30% and 52%, respectively. Shortly after this, Allmann and Gibson (2) discovered that the induction of hepatic lipogenesis associated with carbohydrate feeding could be inhibited by the inclusion of 18:2(n-6) in the diet but not by the addition of 16:0. Since the initial reports by these investigators (1,2), numerous studies have shown that the ingestion of fats rich in n-6 and n-3 fatty acids suppress hepatic *de novo* fatty acid biosynthesis, triglyceride production, and triglyceride secretion, while they enhance hepatic and skeletal muscle fatty acid oxidation (3–16). These metabolic effects cannot be mimicked by dietary monounsaturated and saturated fatty acids or by polyunsaturated fatty acids not derived from the n-6 and n-3 fatty acid families (for example, 5,13-22:2) (15). Moreover, in order for 18:2(n-6) and 18:3(n-3) to influence gene expression and metabolism, they must be metabolized to highly unsaturated fatty acid (HUFA) products of the Δ-6 desaturase pathway (16).

In addition to regulating lipid metabolism, HUFA may improve insulin sensitivity and increase nonoxidative glucose metabolism (17,18), but this promising action appears to be highly dependent upon the amount of HUFA consumed. For example, we have recently discovered that rats fed a diet containing 40% fish oil energy showed a pattern of muscle gene expression that suggested a severe impairment of glucose metabolism (a 20-fold increase in the expression of pyruvate dehydrogenase kinase 4). Interestingly, feeding rats a 40% olive oil diet [rich in 18:1(n-9)] resulted in a muscle gene expression pattern suggesting minimal interference with skeletal muscle glycolysis, an appreciable increase in glucose conversion to glycogen, and significantly reduced DNA damage (Nelson CM, Heird W, Clarke SD, unpublished data).

A key component of the HUFA repartitioning mechanism involves an HUFA-dependent reduction in the hepatic activity of several glycolytic and lipogenic enzymes, including glucokinase, pyruvate kinase, acetyl-CoA carboxylase, fatty acid synthase, citrate lyase, malic enzyme, glucose-6-phosphate dehydrogenase, pyruvate dehydrogenase, and diacylglycerol acyl transferase (2–15,19). Expression of these enzymes is very sensitive to the suppressive effects of HUFA. Diets that contain as little as 2% to 4% of their dietary energy as HUFAs are sufficient to reduce hepatic fatty acid synthase activity by 25% to 40%, but maximum inhibition requires a diet that derives 10% of its energy from HUFA (15).

Paralleling the HUFA-mediated decrease in hepatic lipogenesis is an increase in the rate of hepatic fatty acid oxidation and ketogenesis (12–14). HUFAs enhance fatty acid oxidation by coordinately increasing the expression of mitochondrial L-type carnitine palmitoyltransferase and peroxisomal acyl-CoA oxidase, by decreasing the synthesis of malonyl-CoA, and by decreasing the sensitivity of carnitine palmitoytransferase to malonyl-CoA inhibition (20). Carnitine palmitoyltransferase regulates the entry of fatty acids into the mitochondria, and acyl-CoA oxidase catalyzes the rate-limiting step in peroxisomal fatty acid oxidation. The contribution of hepatic peroxisomal fatty acid oxidation to overall fatty acid oxidation is unclear but has been reported to be as high as 20%. Dietary HUFAs readily induce the expression of both liver-type carnitine palmitoyltransferase and acyl-CoA oxidase. We have found that a diet providing as little as 7% of energy as HUFA will significantly increase the hepatic abundance of both transcripts [(21) and Nelson CM, Heird W, Clarke SD, unpublished data].

In addition to the HUFA-dependent up-regulation in hepatic capacity for fatty acid oxidation, metabolite regulation of fatty acid entry into the mitochondria is an important determinant of fatty acid oxidation (22–24). In this respect, malonyl-CoA has an important negative influence on carnitine palmitoyltransferase activity (22–24). The concentration of malonyl-CoA is highly dependent upon the rate of substrate flux through the fatty acid biosynthetic pathway (22–24). Hepatic concentrations of malonyl-CoA increase 10- to 20-fold when fasted rats are fed a high-carbohydrate meal. Several years ago we showed that if 18:2(n-6) was supplemented to the carbohydrate meal, malonyl-CoA synthesis was inhibited (20). This is one of the earliest metabolic changes detectable with HUFA feeding. Thus the fatty acid biosynthetic pathway (and its inhibition by dietary HUFA) is not only important for governing the rate of conversion of dietary carbohydrate to fatty acids and triglycerides, but is also a key determinant of the rate of fatty acid flux through the pathway of β oxidation. This latter function may be most important in humans, where dietary HUFAs are known to enhance fatty acid oxidation and lower hepatic triglyceride output but where net fatty acid synthesis is reportedly very low (25,26).

In addition to suppressing the production of malonyl-CoA, HUFAs also reduce the sensitivity of carnitine palmitoyltransferase to malonyl-CoA inhibition (27). In this respect, 20- and 22-carbon (n-3) HUFA appear to be more potent than 18:2(n-6) (27).

Thus, dietary HUFAs regulate hepatic lipid metabolism at three different levels: gene expression and enzyme abundance, malonyl-CoA synthesis, and membrane fluidity (27,28).

Malonyl-CoA is also a key regulator of fatty acid oxidation in skeletal muscle (23,24). Like the liver, the concentration of malonyl-CoA in soleus and gastrocnemius is highly dependent upon nutritional state. For example, feeding fasted rats a high-carbohydrate diet increased skeletal muscle malonyl-CoA levels by 50% to 250%, and, as in the hepatic system, the rate of fatty acid oxidation decreased as the level of skeletal muscle malonyl-CoA increased (23). However, unlike the liver, the production of malonyl-CoA in gastrocnemius muscle was not dependent upon the amount of muscle-type acetyl-CoA carboxylase protein found in the muscle (23).

The expression of muscle-type acetyl-CoA carboxylase is also unaffected by the type or amount of dietary HUFA (Nelson CM, Heird W, Clarke SD, unpublished data). Moreover, unlike the situation in the liver, muscle-type carnitine palmitoyltransferase and peroxisomal acyl-CoA oxidase gene expression is very resistant to up-regulation by dietary HUFA. In fact, achievement of a modest 50% to 100% increase in muscle-type carnitine palmitoyltransferase activity requires a diet containing less than 35% of HUFA energy (12–14), and even under these conditions peroxisomal acyl-CoA oxidase expression in gastrocnemius muscle remains virtually unchanged (Nelson CM, Heird W, Clarke SD, unpublished data).

A close examination of published reports reveals that investigators have only used high HUFA diets or cells in culture to investigate the influence of fatty acid composition on muscle fatty acid oxidation (12–14,21,29,30). Thus, although HUFAs may increase mitochondrial membrane fluidity and reduce the malonyl-CoA sensitivity of muscle-type carnitine palmitoyltransferase, the impact of physiologic intakes of HUFA on skeletal muscle fatty acid oxidation remains an open question. If increases in skeletal muscle fatty acid oxidation are to be seriously considered in formulating dietary recommendations for improving glucose sensitivity and for lowering blood triglycerides (17,18,31), the influence of HUFA on skeletal muscle metabolism must be examined at lower, more meaningful dietary levels.

A GENOMIC MECHANISM FOR THE POLYUNSATURATED FATTY ACID REGULATION OF LIPID METABOLISM

The Role of PPAR in PUFA Fatty Acid Oxidation and Synthesis

HUFA inhibition of gene expression occurs too quickly (< 90 minutes) to be explained simply by modification of hormone receptor signaling by HUFA (7,30–33). Rather, the changes are more consistent with the idea that HUFAs regulate the activity or abundance of a nuclear transcription factor directly. The 1990 discovery of a novel lipid activated transcription factor, peroxisome proliferator activated receptor α (PPARα), provided the first evidence that the nucleus did in fact contain transcription factors that were dependent on fatty acid ligands (34). PPARα is a member of the steroid receptor superfamily and like other steroid receptors, it possesses a DNA binding domain and a ligand binding domain (34–38). Interaction of PPARα with its

DNA recognition site is markedly enhanced by ligands such as the hypotriglyceridemic fibrate drugs, conjugated linoleic acid, and HUFA (34–38). In general, PPARα activation leads to the induction of several genes encoding proteins involved in lipid transport, oxidation, and thermogenesis (Fig. 2) (29,39–42).

The importance of PPARα in overall glucose and fatty acid homeostasis was shown by the observation that PPARα −/− mice lack the ability to increase rates of fatty acid oxidation during periods of food deprivation, and they develop characteristics of adult onset diabetes including fatty liver, raised blood triglycerides, and hyperglycemia (43). The fact that PPARα is essential for lipid oxidation was further underscored when hyperglycemia was found to suppress PPARα expression, induce PPARγ expression, increase β cell and cardiomyocyte lipids, and accelerate cell death (44,45).

That PPARα may play a pivotal role in both fatty acid and glucose metabolism quickly led to the idea that PPARα was a "master switch" transcription factor that was targeted by HUFAs to coordinate the suppression of genes encoding proteins of lipid synthesis and to induce genes encoding proteins of fatty acid oxidation and thermogenesis. More important, PPARα regulation by HUFA—particularly (n-3) HUFA and possibly conjugated linoleic acid—offered an explanation for the reported benefits of these fatty acids in protecting individuals from developing the detrimental characteristics of non–insulin-dependent diabetes (18). This attractive hypothesis was strengthened by reports that potent pharmacologic activators of PPARα modestly reduced lipogenic gene transcription (33). However, PPARα does not appear to interact with HUFA response sequences of lipogenic genes (10). Moreover, HUFA

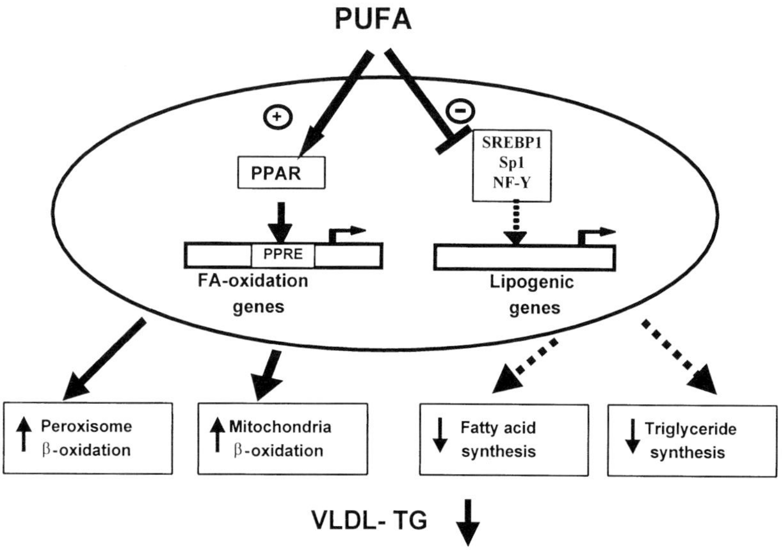

FIG. 2. Nuclear mechanism for polyunsaturated fatty acids (PUFA) regulation of gene expression. FA, fatty acids; NF-Y, nuclear factor Y; PPAR, peroxisome proliferator-activated receptor; PPRE, peroxisome proliferator-activated receptor response element; Sp1, stimulatory protein 1; SREBP-1, sterol regulatory element binding protein-1; TG, triglycerides.

continue to suppress the transcription of hepatic lipogenic genes in PPARα −/− mice (46). Thus the HUFA inhibition of lipogenic gene expression is independent of PPARα and may simply reflect the increase in HUFA synthesis resulting from the PPARα-dependent induction of the Δ-6 desaturase pathway (Fig. 2) (16,47).

HUFA Regulation of Transcription Factors in the Insulin Response Element and Lipogenic Gene Transcription

Dietary HUFAs inhibit hepatic lipogenesis by suppressing the transcription of various hepatic enzymes involved in glucose metabolism and fatty acid biosynthesis, including glucokinase, pyruvate kinase, acetyl-CoA carboxylase, fatty acid synthase, stearoyl-CoA desaturase, and the Δ-6 and Δ-5 desaturases (2–11,33). HUFA response sequences have been well characterized in only three genes: fatty acid synthase, S14, and L-type pyruvate kinase. The rat fatty acid synthase gene contains two independent HUFA regulatory sequences that are located between −118 and −43 and between −7250 and −7035 (Teran-Garcia M, Clarke SD, unpublished data). Approximately 65% and 35%, respectively, of the HUFA control of the fatty acid synthase gene can be attributed to the proximal and distal elements. Interestingly, the proximal HUFA response region of the fatty acid synthase gene has characteristics that are very similar to the HUFA response region of the S14 gene (−220 to −80), while the distal HUFA response region of the fatty acid synthase has similarities to the L-type pyruvate kinase HUFA response region (−160 to (−97) (10).

The proximal HUFA response region of the fatty acid synthase gene imparts insulin responsiveness to the gene and contains DNA binding sites for sterol regulatory element binding protein 1 (SREBP-1), USF (upstream stimulatory factor), Sp1 (specific transcription factor 1), and nuclear factor Y (NF-Y) (48–50). The nuclear abundance of USF is unaffected by dietary HUFA (33). In contrast, HUFAs rapidly reduce the nuclear content of hepatic SREBP-1, and this reduction is associated with a decreased rate of fatty acid synthase and S14 gene transcription (33). SREBPs are a family of transcription factors (SREBP-1a, -1c, and -2) that were first isolated as a result of their property of binding to the sterol regulatory element. SREBP-2 is a regulator of genes encoding proteins involved in cholesterol metabolism (48–53). SREBP-1 exists in two forms, 1a and 1c. SREBP-1a is the dominant form in cell lines and is a regulator of genes encoding proteins involved in both lipogenesis and cholesterologenesis. SREBP-1c constitutes 90% of the SREBP-1 found *in vivo* and is a determinant of lipogenic gene transcription (54,55).

SREBP-1 is synthesized as a 125-kDa precursor protein that is anchored in the endoplasmic reticulum membrane (48,53). Proteolytic release of the 68-kDa mature SREBP-1 occurs in the Golgi system, and movement of SREBP-1 from the endoplasmic reticulum to the Golgi apparatus requires the trafficking protein, SREBP cleavage activating protein (SCAP) (53,56). Once released, mature SREBP-1 translocates to the nucleus and binds both to the classical sterol response element and to a palindrome CATG sequence. In the case of fatty acid synthase, SREBP-1 interacts with a CATG palindrome that also functions as an insulin response element (48).

Overexpression of mature SREBP-1a in the liver is associated with high rates of fatty acid biosynthesis and the development of a fatty liver (57). In contrast, the ablation of the SREBP-1 gene results in low expression of lipogenic genes (58). These observations led us to hypothesize that HUFAs inhibit lipogenic gene transcription by impairing the proteolytic release of SREBP-1c or by suppressing SREBP-1c gene expression, or both. Diets rich in 18:2 (n-6) or 20:5 and 22:6 (n-3) were found to reduce the hepatic nuclear and precursor content of mature SREBP-1 65% and 90%, and 60% and 75%, respectively (33). The decrease in SREBP-1c was accompanied by a comparable decrease in the transcription rate of hepatic fatty acid synthase (33).

Unlike HUFAs, saturated and monounsaturated fats have no effect on the nuclear content or precursor content of SREBP-1 or on lipogenic gene expression (33). The HUFA-dependent reduction in hepatic content of SREBP-1 may explain how HUFAs inhibit the transcription of several genes encoding proteins that are involved in hepatic glucose metabolism and fatty acid biosynthesis, including glucokinase, acetyl-CoA carboxylase, stearoyl-CoA desaturase, and the Δ-6 and Δ-5 desaturases (59). Interestingly, the inhibition of lipogenic gene expression that reportedly occurs in adipose tissue with the ingestion of fish oil does not involve an SREBP-1–dependent mechanism (51).

HUFAs reduce the nuclear content of SREBP-1 by a two-phase mechanism. The first phase is a rapid (< 60 minutes) inhibition of the proteolytic release process (54,56,60). The second phase involves an adaptive reduction (taking about 48 hours) in the hepatic content of SREBP-1 mRNA, which is subsequently followed by a reduction in the amount of precursor SREBP-1 protein (33). The mechanism whereby HUFAs acutely inhibit the proteolytic processes is unknown. However, nuclear run-on assays suggest that HUFAs reduce the hepatic content of SREBP-1 mRNA by posttranscriptional mechanisms (33). Using rat liver cells in primary culture, we have determined that HUFAs reduce the half-life of SREBP-1c mRNA from 11 hours to less than 5 hours (54). The mechanism by which HUFAs control the half-life of SREBP-1 is unknown, but it may require the synthesis of a rapidly turning-over HUFA-dependent protein (54).

Recently, we found that feeding rats the PPARα-specific activator WY 14,643 induced hepatic Δ-6 and Δ-5 desaturase gene expression severalfold. The induction in Δ-6 and Δ-5 desaturase gene transcription was paralleled by an increase in the hepatic abundance of the respective transcripts and by a rise in enzymatic activity (Tang Z, Cho H, Clarke SD, unpublished data). This apparent increase in the rate of conversion of 18-carbon HUFA to 20- and 22-carbon products was paralleled by a 50% reduction in the expression of SREBP-1 and its target gene, fatty acid synthase (33). Thus it appears that nonlipid ligand activators of PPARα reduce hepatic lipogenic gene transcription by inducing the activity of the Δ-6 and Δ-5 desaturase pathway. This suggests that PPARα activators (for example fibrates) lower hepatic triglyceride secretion and hence blood triglycerides in two ways: first, by stimulating the hepatic oxidation of fatty acids and thereby diverting fatty acids away from triglyceride synthesis; and second, by inducing the Δ-6 desaturase pathway and the production of 20- and 22-carbon fatty acids, which in turn lower the SREBP-1, thereby decreasing lipogenic

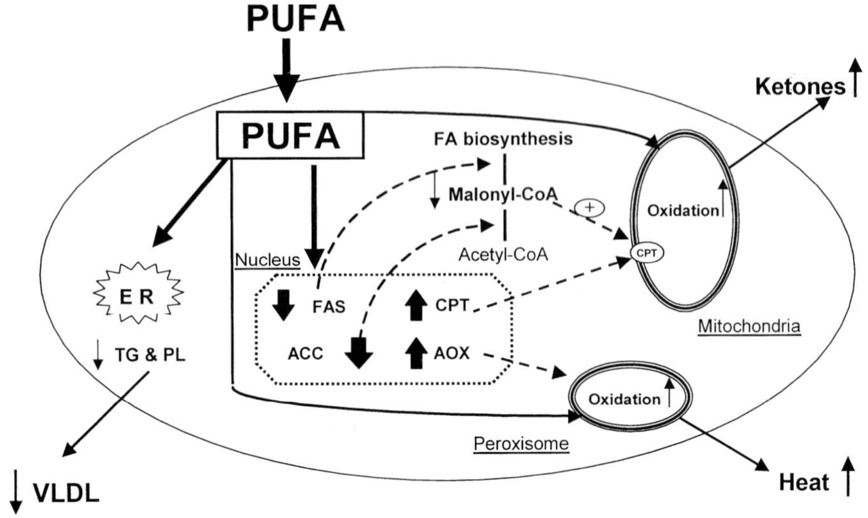

FIG. 3. Schematic mechanism of polyunsaturated fatty acid (PUFA) regulation of fuel partitioning. FAS, fatty acid synthase; PL, phospholipids; PPAR, peroxisome proliferator activated receptor; SREBP, sterol response element-binding protein; thiazole, thiazolidinediones; UCPs, mitochondrial uncoupling proteins; +, activation or induction; −, suppression.

gene expression and the synthesis of malonyl-CoA for fatty acid biosynthesis and carnitine palmitoyltransferase regulation (Fig. 3). It remains to be determined whether the Δ-6 and Δ-5 desaturase pathway is enhanced by ligands for other PPARs such as PPARγ and PPARδ. However, it is interesting that enrichment of skeletal muscle phospholipids with 20- and 22-carbon HUFAs improves glucose uptake (17,18), and that PPARγ enhances skeletal muscle insulin sensitivity (61).

SREBP-1c by itself possesses weak *trans* activating power, but the binding of SREBP-1c to its recognition sequence enhances the upstream DNA binding of NF-Y and Sp1 and this in turn amplifies the *trans* activating activities of the three transcription factors (48). NF-Y is a heterotrimeric nuclear protein that reportedly plays a role in regulating chromatin structure by way of its interaction with histone acetyl transferases. The binding sites for NF-Y are essential for fatty acid synthase (49) and S14 promoter activity (62). Mutations within the Y box region of −104 to −99 of the S14 gene eliminate promoter activity by preventing NF-Y from interacting with upstream T3 (−2800 to −2500) and carbohydrate response (−1600 to −1400) regions (62). Similarly mutating the Y box motif of the rat fatty acid synthase gene eliminated 80% of the promoter activity (Teran-Garcia M, Clarke SD, unpublished data). In contrast, eliminating the SREBP-1 site (−67 to −53) reduced fatty acid synthase promoter activity by only 40%. More important, only 25% of the HUFA inhibition of fatty acid synthase promoter activity was lost with the SREBP-1 site mutation, whereas 55% was lost with the NF-Y site mutation. Apparently, HUFAs regulate fatty acid synthase gene transcription by governing the nuclear content of SREBP-1 and by interfering with the enhancer activity of NF-Y.

HUFA Regulation of the Carbohydrate Response Element and Lipogenic Gene Transcription

The insulin response region and its associated transcription factors (SREBP-1, NF-Y, and Sp1) are not the only nuclear factors regulated by HUFA. Transfection-reporter analyses indicate that HUFAs exert a negative influence on the carbohydrate response element of the L-type pyruvate kinase (10,63) and fatty acid synthase genes (Teran-Garcia M, Clarke SD, unpublished data). The nature of the transcription factors and the mechanism by which HUFAs regulate them are not well defined. One hepatic protein that may be an HUFA target is hepatic nuclear factor 4 (HFN-4) (64). HNF-4 is a member of the steroid receptor superfamily. It enhances the glucose/insulin induction of L-type pyruvate kinase transcription by binding as a homo-dimer to a direct repeat-1 motif (63). Like PPARα, HNF-4 has a ligand binding domain that interacts with acyl-CoA esters, but unlike PPARα, fatty acyl-CoA binding to HNF-4 decreases its DNA binding activity (63). This suggests that HUFA may exert part of its negative influence on gene transcription by reducing HNF-4 DNA binding activity. Linker scanner mutations through the carbohydrate response region of the L-type pyruvate kinase promoter (that is, −183 to −97) did in fact reveal that the HNF-4 recognition elements were essential for HUFA suppression of the promoter (63). Recently we found that sequences between −7242 and −7150 of the fatty acid synthase gene were required for glucose to induce fatty acid synthase gene transcription (65,66). Subsequent studies have shown that −7242 to −7150 contains DNA recognition sites for HNF-4 and a novel carbohydrate response factor (66). Moreover, deleting this sequence eliminated 30% to 40% of the total HUFA suppression of the fatty acid synthase promoter (Teran-Garcia M, Clarke SD, unpublished data). Thus, HUFA suppression of lipogenic gene expression may involve two mechanisms: interference with the insulin regulation of transcription factors interaction with the insulin response element; and interference with the glucose signal governing the transcription factors affecting the carbohydrate response element.

HUFA Regulation of mRNA Processing

One of the first transcripts shown to be reduced by dietary HUFA was glucose-6-phosphate dehydrogenase mRNA. Interestingly, nuclear run-on assays revealed that this did not reflect an inhibition of gene transcription (67,68). Recently, Salati and coworkers have shown that HUFA govern the hepatic abundance of glucose-6-phosphate dehydrogenase mRNA by inhibiting the processing events involved in mRNA maturation (67,68). HUFA regulation of transcript processing may be unique for glucose-6-phosphate dehydrogenase, because transcription studies indicate that HUFAs exert their inhibitory influence primarily by suppressing the rate of lipogenic gene transcription. Nevertheless, these results show that the dietary HUFA influence lipid synthesis and oxidation at several different regulatory steps including transcription, mRNA processing, mRNA decay, and post-translational modifications.

ARE Ω-3 FATTY ACIDS BETTER FUEL REPARTITIONERS THAN Ω-6?

To answer the relative potency question requires first, a knowledge of the structural requirements for a HUFA to possess inhibitory activity, and second, an appreciation that much of the work evaluating the effectiveness of ω-6 and ω-3 fatty acids has compared vegetable oils with marine oils. Such a comparison not only mixes the effects of n-6 with n-3 fatty acids, but also in essence compares Δ-6 desaturase substrates [for example, 18:2(n-6)] with Δ-6 desaturase products [for example, 20:5(n-3)].

To begin with, an inhibitory HUFA must consist of 18 carbons and contain at least two conjugated double bonds in the 9,12 position. The loss of one of the double bonds by hydroxylation renders the fatty acid inactive, but one of the double bonds may be in a *trans* configuration (15). A fatty acid may contain additional double bonds and retain inhibitory potency (for example, columbinic acid, 5t,9c,12c-18:3) (15). Most important, in order for an HUFA to influence fuel repartitioning it must be a product of the Δ-6 desaturase pathway (16). These fatty acid structural requirements appear also to be instrumental in the HUFA regulation of skeletal muscle metabolism, but unequivocal data establishing these requirements remain to be acquired.

Δ-6 Desaturase has a two- to three-fold greater affinity for 18:3(n-3) than for 18:2(n-6). Consequently, 18:3(n-3) is more effective than 18:2(n-6) as a regulator of lipid synthesis and oxidation. However, only a small amount of 18:2(n-6) and 18:3(n-3) undergoes Δ-6 desaturation ($< 1\%$) (69). Thus it is not surprising that dietary oils rich in fatty acid products of Δ-6 desaturase are approximately three-fold more effective as suppressors of hepatic lipogenesis and inducers of fatty acid oxidation than the 18-carbon HUFA substrates (15). Moreover, when n-6 and n-3 HUFAs are compared on an equal carbon basis, n-6 and n-3 HUFAs appear to be equipotent regulators of lipid metabolism. For example, when 18:3(n-6) and 20:4(n-6) were compared with fish oil, those fatty acids were equipotent inhibitors of hepatic fatty acid biosynthesis (15). Thus, the greater effectiveness of fish oil may have less to do with its n-3 fatty acid content than with the fact that fish oils are rich in 20- and 22-carbon HUFA, which bypass the regulated and required steps of desaturation and elongation. However, that said, one cannot overlook the fact that 20:4(n-6) may increase the production of eicosanoids and thereby offset the health benefits of HUFA (for example, by enhancement of inflammatory responses). On the other hand, 20:5(n-3) would give rise to eicosanoids with low bioactivity, and in this way, would provide benefits for both lipid metabolism and eicosanoid target diseases.

Finally, although n-6 and n-3 fatty acid products of Δ-6 desaturase are modest inducers of fatty acid oxidation, both fatty acid groups are relatively weak ligand activators of PPARα. In contrast, eicosanoid products of n-6 and n-3 HUFAs have two- to three-fold greater affinity for PPARα and hence are more effective activators of PPAR-responsive genes (70). This raises the possibility that the bioactive n-6 or n-3 fatty acid that governs lipid metabolism may be a prostaglandin, leukotriene, or hydroxylated fatty acid. While this question remains unresolved, numerous studies using a wide array of eicosanoid synthesis inhibitors have failed to detect eicosanoid involvement in the regulation of hepatic lipid and glucose metabolism (5,30).

Prostaglandins have been implicated in the inhibition of lipogenic gene expression in 3T3-L1 adipocytes (71), but their involvement in the *in vivo* regulation of lipogenesis remains to be established. In general, there are few data to support the argument that dietary HUFAs exert their influence on hepatic lipogenesis by enhancing eicosanoid production. In fact, the signaling mechanism whereby HUFAs govern lipogenesis and lipogenic gene expression remains a mystery.

ACKNOWLEDGMENTS

Supported by National Institutes of Health grants DK-53872 (S.D.C.) and HD-37133 (W.C.H. and S.D.C.), and by the sponsors of the M. M. Love Chair of Nutritional, Cellular, and Molecular Sciences.

REFERENCES

1. Nichaman MZ, Olson RE, Sweely CL. Metabolism of linoleic acid-1-14-C in normolipemic and hyperlipemic humans fed linoleate diets. *Am J Clin Nutr* 1967; 20: 1070–108.
2. Allman DW, Gibson DW. Fatty acid synthesis during early linoleic acid deficiency in the mouse. *J Lipid Res* 1969; 6: 51–62.
3. Musch K, Ojakian MA, Williams MA. Comparison of alpha-linolenate and oleate in lowering activity of lipogenic enzymes in rat liver: evidence for a greater effect of dietary linolenate independent of food and carbohydrate intake. *Biochim Biophys Acta* 1974; 37: 343–8.
4. Clarke SD, Romsos DR, Leveille GA. Differential effects of dietary methyl esters of long-chain saturated and polyunsaturated fatty acids on rat liver and adipose tissue lipogenesis. *J Nutr* 1977; 107: 1170–81.
5. Flick PK, Chen J, Vagelos PR. Effect of dietary linoleate on synthesis and degradation of fatty acid synthetase from rat liver. *J Biol Chem* 1977; 252: 4242–7.
6. Toussant MJ, Wilson MD, Clarke SD. Coordinate suppression of liver acetyl-CoA carboxylase and fatty acid synthetase by polyunsaturated fat. *J Nutr* 1981; 111: 146–53.
7. Clarke SD, Armstrong MK, Jump DB. Dietary polyunsaturated fats uniquely suppress rat liver fatty acid synthase and S14 mRNA content. *J Nutr* 1990; 120: 225–32.
8. Clarke SD, Abraham S. Gene expression: nutrient control of pre- and posttranscriptional events. *FASEB J* 1992; 6: 3146–52.
9. Clarke SD, Jump DB. Dietary polyunsaturated fatty acid regulation of gene transcription. *Annu Rev Nutr* 1994; 14: 83–98.
10. Jump DB, Clarke SD. Regulation of gene expression by dietary fat. *Annu Rev Nutr* 1999; 19: 63–90.
11. Clarke SD. Polyunsaturated fatty acid regulation of gene transcription: a molecular mechanism to improve the metabolic syndrome. *J Nutr* 2001; 131: 1129–32.
12. Ide T, Kobayashi H, Ashakumary L, *et al.* Comparative effects of perilla and fish oils on the activity and gene expression of fatty acid oxidation enzymes in rat liver. *Biochim Biophys Acta* 2000; 1485: 23–35.
13. Power GW, Newsholme EA. Dietary fatty acids influence the activity and metabolic control of mitochondrial carnitine palmitoyltransferase I in rat heart and skeletal muscle. *J Nutr* 1997; 127: 2142–50.
14. Takada R, Saitoh M, Mori T. Dietary gamma-linolenic acid-enriched oil reduces body fat content and induces liver enzyme activities relating to fatty acid beta-oxidation in rats. *J Nutr* 1994; 124: 469–74.
15. Clarke SD, Jump DB. *Nutrition and gene expression*. Boca Raton: CRC Press, 1993.
16. Nakamura MT, Cho HP, Clarke SD. Regulation of hepatic delta-6 desaturase expression and its role in the polyunsaturated fatty acid inhibition of fatty acid synthase gene expression in mice. *J Nutr* 2000; 130: 1561–5.
17. Storlien LH, Kraegen EW, Chisholm DJ, Ford GL, Bruce DG, Pascoe WS. Fish oil prevents insulin resistance induced by high-fat feeding in rats. *Science* 1987; 237: 885–8.
18. Baur A, O'Connor J, Pan DA, Storlien LH. Relationships between maternal risk of insulin resistance and the child's muscle membrane fatty acid composition. *Diabetes* 1999; 48: 112–16.
19. Odin RS, Adkins-Finke BA, Blake WL, Phinney SD, Clarke SD. Modification of fatty acid composition of membrane phospholipid in hepatocyte monolayer with n-3, n-6 and n-9 fatty acids and its relationship to triacylglycerol production. *Biochim Biophys Acta* 1987; 921: 378–91.

20. Wilson MD, Blake WL, Salati LM, Clarke SD. Potency of polyunsaturated and saturated fats as short-term inhibitors of hepatic lipogenesis in rats. *J Nutr* 1990; 120: 544–54.
21. Baillie RA, Takada R, Nakamura M, Clarke, SD. Coordinate induction of peroxisomal acyl-CoA oxidase and UCP-3 by dietary fish oil: a mechanism for decreased body fat deposition. *Prostaglandins Leukot Essent Fatty Acids* 1999; 60: 351–6.
22. McGarry JD, Brown NF. The mitochondrial carnitine palmitoyltransferase system. From concept to molecular analysis. *Eur J Biochem* 1997; 244: 1–8.
23. Chien D, Dean D, Saha AK, Flatt JP, Ruderman NB. Malonyl-CoA content and fatty acid oxidation in rat muscle and liver in vivo. *Am J Physiol Endocrinol Metab* 2000; 279: E259–65.
24. Abu-Elheiga L, Matzuk MM, Abo-Hashema AH, Wakil SJ. Continuous fatty acid oxidation and reduced fat storage in mice lacking acetyl-CoA carboxylase 2. *Science* 2001; 91: 2613–16.
25. Hellerstein MK. Methods for measurement of fatty acid and cholesterol metabolism *Curr Opin Lipidol* 1995; 6: 172–81.
26. Couet C, Delarue J, Ritz P, Antoine JM, Lamisse F. Effect of dietary fish oil on body fat mass and basal fat oxidation in healthy adults. *Int J Obes* 1997; 21: 637–43.
27. Wong SH, Nestel PJ, Trimble RP, Storer GB, Illman RJ, Topping DL. The adaptive effects of dietary fish and safflower oil on lipid and lipoprotein metabolism in perfused rat liver. *Biochim Biophys Acta* 1984; 792: 103–9.
28. Broadway NM, Saggerson ED. Effect of membrane environment on the activity and inhibitability by malonyl-CoA of the carnitine acyltransferase of hepatic microsomal membranes. *Biochem J* 1997; 322: 435–40.
29. Brandt JM, Djouadi F, Kelly DP. Fatty acids activate transcription of the muscle carnitine palmitoyltransferase I gene in cardiac myocytes via the peroxisome proliferator-activated receptor alpha. *J Biol Chem* 1998; 273: 23786–92.
30. Clarke SD, Turini M, Jump DB, Abraham S, Reedy M. Polyunsaturated fatty acid inhibition of fatty acid synthase transcription is independent of PPAR activation. *Z Ernahrungswiss* 1998; 37: 14–20.
31. Malasanos TH, Stacpoole PW. Biological effects of omega-3 fatty acids in diabetes mellitus. *Diabetes Care* 1991; 14: 1160–79.
32. Clarke SD, Romsos DR, Leveille GA. Influence of dietary fatty acids on liver and adipose tissue lipogenesis and on liver metabolites in meal-fed rats. *J Nutr* 1977; 107: 1277–87.
33. Xu J, Nakamura MT, Cho HP, Clarke SD. Sterol regulatory element binding protein-1 expression is suppressed by dietary polyunsaturated fatty acids. A mechanism for the coordinate suppression of lipogenic genes by polyunsaturated fats. *J Biol Chem* 1999; 274: 23577–83.
34. Issemann I, Green S. Activation of a member of the steroid hormone receptor superfamily by peroxisome proliferators. *Nature* 1990; 347: 645–50.
35. Kliewer SA, Sundseth SS, Jones SA, *et al.* Fatty acids and eicosanoids regulate gene expression through direct interactions with peroxisome proliferator-activated receptors alpha and gamma. *Proc Natl Acad Sci USA* 1997; 94: 4318–23.
36. Krey G, Braissant O, L'Horset F, *et al.* Fatty acids, eicosanoids, and hypolipidemic agents identified as ligands of peroxisome proliferator-activated receptors by coactivator-dependent receptor ligand assay. *Mol Endocrinol* 1997; 11: 779–91.
37. Moya-Camarena SY, Vanden Heuvel JP, Blanchard SG, Leesnitzer LA, Belury MA. Conjugated linoleic acid is a potent naturally occurring ligand and activator of PPARalpha. *J Lipid Res* 1999; 40: 1426–33.
38. Kersten S, Seydoux J, Peters JM, Gonzalez FJ, Desvergne B, Wahli W. Peroxisome proliferator-activated receptor alpha mediates the adaptive response to fasting. *J Clin Invest* 1999; 103: 1489–98.
39. Mascaro C, Acosta E, Ortiz JA, Haro D. Control of human muscle-type carnitine palmitoyltransferase I gene transcription by peroxisome proliferator-activated receptor. *J Biol Chem* 1998; 273: 8560–63.
40. Rodriguez JC, Gil-Gomez B, Hegradt FG, Haro D. Peroxisome proliferator-activated receptor mediates induction of the mitochondrial 3-hydroxy-3-methylglutaryl-CoA synthase gene by fatty acids. *J Biol Chem* 1994; 269: 18767–72.
41. Clarke SD, Thuillier P, Baillie RA, Sha X. Peroxisome proliferator-activated receptors: a family of lipid-activated transcription factors. *Am J Clin Nutr* 1999; 70: 566–77.
42. Varanasi U, Chu R, Huang Q, Castellon R, Yeldani AV, Reddy JK. Identification of a peroxisome proliferator-responsive element upstream of the human peroxisomal fatty acyl coenzyme A oxidase gene. *J Biol Chem* 1996; 271: 2147–55.
43. Aoyama T, Peters JM, Iritani N, *et al.* Altered constitutive expression of fatty acid-metabolizing enzymes in mice lacking the peroxisome proliferator-activated receptor alpha (PPARalpha). *J Biol Chem* 1998; 273: 5678–84.

44. Kakuma T, Lee Y, Higa M, *et al.* Leptin, troglitazone, and the expression of sterol regulatory element binding proteins in liver and pancreatic islets. *Proc Natl Acad Sci USA* 2000; 97: 8536–41.
45. Zhou YT, Grayburn P, Karim A, *et al.* Lipotoxic heart disease in obese rats: implications for human obesity. *Proc Natl Acad Sci USA* 2000; 97: 1784–9.
46. Ren B, Thelen AP, Peters JM, Gonzalez FJ, Jump DB. Polyunsaturated fatty acid suppression of hepatic fatty acid synthase and S14 gene expression does not require peroxisome proliferator-activated receptor alpha. *J Biol Chem* 1997; 272: 26827–32.
47. Matsui H, Okumura K, Kawakami K, Hibino M, Toki Y, Ito T. Improved insulin sensitivity by bezafibrate in rats: relationship to fatty acid composition of skeletal-muscle triglycerides. *Diabetes* 1997; 46: 348–53.
48. Osborne TF. Sterol regulatory element-binding proteins (SREBPs): key regulators of nutritional homeostasis and insulin action. *J Biol Chem* 2000; 275: 32379–82.
49. Magana MM, Koo SH, Towle HC, Osborne TF. Different sterol regulatory element-binding protein-1 isoforms utilize distinct co-regulatory factors to activate the promoter for fatty acid synthase. *J Biol Chem* 2000; 275: 4726–33.
50. Xiong S, Chirala SS, Wakil SJ. Sterol regulation of human fatty acid synthase promoter I requires nuclear factor-Y- and Sp-1-binding sites. *Proc Natl Acad Sci USA* 2000; 97: 3948–53.
51. Mater MK, Thelen AP, Pan DA, Jump DB. Sterol response element-binding protein 1c (SREBP1c) is involved in the polyunsaturated fatty acid suppression of hepatic S14 gene transcription. *J Biol Chem* 1999; 274: 32725–32.
52. Yahagi N, Shimano H, Hasty AH, *et al.* A crucial role of sterol regulatory element-binding protein-1 in the regulation of lipogenic gene expression by polyunsaturated fatty acids. *J Biol Chem* 1999; 274: 35840–4.
53. Brown M, Goldstein JL. A proteolytic pathway that controls the cholesterol content of membranes, cells, and blood. *Proc Natl Acad Sci USA* 1999; 96: 11041–8.
54. Xu J, Teran-Garcia M, Park JH, Nakamura MT, Clarke SD. Polyunsaturated fatty acids suppress hepatic sterol regulatory element-binding protein-1 expression by accelerating transcript decay. *J Biol Chem* 2001; 276: 9800–7.
55. Shimomura I, Bashmakov Y, Ikemoto S, Horton JD, Brown MS, Goldstein JL. Insulin selectively increases SREBP-1c mRNA in the livers of rats with streptozotocin-induced diabetes. *Proc Natl Acad Sci USA* 1999; 96: 13656–61.
56. Hannah VC, Ou J, Luong A, Goldstein JL, Brown MS. Unsaturated fatty acids down-regulate srebp isoforms 1a and 1c by two mechanisms in HEK-293 cells. *J Biol Chem* 2001; 276: 4365–72.
57. Shimoura I, Hammer R, Richardson JA, *et al.* Insulin resistance and diabetes mellitus in transgenic mice expressing nuclear SREBP-1c in adipose tissue: model for congenital generalized lipodystrophy. *Genes Dev* 1998; 12: 3182–94.
58. Shimano H, Yahagi N, Amemiya-Kufo M, *et al.* Sterol regulatory element-binding protein-1 as a key transcription factor for nutritional induction of lipogenic enzyme genes. *J Biol Chem* 1999; 274: 35832–9.
59. Sessler AM, Ntambi JM. Polyunsaturated fatty acid regulation of gene expression. *J Nutr* 1998; 128: 923–6.
60. Xu J, Garcia-Teran M, Cho HP, Clarke SD. Polyunsaturated fatty acids (PUFA) regulate the action of sterol regulatory element binding protein-1 (SREBP-1) by inhibiting SREBP-1 nuclear localization and by enhancing SREBP-1 mRNA decay [abstract]. *FASEB J* 2000; 14: A210.
61. Kletzien RF, Foellmi LA, Harris PKW, Wyse BM, Clarke SD. Adipocyte fatty acid-binding protein: regulation of gene expression in vivo and in vitro by an insulin-sensitizing agent. *Mol Pharmacol* 1992; 42: 558–62.
62. Jump DB, Badin MV, Thelen A. The CCAAT box binding factor, NF-Y, is required for thyroid hormone regulation of rat liver S14 gene transcription. *J Biol Chem* 1997; 272: 27778–86.
63. Liimatta M, Towle HC, Clarke SD, Jump DB. Dietary polyunsaturated fatty acids interfere with the insulin/glucose activation of L-type pyruvate kinase gene transcription. *Mol Endocrinol* 1994; 8: 1147–53.
64. Hertz R, Magenheim J, Berman I, Bar-Tana J. Fatty acyl-CoA thioesters are ligands of hepatic nuclear factor-4alpha. *Nature* 1998; 392: 512–16.
65. Rufo C, Gasperikova D, Clarke SD, Teran-Garcia M, Nakamura MT. Identification of a novel enhancer sequence in the distal promoter of the rat fatty acid synthase gene. *Biochem Biophys Res Commun* 1999; 261: 400–5.
66. Rufo C, Teran-Garcia M, Nakamura MT, Koo SH, Towle HC, Clarke SD. Involvement of a unique carbohydrate-responsive factor in the glucose regulation of rat liver fatty-acid synthase gene transcription. *J Biol Chem* 2001; 276; 21969–75.

67. Stabile LP, Klautky SA, Minor SM, Salati LM. Polyunsaturated fatty acids inhibit the expression of the glucose-6-phosphate dehydrogenase gene in primary rat hepatocytes by a nuclear posttranscriptional mechanism. *J Lipid Res* 1998; 39: 1951–63.
68. Amir-Ahmady B, Salati LM. Regulation of the processing of glucose-6-phosphate dehydrogenase mRNA by nutritional status. *J Biol Chem* 2001; 276: 10514–23.
69. Pawlosky R J, Hibbeln JR, Novotny JA, Salem N. Physiological compartmental analysis of alpha-linolenic acid metabolism in adult humans. *J Lipid Res* 2001; 42: 1257–65.
70. Devchand PR, Keller H, Peters JM, Vaquez M, Gonzalez FM, Wahli W. The PPARalpha-leukotriene B4 pathway to inflammation control. *Nature* 1996; 384: 39–43.
71. Mater MK, Pan D, Bergen WG, Jump DB. Arachidonic acid inhibits lipogenic gene expression in 3T3-L1 adipocytes through a prostanoid pathway. *J Lipid Res* 1998; 39: 1327–34.

DISCUSSION

Dr. Kon: What about the antioxidant effects of ω-3 fatty acids? You spoke mainly about oxidation, but there are data on the antioxidant effects of fish oil supplements, in woman particularly.

Dr. Clarke: I removed a data set because I needed to shorten my presentation. We've finished the DNA arrays and measured 10,000 gene products. Interestingly, a high olive oil and high fish oil diet decreased DNA repair enzymes, high olive oil having the greatest effect. Therefore, the implication is that there's a decrease in apoptosis, perhaps due to the antioxidant components of olive oil or perhaps due to the 18-1 itself. We also saw some changes in skeletal muscle inflammatory responses, but they didn't go the way you'd expect them to—that is, the olive oil had some effects but fish oil did not, and it didn't enhance inflammation. So it doesn't look as though fish oil is going to cause any additional oxidative stress in skeletal muscle. We saw more positive effects with olive oil than we did with fish oil. Peroxisomal proliferation occurs, which means there will be increased hydrogen peroxide production as well, but that's about as much as we've done in terms of looking at antioxidant gene expression.

Dr. Scott: I have a philosophical question. If you go along with the idea that the genetic blueprint was determined thousands of years ago when men were hunter–gatherers and presumably living off vegetables and animal foodstuffs, where did all this fish come from?

Dr. Clarke: Fish oil happens to be an effective tool. Arachidonic acid or GLA, its precursor, will have the same inhibitory effects as fish oil fatty acids. Anything that's a product of the Δ6 desaturase that's derived from n-6, n-3 family will inhibit lipogenic gene expression and decrease triglyceride synthesis. Now, why did that evolve? Probably because too much oleic acid increases the essential fatty acid requirement. When we used oleic acid in our hepatocyte preparation, we always got increased fatty acid synthase expression. So, my guess is that this evolved as a regulatory mechanism so that the liver would not accumulate excessive amount of triglyceride or synthesize excessive 18-0 and 18-1, while being able to conserve its precursors for essential fatty acids.

Dr. Böhles: In classical biochemistry, the shorter the fatty acid and the less unsaturated, the greater the degree of self-oxidation. From your research, would you say that PUFAs are only signal-providing substances or are they also oxidized themselves?

Dr. Clarke: They are oxidized, but depending on length. Their rate of oxidation depends greatly upon the number of double bonds in the carbon link. The first four steps in 20-carbon polyunsaturate oxidation take place in the peroxisome, which is probably why we get peroxisomal proliferation. After that, the mitochondria take over. Therefore, something like DHA is oxidized very slowly. Linoleate is readily oxidized, but probably not more rapidly than palmitate or oleate.

Dr. Wanders: What always strikes me about these nuclear hormone receptors is that the most powerful ligands are the free fatty acids, but in the cell I would be very surprised if the ligands were present as free acids. Could you comment?

Dr. Clarke: I agree with you that there is very little free acid in the cell. My guess is that the ligands for these fatty acid binding receptors are not truly fatty acids. The assays are set up to make it look as though fatty acids are the activators, but any lipophilic compound will bind and activate. You can't get the type of separation of function between saturates, monounsaturates, and polyunsaturates if your investigation is solely based upon binding constants of lipid to proteins—they are just not large enough to give you adequate levels of difference in biologic responses. In order to achieve that level of selectivity, you must arrange for some sort of metabolic transformation of the fatty acid, probably to an oxide or a hydroxide. In that way, you can get orders of magnitude difference in binding activity. I'm sure that it's not directly the result of the fatty acid.

Dr. Bachmann: How are the PUFAs transported across the plasma membrane, and how are they exchanged in plasma?

Dr. Clarke: I think there is good evidence that CD36 is a fatty acid transport protein in certain cell types, and there has been a lot of work showing impaired fatty acid utilization in knockout mice (1). So there is involvement of a transporter for fatty acids. The question is whether that transporter has selectivity for polyunsaturates *versus* saturates. With knockout mice and using genetic manipulations such as overexpression, we should be able to determine eventually whether one fatty acid or another is preferred.

Dr. Bier: What was called the nuclear hormone super receptor family was discovered because when individual examples of the class were identified people started looking for other receptors and other members of the class, using molecular characteristics and concentrations. Thus, they completely missed a whole family of receptors because the concentration differences and characteristics were different. Now that we've dropped the "hormone" out of the name of the class, we're discovering a new family of orphan receptors or potential ligands. These studies may provide us with information about the dietary ligands that regulate these receptor families.

Dr. Clarke: In my opinion, nutrients were the first hormones. The cell had to respond to its nutrient environment and be able to survive, and hormones represented a more sophisticated system to ensure this. I think the next hot target will be the LXR receptor. This is an orphan receptor that apparently binds oxysterols and may also bind polyunsaturates. Thus, it is another member of the steroid receptor superfamily and has some very interesting characteristics. There are several members of this family of orphan receptors that are being found to bind metabolites of lipids.

Dr. Daniel: I would like to comment on the fatty acid transport proteins. As far as I know, six genes have now been isolated. These proteins are expressed at different levels in different tissues but all the functional data I'm aware of show that they have a very high preference for saturated fatty acids and show only weak interaction with polyunsaturated fatty acids.

Dr. Clarke: That could well be. Although I'm not a lipid biochemist, I think that there are many problems with fatty acid transport studies, and I don't have a lot of confidence that we can do them correctly. There was a time when it was thought that fatty acids inhibited virtually every enzyme in glycolysis and fatty acid biosynthesis, but it turned out that this was a denaturing effect and had no biological significance at all. Therefore, I'm a skeptic when it comes to those kinds of studies.

Dr. Daniel: You said that PUFAs could inhibit fatty acid synthase, or reduce its expression, and so diminish triglyceride synthesis in liver, while at the same time improving β oxidation.

What is the mechanism behind that? Just reducing synthesis of triglycerides does not necessarily mean that the fatty acids go to β oxidation, at least not in humans.

Dr. Clarke: In the rat, the mechanism is that we inhibit the lipogenic side, but at the same time we stimulate the oxidative side and lower malonyl-CoA concentrations, which allows the fatty acid to enter the mitochondria more readily. So you have polyunsaturates working in both ways: down-regulating synthesis and up-regulating oxidation. That happens in the liver and probably in the β cell; it also definitely happens in the heart. Whether or not it happens in skeletal muscle seems to be an open question. All the studies I'm aware of in skeletal muscle that have claimed that polyunsaturates stimulate fatty acid oxidation in the muscle have used diets that are very high in fat—40% polyunsaturated or 40% fish oil, which are not meaningful levels. When you look at lower levels, you don't see the kind of gene expression control that one would predict from the high-level studies. I think the increased oxidation of fatty acid in skeletal muscle—if it occurs at all—will most probably occur through a metabolite control mechanism rather than through changes in enzyme amount and gene expression.

Dr. Borum: Is the metabolic syndrome–like phenotype that is seen in HIV patients treated with some antiretroviral agents associated with the story you have recounted? In addition, could you utilize any of this information for the benefit of such patients?

Dr. Clarke: My familiarity with this is not great. My understanding of this condition is that it resembles a lipid dystrophy, so the problem is a different one in that there is an inability to get glucose into the fat cells. Here we are talking about a high correlation between lipid accumulation and cell death, and then the development of insulin resistance, so I think the scenarios are different. In adipose tissue, you don't see this polyunsaturate control of the fat.

Dr. Böhles: If you take the situation where malonyl-CoA influences CPT-1 (carnitine palmitoyltransferase 1), have you any idea how polyunsaturated fatty acids could influence the KI with respect to infinity of fatty acids for CPT?

Dr. Clarke: There have been good data in the last year or so showing that making the mitochondrial membranes more unsaturated increases the Vmax for CPT-1 (2). Therefore, that is a case where you do have changes in kinetic behavior that accelerate fatty acid oxidation by CPT-1 transport. We have no data yet with respect to KIs for polyunsaturates in lowering malonyl-CoA. Several years ago, we did show the decrease in malonyl-CoA in the liver, but we didn't look at KIs. The evolving malonyl-CoA decarboxylase story opens another door in determining whether polyunsaturates accelerate decarboxylation of malonyl-CoA, and I think that is going to be important. The polyunsaturates we're talking about here must be products of the Δ6-desaturase. If we knock out the Δ6-desaturase, we do not see any of this gene expression control on the lipogenic side.

Dr. Bachmann: In relation to PUFAs, are the differences between the two chain lengths, 20 and 22?

Dr. Clarke: Probably not in the studies we're looking at in terms of lipogenic control and fat synthesis. In terms of fatty acid oxidation, I would say that DHA is probably going to be more potent. There aren't many dose–response studies available, however.

Dr. Bachmann: You also mentioned the monounsaturated fatty acids, and you said at the end of your presentation that you should use four parts of olive oil, one part of fish oil, and one part of vegetable oil. Can you comment on that?

Dr. Clarke: That's based on our micro arrays. These have turned up at least six DNA repair enzymes that are down-regulated with olive oil. This is a significant change and indicates that the olive oil is protecting the muscle DNA against damage. The antioxidant components of olive oil may be responsible, but we don't know that yet. My recommendation may be somewhat provocative, but we can accomplish the triglyceride effects—the lowering of blood and tissue triglycerides, in the liver particularly—with about 10% of dietary energy from fish oil type fatty acids. In respect of the safflower oil, we need certain amounts of essential fatty acids for growth, but the effects on gene expression don't saturate until you reach about 10% of energy. So for both the n-3 and the n-6 fatty acids about 10% of dietary energy is needed. Then based upon the marked changes in reducing DNA repair enzymes, I would provide the balance as olive oil. That's where those suggestions came from.

REFERENCES

1. Coburn CT, Knapp FF, Febbraio M, *et al.* Defective uptake and utilization of long chain fatty acids in muscle and adipose tissues of CD 36 knockout mice. *J Biol Chem* 2000; 275: 32523–9.
2. Power GW, Yoqqob P, Harvey DJ, Newsholme EA, Calder PC. The effect of dietary lipid manipulation on hepatic mitochondrial phospholipid fatty acid composition and carnitine palmitoyltransferase I activity. *Biochem Molec Biol Intern* 1994; 34: 671–84.

Genetic Expression and Nutrition, edited by Claude Bachmann and Berthold Koletzko. Nestlé Nutrition Workshop Series, Pediatric Program, Vol. 50. Nestec Ltd., Vevey/Lippincott Williams & Wilkins, Philadelphia, © 2003.

Genetic Disorders of Mitochondrial and Peroxisomal Fatty Acid Oxidation and Peroxisome Proliferator-Activated Receptors

Ronald J. A. Wanders

Department of Pediatrics, Academic Medical Center; Department of Clinical Chemistry, University of Amsterdam, Amsterdam, The Netherlands

Fatty acids are an important source of energy in humans, especially during fasting. Most tissues are able to degrade fatty acids to carbon dioxide and water, but in addition, some organs—notably the liver—have the capacity to convert the acetyl-CoA units produced during β oxidation into the ketone bodies acetoacetate and 3-hydroxybutyrate. These are important fuels for certain organs, especially the brain. Fatty acids are the main source of energy in the heart. Indeed, under normal conditions 60% to 70% of the energy requirements of the heart are provided by fatty acid oxidation, and even more under certain conditions such as fasting and diabetes.

In principle, fatty acids can be degraded by several mechanisms, including α, β, and ω oxidation. However, most fatty acids are degraded by β oxidation, which can occur in the mitochondria or in the peroxisomes. It is generally agreed that the mitochondria form the major site of fatty acid β oxidation, catalyzing the oxidation of all major dietary fatty acids. In recent years, tremendous progress has been made in the identification of patients affected by a genetic deficiency in either the mitochondrial or the peroxisomal β oxidation system, and defects in virtually all the steps involved have now been identified. Thanks to these efforts, reliable laboratory methods have now become available to identify patients with these conditions. Much remains to be learned, however, about the pathophysiologic mechanisms behind these disorders. In this respect, the role of peroxisome proliferator-activated receptors (PPARs) is intriguing, especially since the metabolites accumulating as a result of a block in mitochondrial or peroxisomal β oxidation are PPAR activators.

In this chapter, I will describe our current state of knowledge about mitochondrial and peroxisomal fatty acid oxidation, with particular emphasis on the various inherited diseases and the role of PPARs.

FATTY ACID OXIDATION

Fatty acids may undergo α, β, or ω oxidation.

Alpha-Oxidation

α-oxidation is required for fatty acids with a methyl group at the 3-position, as 3-methyl branched chain fatty acids cannot undergo direct β oxidation. The mechanism of fatty acid α oxidation has remained mysterious ever since its discovery in the early 1960s, when the accumulation of the 3-methyl branched chain fatty acid phytanic acid (3,7,11,14-tetramethylhexadecanoic acid) was identified in patients suffering from a rare genetic disease known as Refsum disease. Patients with Refsum disease show various abnormalities including retinitis pigmentosa, cerebellar ataxia, peripheral neuropathy, anosmia, and cardiac abnormalities (1). The finding of markedly raised phytanic acid concentrations has led to detailed studies of phytanic acid oxidation. These studies showed that phytanic acid undergoes oxidative decarboxylation, with pristanic acid (2,4,6,10-tetramethylpentadecanoic acid) and CO_2 as the end products, and that patients with Refsum disease cannot convert phytanic acid to pristanic acid and CO_2. Despite intense efforts, the mechanism of phytanic acid α oxidation remained unclear until very recently. However, work from our laboratory and others has now established the sequence of events (Fig. 1A), and most of the enzymes involved have been characterized, purified, and studied at the molecular level (2). This led to the identification of the enzyme defect in Refsum disease—at the level of phytanoyl-CoA hydroxylase (3)—and later to the resolution of the molecular basis of Refsum disease (4). Interestingly, phytanic acid has been found to be a powerful ligand for some nuclear receptors, including PPARα (5–8), as discussed below.

Omega-Oxidation

Ω Oxidation is a relatively minor pathway, like α oxidation. The initial step occurs in the smooth endoplasmic reticulum by ω hydroxylases belonging to the CYP4A subfamily, with a ω-hydroxy fatty acid as the product. The CYP4A family of cytochrome P450 enzymes, which are constitutively expressed in liver and kidney, catalyze both the ω and the (ω-1) hydroxylation of long-chain fatty acids (9). The major ω-hydroxylase expressed in human liver and kidney is CYP4A11, which is functionally similar to the rat CYP4A1. The ω-hydroxy fatty acids produced by the initial hydroxylation reaction then undergo two dehydrogenations to generate the corresponding ω-keto fatty acids and ω-carboxy fatty acids. These reactions are catalyzed by alcohol and aldehyde dehydrogenases in the cytosol. Subsequently, the dicarboxylic acids are activated to their CoA esters by an acyl-CoA synthetase present in the endoplasmic reticulum (10). The long-chain dicarboxylyl-CoA esters produced then appear to undergo several β oxidation cycles in the peroxisome to produce medium-chain dicarboxylyl-CoA esters that undergo final oxidation to CO_2 and H_2O in the mitochondrion (Fig. 1B).

Beta-Oxidation

β Oxidation is the main mechanism whereby fatty acids are broken down, and involves a sequence of four reactions including dehydrogenation, hydration, dehydrogenation again, and thiolytic cleavage (Fig. 2). Both mitochondria and peroxisomes

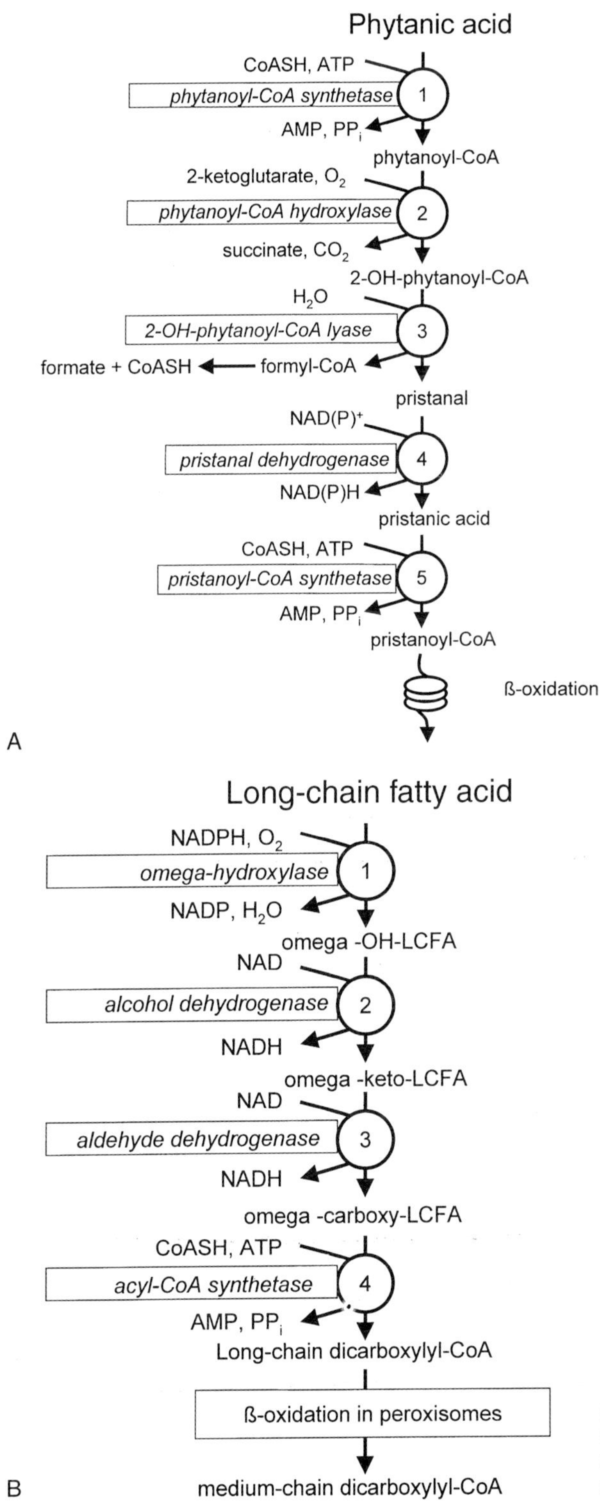

FIG. 1. Schematic representation of the fatty acid α oxidation system **(A)** and the ω oxidation system **(B)**.

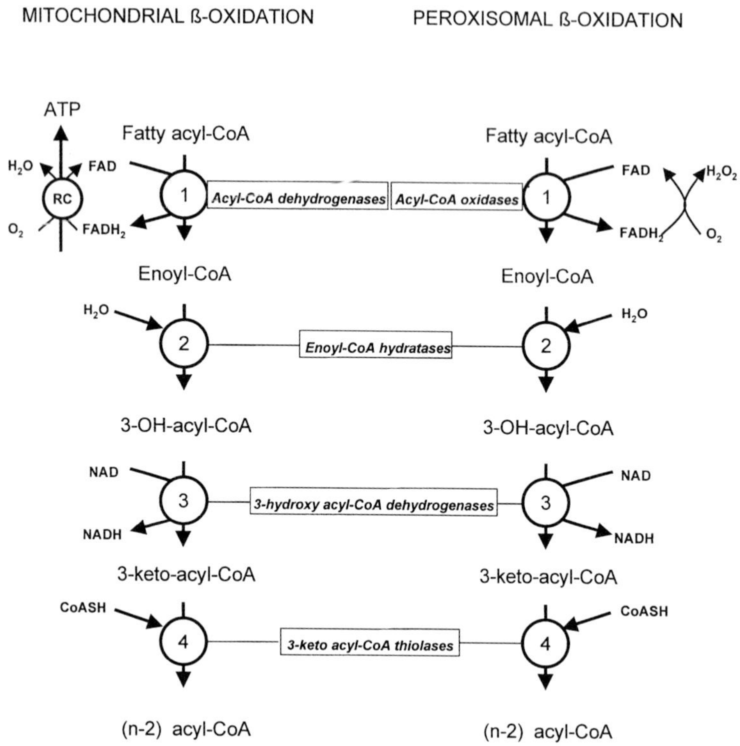

FIG. 2. The mitochondrial and peroxisomal β oxidation of fatty acids. The first step in the β oxidation of fatty acids in mitochondria is catalyzed by acyl-CoA dehydrogenases but in peroxisomes by acyl-CoA oxidases. Both acyl-CoA dehydrogenases and acyl-CoA oxidases are flavoproteins which use different mechanisms for the reoxidation of enzyme-bound $FADH_2$. In case of mitochondrial acyl-CoA dehydrogenases, enzyme-bound E-$FADH_2$ is reoxidized by electron transfer flavoprotein (ETF) followed by reoxidation of reduced ETF by ETF-dehydrogenase, so that the electrons ultimately enter the respiratory chain at the level of ubiquinone. In case of acyl-CoA oxidases, enzyme-bound $FADH_2$ is directly reoxidized by molecular oxygen to produce H_2O_2. Steps 2, 3, and 4 are catalyzed by enoyl-CoA hydratases, 3-hydroxyacyl-CoA dehydrogenases, and 3-ketothiolases, which catalyze identical reactions. RC, respiratory chain.

use the same mechanism for β oxidation of fatty acids, although the two systems differ in many respects as summarized below.

Oxidation of Fatty Acids to CO_2 and H_2O in Mitochondria Versus Peroxisomes

Peroxisomes lack a citric acid cycle and thus cannot degrade the acetyl-CoA units produced during fatty acid oxidation to CO_2 and H_2O. Available evidence indicates that the acetyl-CoA units produced in peroxisomes are transferred to mitochondria in the form of acetylcarnitine (11,12).

Peroxisomal β Oxidation Versus Mitochondrial β Oxidation

In mitochondria the first step in the β oxidation of fatty acids is catalyzed by various acyl-CoA dehydrogenases, all FAD-linked, which donate their electrons to the respiratory chain, thus generating adenosine triphosphate (ATP). In peroxisomes, however, the acyl-CoA oxidases—which are also FAD linked—donate their electrons directly to molecular oxygen to produce H_2O_2 which is subsequently decomposed to H_2O and O_2. Consequently, one cycle of β oxidation in peroxisomes is at most half as efficient as β oxidation in mitochondria in terms of ATP production.

Role of Carnitine in Peroxisomal and Mitochondrial β Oxidation

Carnitine plays an indispensable role in both mitochondrial and peroxisomal β oxidation, but at different levels. In the mitochondria, carnitine is involved in the transfer of long-chain fatty acids across the mitochondrial inner membrane by the concerted action of carnitine palmitoyltransferase 1 (CPT1), mitochondrial carnitine/acylcarnitine translocase (CACT), and carnitine palmitoyltransferase 2 (CPT2) (13). In peroxisomes, however, carnitine plays no role in fatty acid uptake but does play an indispensable role in the transfer of chain-shortened fatty acids from peroxisomes to mitochondria (12) (Fig. 3).

Regulation of Mitochondrial and Peroxisomal β Oxidation

Mitochondrial but not peroxisomal β oxidation is under rapid short-term control by malonyl-CoA, the key regulator of mitochondrial fatty acid β oxidation at the level of CPT1. On the other hand, despite this difference in short-term control, the long-term control of both systems shares some features, with major effects on both systems exerted by PPARα ligands.

Genetic Basis of Enzymes Involved in Fatty Acid β Oxidation in Peroxisomes and Mitochondria

Although the reactions involved in peroxisomal and mitochondrial β oxidation are identical, these reactions are catalyzed by different enzymes that are in general encoded by distinct genes—with a few exceptions where a single gene codes for a protein directed to both peroxisomes and mitochondria. Examples are carnitine acetyltransferase, Δ3,5,Δ2,4-dienoyl-CoA reductase, and 2-methylacyl-CoA racemase (2).

Substrate Specificities of the Mitochondrial and Peroxisomal β Oxidation Systems

From a physiologic viewpoint, the most important difference between the mitochondrial and the peroxisomal β oxidation systems is that the two systems have different substrate specificities. Realization that the peroxisomal system handles a distinct set of

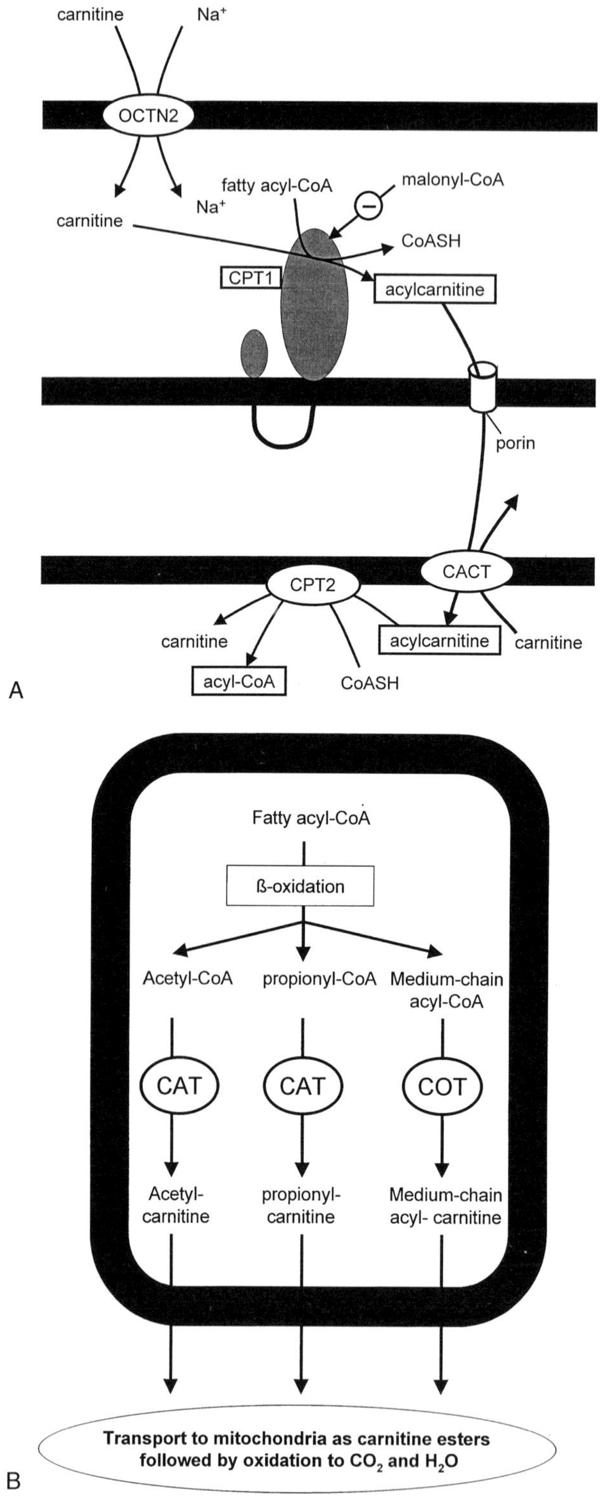

FIG. 3. Illustration of the different roles of carnitine in mitochondrial **(A)** and peroxisomal **(B)** β oxidation. See text for details. CAT, carnitine acetyltransferase; COT, carnitine octanoyltransferase; MIM, mitochondrial inner membrane; MOM, mitochondrial outer membrane.

substrates has largely come from studies on a rare genetic disease—the so-called cerebrohepatorenal syndrome of Zellweger or Zellweger syndrome. In patients with Zellweger, peroxisome biogenesis is defective owing to a genetic defect in one of the many genes involved in peroxisome biogenesis, leading to the complete absence of peroxisomes. Careful studies in plasma and cells from these patients have shown that peroxisomes play an indispensable role in the β oxidation of the following fatty acids (Fig. 4):

- Very long-chain fatty acids, notably hexacosanoic acid (C26:0);
- Certain 2-methyl branched chain fatty acids including pristanic acid [2,4, 6,10-tetramethylpentadecanoic acid], which is in part derived from dietary sources and in part from α oxidation of phytanic;
- Di- and trihydroxycholestanoic acids, as produced in the liver from cholesterol: After activation to their CoA esters, these undergo β oxidation in the peroxisome to produce chenodeoxycholoyl-CoA and choloyl-CoA, respectively, which are

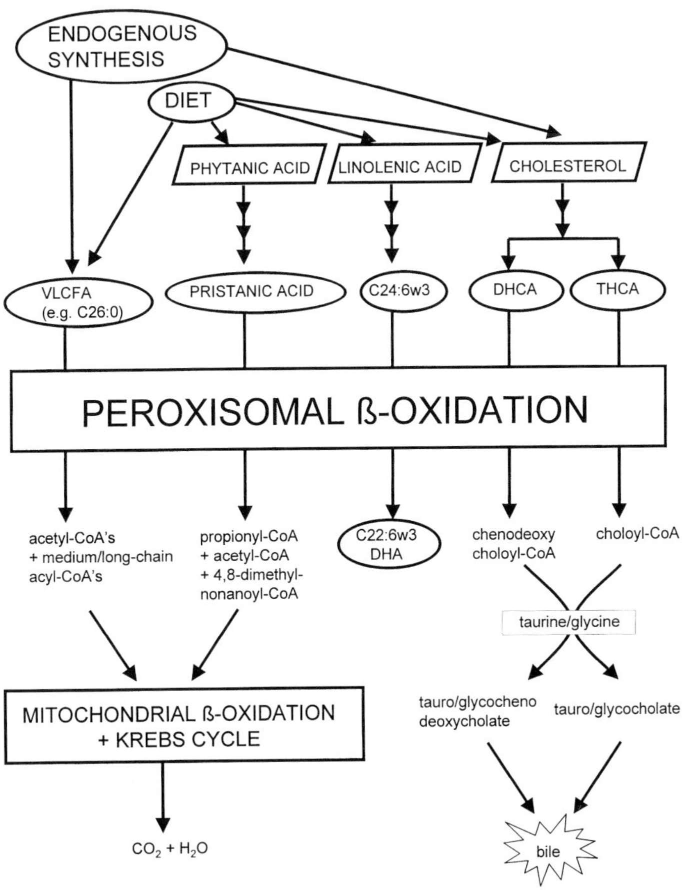

FIG. 4. Simplified scheme depicting the different roles of mitochondria and peroxisomes in cellular fatty acid β oxidation.

then converted into the corresponding tauro- and/or glycoconjugates with tauro/glycochenodeoxycholate, and tauro/glycocholate as end products;
- Peroxisomes play a major role in the β oxidation of a series of mono- and polyunsaturated fatty acids; recent studies have shown that the conversion of C24:6 into C22:6, the last step in the synthesis of docosahexaenoic acid (DHA, C22:6) from linolenic acid, is mediated by the peroxisomal β oxidation system and not by the mitochondrial system (Fig. 4).

In addition to the substrate listed above, peroxisomes also catalyze the β oxidation of a range of additional fatty acids and fatty acid derivatives (14).

FATTY ACID TRANSPORT

Before fatty acids can undergo β oxidation in the mitochondrion or peroxisome, transport across the plasma membrane must occur. Although it was long believed that this was by simple diffusion, increasing evidence favors the involvement of a carrier-mediated process, although the precise mechanism is still to be worked out. Three different proteins seem to be involved in transmembrane translocation of long-chain fatty acids (LCFA): FARPpm, FAT/CD36, and FATP. All three proteins probably play a role in LCFA transport, with FAT/CD36 as the best candidate for the actual transmembrane translocation (15). Following transport, fatty acids and fatty acyl-CoA esters are further carried through the cytosol to their actual site of action or metabolism by the proteins FABP and ACBP, respectively.

Although the mechanism of transport of fatty acids across the mitochondrial membrane has been resolved in detail (Fig. 3A), much remains to be learned about the transport across the peroxisomal membrane, although there is increasing evidence to suggest the involvement of members of the family of ATP binding cassette (ABC) proteins. Four of these proteins have been identified in peroxisomal membranes. These are ALDP, ALDRP, PMP70, and PMP69 (2).

ENZYMOLOGY OF THE MITOCHONDRIAL AND PEROXISOMAL β OXIDATION SYSTEMS

Mitochondrial β Oxidation Enzymes

Studies over the years have clearly shown that the four reactions of the β oxidation spiral are not catalyzed by single enzymes covering the whole spectrum of different substrates. Instead, each reaction is catalyzed by multiple enzymes, each having a certain chain length specificity (16). Full oxidation of a LCFA such as palmitate to its 8 acetyl-CoA units requires the active participation of the following:

- Three acyl-CoA dehydrogenases [very long-chain (VLCAD), medium-chain (MCAD), and short-chain (SCAD) acyl-CoA dehydrogenase];
- Two enoyl-CoA hydratases [short-chain (SCEH) and long-chain (LCEH) enoyl-CoA hydratase]; it should be noted that LCEH is not a monofunctional enzyme

but part of the mitochondrial trifunctional protein (MTP), an octamer of 4α and 4β subunits with long-chain enoyl-CoA hydratase (LCEH), long-chain 3-hydroxyacyl-CoA dehydrogenase (LCHAD), and long-chain 3-ketothiolase (LCKT) activity;

- Two 3-hydroxyacyl-CoA dehydrogenases, including short-chain 3-hydroxyacyl-CoA dehydrogenase (SCHAD) and long-chain 3-hydroxyacyl-CoA dehydrogenase, again part of MTP;
- Two 3-kethothiolases, including medium-chain 3-ketothiolase (MCKT) and long-chain thiolase, again part of MTP.

The fact that the mitochondrial enzymes involved in long-chain fatty acid β oxidation (VLCAD and MTP) are membrane-bound, whereas the other enzymes are soluble enzymes localized in the mitochondrial matrix, suggests a different spatial organization, as shown schematically in Fig. 5.

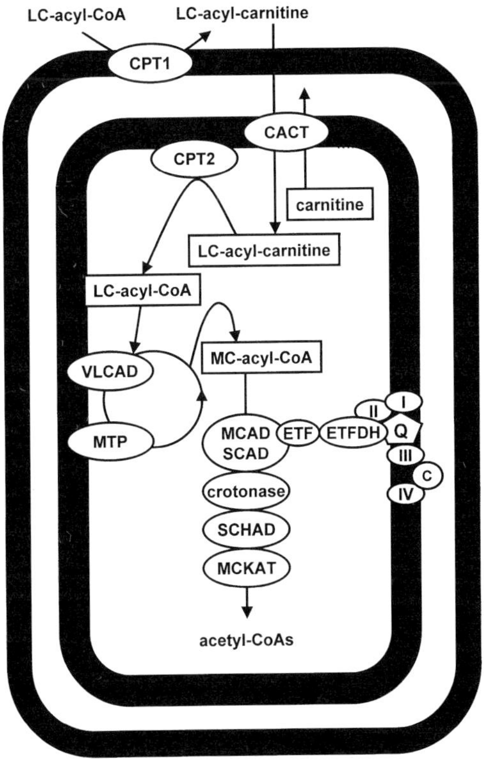

FIG. 5. Intramitochondrial organization of the mitochondrial fatty acid β oxidation machinery. The fact that carnitine palmitoyltransferase 2, VLCAD, and MTP are all membrane-bound enzymes, in contrast to the other enzymes involved in the complete oxidation of long-chain fatty acids, suggests that the β oxidation of long-chain and medium-/short-chain acyl-CoAs occurs in different submitochondrial compartments.

Peroxisomal β Oxidation Enzymes

Like the mitochondria, the peroxisomes contain multiple enzyme proteins to catalyze the oxidative chain shortening of the various acyl-CoA esters, including two acyl-CoA oxidases, two so-called bifunctional proteins with enoyl-CoA hydratase and 3-hydroxyacyl-CoA dehydrogenase activities, and two 3-ketothiolases (14).

Acyl-CoA Oxidases

Human peroxisomes contain two acyl-CoA oxidases with different substrate specificities. The first oxidase (ACOX1) is the human equivalent of the clofibrate inducible rat enzyme identified by Hashimoto and coworkers, and handles straight chain acyl-CoAs. The enzyme is *not* reactive with 2-methyl branched chain acyl-CoAs, like pristanoyl-CoA and trihydroxycholestanoyl-CoA, which are handled by the second oxidase (ACOX2), also known as branched chain acyl-CoA oxidase.

Bifunctional Proteins

Human peroxisomes contain two bifunctional proteins. Both enzymes catalyze the conversion of enoyl-CoA esters into the respective 3-keto esters, but they do this in different ways through an L-3-hydroxy and a D-3-hydroxy intermediate, respectively. Different names have been given to these proteins, including L-BP/D-BP, L-PBE/D-PBE, MFEI/MFEII, and MFP1/MFP2. Studies in humans (17–19) and in mutant

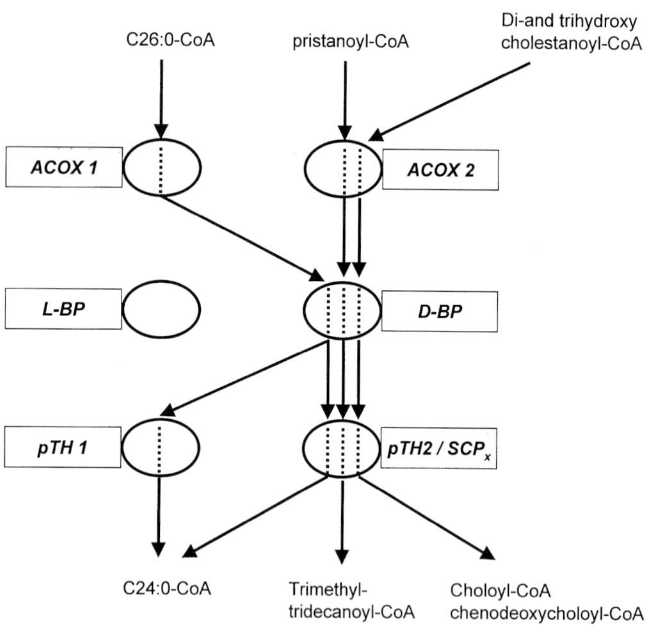

FIG. 6. Enzymology of the peroxisomal fatty acid β oxidation system. See text.

mice (20) have shown that D-BP is the enzyme involved in the oxidation of C26:0, pristanic acid, and di- and trihydroxycholestanoic acids. The function of L-BP remains to be established. Clofibrate feeding in rats leads to a large induction of L-BP but not of D-BP.

3-Ketothiolases

Human peroxisomes contain two thiolases, one of which is the human counterpart of the clofibrate inducible thiolase identified by Hashimoto (21). The second thiolase is the 58-kDa sterol carrier protein/3-ketothiolase identified by Seedorf *et al.* (22), which goes by the name SCPx or alternatively pTH2 (2). SCPx/pTH2 plays an indispensable role in the β oxidation of 2-methyl branched chain fatty acids such as pristanic acid and di- and trihydroxycholestanoic acids, as revealed by several studies, including ones carried out in SCPx (−/−) mice (23) (Fig. 6).

INHERITED DISORDERS OF MITOCHONDRIAL AND PEROXISOMAL OXIDATION

As described above, much has been learned about inherited disorders of mitochondrial and peroxisomal β oxidation, and defects in almost all the enzymes and transporters involved have been identified recently. The introduction of tandem mass spectrometry for the analysis of plasma acylcarnitine species has greatly stimulated progress in the field of mitochondrial fatty acid oxidation disorders, while detailed studies on patients with Zellweger syndrome have paved the way for the identification of peroxisomal fatty acid oxidation disorders.

Mitochondrial Fatty Acid Oxidation Disorders

Table 1 lists the various mitochondrial fatty acid oxidation defects. The clinical manifestations of these defects are diverse and depend on the nature of the enzyme block. The main organs directly affected by a defect in mitochondrial fatty acid oxidation are the liver (fatty change, microvesicular steatosis, liver failure), the heart (acute heart block, progressive cardiomyopathy, arrhythmia, tachycardia), skeletal muscle (rhabdomyolysis), brain (energy deficit), and the kidneys (renal failure, renal tubular acidosis). Depending on the primary organ involved, the clinical manifestations of the fatty acid oxidation defects may range from a pure hepatic or cardiac presentation to one dominated by skeletal or kidney involvement. In many cases, these clinical manifestations occur in different combinations and these may vary with the age of the affected individual. In general, patients affected by a defect in the oxidation of long-chain fatty acids—such as in mitochondrial carnitine/acylcarnitine translocase (CACT) deficiency, carnitine palmitoyltransferase 2 deficiency, VLCAD deficiency, and LCHAD/MTP deficiency—have a bad prognosis, usually with early death from cardiac abnormalities, especially in cases of severe enzyme deficiency.

TABLE 1. *Disorders of mitochondrial and peroxisomal fatty acid β oxidation*

Mitochondrial FAO disorder	Gene involved
1. Plasma membrane Na^+/carnitine transport defect	*OCTN2*
2. Carnitine palmitoyl transferase 1 deficiency (CPT1)	*M-CPT1*
3. Mitochondrial carnitine/acylcarnitine translocase (CACT) deficiency	*CACT*
4. Carnitine palmitoyl transferase 2 deficiency (CPT2)	*CPT2*
5. Very-long-chain acyl-CoA dehydrogenase (VLCAD) deficiency	*VLCAD*
6. Medium-chain acyl-CoA dehydrogenase (MCAD) deficiency	*MCAD*
7. Short-chain acyl-CoA dehydrogenase (SCAD) deficiency	*SCAD*
8. Long-chain 3-hydroxyacyl-CoA dehydrogenase (LCHAD) deficiency	*MTP*
9. Mitochondrial trifunctional protein (MTP) deficiency	*MTP*
10. Medium-chain 3-ketothiolase (MCKT) deficiency	*MCKT*
Peroxisomal FAO disorder	
1. X-linked adrenoleukodystrophy	*ABCD1*
2. Acyl-CoA oxidase 1 (ACOX1) deficiency	*ACOX1*
3. D-Bifunctional protein (D-BP) deficiency	*D-BP*
4. Peroxisomal 3-ketothiolase 1 (pTH1) deficiency	*pTH1?*[a]
5. 2-Methylacyl-CoA racemase (AMACR) deficiency	*AMACR*

[a] Molecular defect as yet unresolved.

Hepatic Presentation

This presentation, consisting of acute attacks of life-threatening coma associated with impaired hepatic ketone body production, is most commonly observed in infants and young children. The first attacks may occur in the newborn period as a result of prolonged fasting associated with unsuccessful attempts to breastfeed. Usually, however, attacks occur later in life at around 3 to 6 months of age, when the frequency of feeding declines. In the older infant or child, intercurrent infections or missed morning meals are events that typically provoke hypoketotic hypoglycemic attacks. An example of a fatty acid oxidation disorder often associated with the hepatic presentation is medium-chain acyl-CoA dehydrogenase (MCAD) deficiency. A large retrospective analysis of 120 MCAD patients by Lafolla *et al.* (24) showed a median age of presentation of 12 months, ranging from the newborn period (2 days) to 6.5 years of age. Importantly, 20% of the patients died during their first attack.

Cardiac Presentation

Cardiac involvement in fatty acid oxidation disorders may be associated with either dilated or hypertrophic cardiomyopathy. The cardiomyopathy may resemble endocardial fibroelastosis on echocardiography. In some fatty acid oxidation disorders such as primary carnitine deficiency, cardiac involvement may be acute, with overt heart block early in life. This is especially true for those fatty acid oxidation disorders in which the oxidation of LCFAs is impaired.

Skeletal Muscle Presentation

In early infancy and childhood, muscle weakness may be an isolated feature of some fatty acid oxidation defects, but it may also accompany acute attacks of the hepatic presentation. Muscle weakness is rarely seen in MCAD deficiency, but is common in defects that interfere with the oxidation of LCFAs. An example of pure skeletal muscle involvement is the mild form of CPT2 deficiency first described by DiMauro and DiMauro (25), which presents later in life with exercise-induced myalgia, rhabdomyolysis, myoglobinuria (sometimes leading to renal shut down), and greatly increased creatine kinase concentrations in plasma.

Treatment of Mitochondrial Fatty Acid Oxidation Disorders

Avoidance of Fasting

The most important therapeutic measure in fatty acid oxidation disorders is avoidance of prolonged fasting, especially during an intercurrent illness. Although the need for a late evening meal has not been established in all these disorders, it is recommended for the long-chain defects by most clinicians. This guarantees suppression of lipolysis by ensuring sufficient carbohydrate absorption with high insulin levels. In the case of an infectious disease, which in children is generally accompanied by anorexia and vomiting, the utmost care must be taken to supply the desired amount of carbohydrates by gavage feeding or intravenous glucose. During a severe metabolic derangement with hypoketotic hypoglycemia, intravenous glucose should be given at a rate of at least 7 mg/kg body weight per minute, with careful monitoring of plasma glucose concentrations. A high rate of glucose intake not only normalizes glycemia but also efficiently suppresses lipolysis, thus diminishing the production of toxic long-chain acylcarnitines in the case of a long-chain fatty acid oxidation defect and, probably, the production of other toxic metabolites such as octanoate in the case of medium- or short-chain defects.

Dietary Treatment

The role of a low-fat/high-carbohydrate diet in the management of fatty acid oxidation disorders is not well established. In long-chain disorders, treatment with medium-chain triglycerides (MCT) and a diet containing very small amounts of long-chain triglycerides (LCT) is generally considered beneficial. MCT supplementation should be avoided in patients with medium- or short-chain disorders. For a more prolonged glycemic effect, the use of uncooked cornstarch in the late evening meal is recommended by some investigators. In patients treated with a diet low in LCT, one has to be aware that this may easily result in a deficiency of essential fatty acids, making supplementation necessary.

L-Carnitine Supplementation

Primary carnitine deficiency will benefit greatly from oral carnitine supplementation. In most fatty acid oxidation disorders, low plasma levels of free carnitine are observed, owing to the accumulation of acylcarnitines. In medium- and short-chain disorders, total carnitine levels can be very low as a result of urinary losses. L-Carnitine supplementation (50–100 mg/kg/d) is considered beneficial in patients with a short- or medium-chain defect to replenish these urinary losses. In addition, L-carnitine may work as a detoxifying agent by binding toxic intermediates such as octanoate, facilitating urinary excretion. During an acute metabolic derangement, intravenous L-carnitine can be given in doses exceeding 100 mg/kg/d.

It is not clear whether L-carnitine supplementation is justified in patients with a long-chain disorder. In these disorders, it may even be harmful by increasing the levels of toxic long-chain acylcarnitines. However, in some patients with a long-chain defect (VLCAD/LCHAD), L-carnitine supplementation in combination with dietary treatment is thought to ameliorate the cardiac abnormalities.

Riboflavin

While a trial of riboflavin should be started in all patients with a multiple acyl-CoA dehydrogenase defect, its effects in the treatment of true fatty acid β oxidation disorders are disputable. In only a few case reports on MCAD deficiency has a beneficial effect of riboflavin been suggested.

Peroxisomal Fatty Acid Oxidation Disorders

To date, five defined disorders of peroxisomal fatty acid β oxidation have been identified): X linked adrenoleukodystrophy (XALD); acyl-CoA oxidase 1 (ACOX1) deficiency; D-bifunctional protein (D-BP) deficiency; peroxisomal thiolase (pTH1) deficiency; and 2-methylacyl-CoA racemase (AMACR) deficiency.

X-linked Adrenoleukodystrophy

X-linked adrenoleukodystrophy (XALD) is a devastating disease showing marked clinical variability even within families. At least six phenotypic variants can be distinguished (26). The classification of the different phenotypes is somewhat arbitrary and is based on the age of onset and the organs principally involved. The two most frequent phenotypes, accounting for approximately 80% of all cases, are childhood cerebral adrenoleukodystrophy (CCALD) and adrenomyeloneuropathy (AMN). CCALD is characterized by rapidly progressive cerebral demyelination. The age of onset ranges from 3 to 10 years. Frequent early neurologic symptoms are behavioral disturbances, a decline in school performance, deterioration of vision, and impaired auditory discrimination. The course is relentlessly progressive, and seizures, spastic paraplegia, and dementia develop within months. Most patients die within 2 to 3 years after the onset of neurologic symptoms.

Adrenomyeloneuropathy (AMN) presents later in life, with neurologic symptoms usually starting in the third or fourth decade. Neurologic deficits are primarily caused by myelopathy, and to a lesser extent by neuropathy. Patients gradually develop a spastic paraparesis, often in combination with a disturbed vibration sense in the legs and sphincter dysfunction.

The biochemical hallmark of all forms of XALD is the accumulation of VLCFAs, notably C26:0, in plasma owing to a defect in the peroxisomal β oxidation of

VLCFAs. The defective gene was discovered in 1993 (27) and codes for a peroxisomal membrane protein which belongs to the ATP binding cassette (ABC) family of transporters. The protein involved, called ALDP, is a half ABC transporter and forms either homo- or heterodimers. Although this has not been resolved definitively, ALDP is supposed to catalyze the transport of VLCFAs across the peroxisomal membrane (2).

Acyl-CoA Oxidase 1 Deficiency

Acyl-CoA oxidase 1 (ACOX1) deficiency has been reported in a few patients only. All patients reported showed neurologic abnormalities including early onset seizures, hypotonia, hearing impairment, and visual failure resulting from retinopathy. Most patients die early in life (28).

D-Bifunctional Protein Deficiency

D-Bifunctional protein (D-BP) deficiency is the second most common disorder of peroxisomal β oxidation after XALD. The clinical presentation of D-BP patients is usually severe and resembles Zellweger syndrome in many respects, with hypotonia, craniofacial dysmorphia, neonatal seizures, hepatomegaly, developmental delay, and usually early death. A particular feature is that patients with D-BP deficiency often show disordered neuronal migration. The central role of D-BP in the peroxisomal β oxidation of both straight chain and 2-methyl branched chain fatty acids explains why there is accumulation of VLCFAs, pristanic acid, and di- and trihydroxycholestanoic acids in most but not all patients (17–19,28).

Peroxisomal Thiolase 1 Deficiency

Peroxisomal thiolase 1 (pTH1) deficiency has so far been described only in a single patient, with a Zellweger-like phenotype (29).

2-Methylacyl-CoA Racemase Deficiency

2-Methylacyl-CoA racemase (AMACR) deficiency is a newly identified disorder of peroxisomal β oxidation in which only the peroxisomal oxidation of the 2-methyl branched chain fatty acids pristanic acid and di- and trihydroxycholestanoic acids is impaired. Following the description of three patients in our first report (30), we have identified a few additional patients, all of whom show a late onset neuropathy.

Treatment of Peroxisomal Oxidation Disorders

In the past few years, little progress has been made in the treatment of peroxisomal fatty acid oxidation disorders. Initially it was hoped that dietary treatment based on administration of Lorenzo's oil—made from trioleate and trierucate—would be beneficial for XALD patients. Unfortunately, long-term studies have shown no beneficial effect despite the fact that plasma VLCFA levels are reduced on this treatment. Bone marrow transplantation remains the only option for XALD patients. There have been no therapeutic trials in the other peroxisomal fatty acid oxidation disorders.

PPARS AND FATTY ACID OXIDATION DISORDERS

So far, studies in inherited diseases of mitochondrial and peroxisomal fatty acid oxidation have largely concentrated on the identification of patients and the development of methods for detecting the different syndromes. A great deal of work has been done to determine the molecular basis of the defects after cloning the genes that code for the various enzymes and transporters. Much remains to be learned, however, about the consequences of these enzyme defects and the pathophysiologic mechanisms involved. This is especially true for the peroxisomal fatty acid oxidation disorders, including D-bifunctional protein deficiency. Patients suffering from this enzyme defect show severe abnormalities including neonatal hypotonia, dysmorphic features, neonatal seizures, hepatomegaly, developmental delay, and early death.

A block in mitochondrial or peroxisomal fatty acid oxidation will lead to the accumulation of certain fatty acids, determined by the site of the block in the pathway. Although not studied in patients directly, there are clear indications of the involvement of PPARs under these conditions, as shown by *in vitro* studies and by studies in experimental animals.

PPARs and Mitochondrial β Oxidation Deficiencies

Patients suffering from a defect in the mitochondrial oxidation of long-chain fatty acids have increased plasma concentrations of a range of free fatty acids and acylcarnitines, including C18:0, C18:1, C18:2, C16:0, C16:1, C16:2, C14:0, C14:1, and C14:2 in cases of VLCAD deficiency for instance. In case of CPT1 deficiency, there is no accumulation of acylcarnitines in plasma but only of long-chain free fatty acids. Early studies by Keller *et al.* (31) have shown that free fatty acids, especially polyunsaturated fatty acids, are powerful activators of PPARα, with C22:6ω3, C18:3ω3, and C18:2ω6 being about equally effective. To study the role of PPARα under conditions of an impaired mitochondrial oxidation of long-chain fatty acids, Djouadi *et al.* (32) used etomoxir as a pharmacologic inhibitor of CPT1. Administration of etomoxir for 5 days led to a marked increase in the expression of PPARα target genes like acyl-CoA oxidase, CYP4A1, CYP4A3, and medium-chain acyl-CoA dehydrogenase in the liver. Studies in heart from the etomoxir-treated animals revealed that acyl-CoA oxidase and MCAD were also markedly induced in this tissue, which is in line with the fact that PPARα is not only expressed in liver but also in brown adipose tissue (BAT), and to a lesser extent in kidney, skeletal muscle, and heart. All the mice tolerated the perturbation induced by etomoxir. These findings indicate that the metabolic changes induced by etomoxir lead to an activation of PPARα, which subsequently causes an increase in LCFA β oxidation.

When the same experiments were repeated in PPAR(−/−) mice, there were some surprising findings. First, in mice lacking PPARα, inhibition of cellular fatty acid flux by etomoxir caused massive hepatic and cardiac lipid accumulation, hypoglycemia, and death in 100% of male but only 25% of female PPAR(−/−) mice. When male PPAR(−/−) mice were treated for 2 weeks with β-estradiol, virtually all the mice survived. The cause of death of the PPAR(−/−) mice treated with etomoxir turned out to be extreme hypoglycemia, which was not observed in PPAR(−/−) female mice or in male PPAR(−/−) mice pretreated with β-estradiol. These results clearly show the important role of PPARα under conditions where mitochondrial fatty acid oxidation is impaired.

PPARs and Peroxisomal β Oxidation Deficiencies

In recent years, several mouse models have been generated in which one of the peroxisomal β oxidation genes has been disrupted. The acyl-CoA oxidase 1 (ACOX1)–deficient mouse generated by Fan *et al.* (33) has growth retardation, hepatomegaly with severe microvesicular steatohepatitis, and lipogranulomas. These ACOX1(−/−) mice show age-progressive hepatocellular regeneration beginning in the periportal region and extending towards the centrizonal region of the liver lobule. Between 6 and 8 months of age, almost all steatotic hepatocytes in ACOX1(−/−) mice are replaced by regenerated hepatocytes without steatosis. Interestingly, these cells show an abundance of peroxisomes. This spontaneous peroxisome proliferation is associated with the increased expression of genes that are transcriptionally regulated by PPARα. The investigators conclude that the principal function of ACOX1 is to keep PPARα in check by β oxidizing very long-chain fatty acids and a range of other biologic ligands of PPARα. Interestingly, in mice nullizygous for both PPARα and ACOX1 (PPARα(−/−)/ACOX1(−/−) the extensive microvesicular steatohepatitis, spontaneous peroxisome proliferation, and induction of PPARα-regulated genes, as observed in ACOX1(−/−) mice, was not present (34).

In addition to the ACOX1(−/−) mice, several other knock-out mouse models have been made including the L-bifunctional protein (L-BP) (−/−) (35), D-bifunctional protein (D-BP) (−/−) (20), and the sterol carrier protein X (SCPx/pTH2) (−/−) mouse (23). The latter mouse model was initially constructed to try to shed new light on the role of the sterol carrier protein in sterol metabolism. The SCPx/pTH2(−/−)mice, however, were completely normal and no abnormalities in sterol metabolism were found. This led to a shift in interest to the potential role of the 58-kDa SCPx in fatty acid oxidation, as SCPx contains both a thiolase and a sterol carrier protein (SCP) domain. This was especially relevant because our own work (36) and that of Antonenkov *et al.* (37) had shown that the thiolase domain of SCPx reacts with 3-ketoacyl-CoA esters of 2-methyl branched chain fatty acids like pristanic acid, in contrast to the other peroxisomal thiolase, pTH1. The studies done in SCPx(−/−) mice have clearly shown the importance of SCPx for peroxisomal fatty acid oxidation. When phytol was given to the (−/−) mice to put pressure on the system, the mice developed a severe phenotype characterized by lethargy, reduced muscle tone, weight loss, ataxia, peripheral neuropathy (incoordinated movements, unsteady gait, and trembling), and death. Remarkably, inspection of the livers from control and phytol-fed SCPx(−/−) mice revealed an increased number of peroxisomes, and subsequent studies showed that various PPARα target genes were up-regulated. Because SCPx deficiency leads to a block in pristanic acid β oxidation, pristanic and phytanic acid were subsequently tested for their potential to activate PPARα. Ellinghaus *et al.* (7) reported that phytanic acid is indeed a ligand for PPARα. Earlier studies by Lemotte *et al.* (5) had already identified phytanic acid as a ligand for the retinoid X receptor (RXR). Subsequent studies by Zomer *et al.* (8) confirmed that phytanic acid is a ligand for both PPARα and all RXRs. Furthermore, pristanic acid was found to be a more powerful ligand of PPARα than phytanic acid (8). These findings must have important consequences for patients suffering from peroxisomal fatty acid oxidation defects, especially as pristanic and phytanic acid can reach extremely high levels in these patients.

Preliminary studies in liver biopsy specimens from D-BP–deficient patients have shown that PPARα target genes are indeed up-regulated in such patients (Ferdinandusse S, Wanders RIA, August, 2001 unpublished data).

ACKNOWLEDGMENTS

I gratefully acknowledge the help of Mrs. Maddy Festen for her expert preparation of the manuscript and Mr. Jos Ruiter for the artwork.

REFERENCES

1. Wanders RIA, Jakobs C, Skjeldal OH. Refsum disease. In: Scriver CR, Beaudet AL, Sly WS, Valle D, eds. *The metabolic and molecular bases of inherited disease.* New York: McGraw-Hill, 2001: 3303–21.
2. Wanders RJA, Vreken P, Ferdinandusse S, *et al.* Peroxisomal fatty acid alpha- and beta-oxidation in humans: enzymology, peroxisomal metabolite transporters and peroxisomal diseases. *Biochem Soc Trans* 2001; 29: 250–67.
3. Jansen GA, Wanders RJA, Watkins PA, Mihalik SJ. Phytanoyl-coenzyme A hydroxylase deficiency–the enzyme defect in Refsum's disease. *N Engl J Med* 1997; 337: 133–4.
4. Jansen GA, Hogenhout EM, Ferdinandusse S, *et al.* Human phytanoyl-CoA hydroxylase: resolution of the gene structure and the molecular basis of Refsum's disease. *Hum Mol Genet* 2000; 9: 1195–200.
5. Lemotte PK, Keidel S, Apfel CM. Phytanic acid is a retinoid X receptor ligand. *Eur J Biochem* 1996; 236: 328–33.
6. Kitareewan S, Burka LT, Tomer KB, *et al.* Phytol metabolites are circulating dietary factors that activate the nuclear receptor RXR. *Mol Biol Cell* 1996; 7: 1153–66.
7. Ellinghaus P, Wolfrum C, Assmann G, Spener F, Seedorf U. Phytanic acid activates the peroxisome proliferator-activated receptor alfa (PPARalfa) in sterol carrier protein 2-/sterol carrier protein x-deficient mice. *J Biol Chem* 1999; 274: 2766–72.
8. Zomer A WM, van der Burg B, Jansen GA, *et al.* Pristanic acid and phytanic acid. Naturally occurring ligands for the nuclear receptor peroxisome proliferator-activated receptor alpha. *J Lipid Res* 2000; 41: 1801–7.
9. Capdevila JH, Falck JR, Harris RC. Cytochrome P450 and arachidonic acid bioactivation. Molecular and functional properties of the arachidonate mono oxygenase. *J Lipid Res* 2000; 41: 163–81.
10. Vamecq J, de Hoffmann E, Van Hoof F. The microsomal dicarboxylyl-CoA synthetase. *Biochem J* 1985; 230: 683–93.
11. Jakobs BS, Wanders RJA. Fatty acid beta-oxidation in peroxisomes and mitochondria: the first, unequivocal evidence for the involvement of carnitine in shuttling propionyl-CoA from peroxisomes to mitochondria. *Biochem Biophys Res Commun* 1995; 213: 1035–41.

12. Verhoeven NM, Roe DS, Kok RM, *et al.* Phytanic acid and pristanic acid are oxidized by sequential peroxisomal and mitochondrial reactions in cultured fibroblasts. *J Lipid Res* 1998; 39: 66–74.
13. McGarry JD, Brown NF. The mitochondrial carnitine palmitoyltransferase system. From concept to molecular analysis. *Eur J Biochem* 1997; 244: 1–14.
14. Wanders RJA, Tager JM. Lipid metabolism in peroxisomes in relation to human disease. *Mol Aspects Med* 1998; 19: 69–154.
15. Glatz JFC, Storch J. Unraveling the significance of cellular fatty acid-binding proteins. *Curr Opin Lipidol* 2001; 12: 267–74.
16. Wanders RJA, Vreken P, Den Boer MEJ, *et al.* Disorders of mitochondrial fatty acyl-CoA beta-oxidation. *J Inherit Metab Dis* 1999; 22: 442–87.
17. van Grunsven EG, van Berkel E, IJlst L, *et al.* Peroxisomal D-hydroxyacyl-CoA dehydrogenase deficiency: resolution of the enzyme defect and its molecular basis in bifunctional protein deficiency. *Proc Natl Acad Sci USA* 1998; 95: 2128–33.
18. van Grunsven EG, van Berkel E, Mooijer PAW, *et al.* Peroxisomal bifunctional protein deficiency revisited: resolution of its true enzymatic and molecular basis. *Am J Hum Genet* 1999; 64: 99–107.
19. van Grunsven EG, Mooijer PAW, Aubourg P, Wanders RJA. Enoyl-CoA hydratase deficiency: identification of a new type of D-bifunctional protein deficiency. *Hum Mol Genet* 1999; 8: 1509–16.
20. Baes M, Huyghe S, Carmeliet P, *et al.* Inactivation of the peroxisomal multifunctional protein-2 in mice impedes the degradation of not only 2-methyl-branched fatty acids and bile acid intermediates but also of very long-chain fatty acids. *J Biol Chem* 2000; 275: 16329–36.
21. Hashimoto T. Peroxisomal beta-oxidation: enzymology and molecular biology. *Ann N Y Acad Sci* 1996; 804: 86–98.
22. Seedorf U, Brysch P, Engel T, Schrage K, Assmann G. Sterol carrier protein X is peroxisomal 3-oxoacyl coenzyme A thiolase with intrinsic sterol carrier and lipid transfer activity. *J Biol Chem* 1994; 269: 21277–83.
23. Seedorf U, Raabe M, Ellinghaus P, *et al.* Defective peroxisomal catabolism of branched fatty acyl coenzyme A in mice lacking the sterol carrier protein-2/sterol carrier protein-x gene function. *Genes Dev* 1998; 12: 1189–201.
24. Lafolla AK, Thompson RJ, Roe CR. Medium-chain acyl-coenzyme A dehydrogenase deficiency: clinical course in 120 affected children. *J Pediatr* 1994; 124: 409–15.
25. DiMauro S, DiMauro PM. Muscle carnitine palmityltransferase deficiency and myoglobinuria. *Science* 1973; 182: 929–31.
26. Moser HW, Smith KD, Moser AB. X-linked adrenoleukodystrophy. In: Scriver CR, Beaudet AL, Sly WS, Valle D, eds. *The metabolic and molecular bases of inherited disease.* New York: McGraw-Hill, 1995: 2325–49.
27. Mosser J, Douar AM, Sarde CO, *et al.* Putative X-linked adreno1eukodystrophy gene shares unexpected homology with ABC transporters. *Nature* 1993; 361: 726–30.
28. Wanders RJA, Barth PG, Heymans HSA. Single peroxisomal enzyme deficiencies. In: Scriver CR, Beaudet AL, Sly WS, Valle D, eds. *The metabolic and molecular bases of inherited disease.* New York: McGraw-Hill, 2001: 3219–56.
29. Schram AW, Goldfischer S, van Roermund CWT, *et al.* Human peroxisomal 3-oxoacyl-coenzyme A thiolase deficiency. *Proc Natl Acad Sci USA* 1987; 84: 2494–6.
30. Ferdinandusse S, Denis S, Clayton PT, *et al.* Mutations in the gene encoding peroxisomal alpha-methylacyl-CoA racemase cause adult-onset sensory motor neuropathy. *Nat Genet* 2000; 24: 188–91.
31. Keller H, Dreyer C, Merlin I, *et al.* Fatty acids and retinoids control lipid metabolism through activation of peroxisome proliferator-activated receptor-retinoid X receptor heterodimers. *Proc Natl Acad Sci USA* 1993; 90: 2160–4.
32. Djouadi F, Weinheimer CJ, Saffitz JE, *et al.* A gender-related defect in lipid metabolism and glucose homeostasis in peroxisome prolifer. *J Clin Invest* 1998; 102: 1083–91.
33. Fan CY, Pan J, Chu R, *et al.* Hepatocellular and hepatic peroxisomal alterations in mice with a disrupted peroxisomal fatty acyl-coenzyme A oxidase gene. *J Biol Chem* 1996; 271: 24698–710.
34. Hashimoto T, Fujita T, Usuda N, *et al.* Peroxisomal and mitochondrial fatty acid beta-oxidation in mice nullizygous for both peroxisome proliferator-activated receptor alpha and peroxisomal fatty acyl-CoA oxidase. Genotype correlation with fatty liver phenotype. *J Biol Chem* 1999; 274: 19228–36.
35. Qi C, Zhu Y, Pan J, Usuda N, *et al.* Absence of spontaneous peroxisome proliferation in enoyl-CoA hydratase/L-3-hydroxyacyl-CoA dehydrogenase-deficient mouse liver: further support for the role of fatty acyl CoA oxidase in PPARalpha ligand metabolism. *J Biol Chem* 1999; 274: 15775–80.

36. Wanders RJA, Denis S, Wouters F, Wirtz KW, Seedorf U. Sterol carrier protein X (SCPx) is a peroxisomal branched-chain beta-ketothiolase specifically reacting with 3-oxo-pristanoyl-CoA: a new, unique role for SCPx in branched-chain fatty acid metabolism in peroxisomes. *Biochem Biophys Res Commun* 1997; 236: 565–9.
37. Antonenkov VD, Van Veldhoven PP, Waelkens E, Mannaerts GP. Substrate specificities of 3-oxoacyl-CoA thiolase A and sterol carrier protein 2/3-oxoacyl-CoA thiolase purified from normal rat liver peroxisomes: sterol carrier protein 2/3-oxoacyl-CoA thiolase is involved in the metabolism of 2-methyl-branched fatty acids and bile acid intermediates. *J Biol Chem* 1997; 272: 26023–31.

DISCUSSION

Dr. Superti-Furga: In the mouse model of acyl-CoA oxidase deficiency, has it been shown that the hepatocytes that repopulate the liver are genetically corrected, as in tyrosinemia?

Dr. Wanders: It is believed that there is *de novo* synthesis of hepatocytes, and that we are not dealing with old hepatocytes going through some phase and being restored to normality.

Dr. Superti-Furga: But the hepatocytes do not have that defect, because they can make peroxisomes. They must somehow be corrected.

Dr. Wanders: No, they are genetically identical, but for some reason they are repopulating the liver. The situation is not like tyrosinemia.

Dr. Bachmann: Could you clarify the term “PPAR”? It seems misleading, as not all PPARs are important for peroxisomes.

Dr. Wanders: The nomenclature is awful, I agree. This nuclear hormone receptor was discovered through work on clofibrate and other peroxisome proliferators and it was given the name “peroxisome proliferator-activated receptor.” Later on, the β form was discovered and then the γ form. These two forms have very little to do with peroxisome proliferation, but as always, once the name has stuck you can’t get rid of it.

Dr. Kon: What is the frequency of peroxisomal disorders?

Dr. Wanders: The current estimate in our institute and Moser’s, which are the largest centers in this field, is between 1/5000 and 1/6000, of which X-linked adrenoleukodystrophy makes up about half.

Dr. Kon: Are there any reliable clinical tests to diagnose this disorder?

Dr. Wanders: I didn’t go into any detail about this, but the diagnosis of peroxisomal and mitochondrial fatty acid oxidation disorders has improved tremendously over the last few years. For peroxisomal disorders, the initial investigation is to measure the very long-chain fatty acids in the plasma. This does not cover all the disorders, but it will pick up most of them. It is an extremely reliable test if you use a suitable assay. For mitochondrial β oxidation disorders, which used to be very difficult to identify, acylcarnitine analysis has now been available for a couple of years. When we receive a sample, we first run acylcarnitines, which you can do overnight because it’s a very simple straightforward assay, and then we measure the enzyme in the lymphocytes, which we retain. Using these procedures, you can usually reach a diagnosis in 1 or 2 days.

Dr. Yamashiro: Does peroxisomal β oxidation decline with age? The reason I ask is that there has been a report that hexaenoic acid in adults with obesity or diabetes is increased in comparison with normal healthy adults.

Dr. Wanders: That’s true. From animal experiments, we know that peroxisomal β oxidation declines with age. I don’t have any data in humans.

Dr. Scott: As it appears that you have very good biochemical markers, can you tell us whether in the noninborn errors there is a wide range of what you would consider to be normal people? I suspect that in many of these conditions it will turn out that there are polymorphisms of these systems. Obviously, it would be a place to start looking, if some people have intermediate levels of these biochemical markers. Is anyone looking for polymorphisms? I'm sure they will be present.

Dr. Wanders: We are not doing that, and I don't know of any other group that is doing so at this time. In principle, you are right, and it should be studied.

Dr. Steinmann: You mentioned the great intrafamilial variability in adrenoleukodystrophy—for instance, a boy who was severely affected and had a maternal uncle with only benign adrenal insufficiency. Can you speculate on the factors that might be involved in this variability? It's an important question because bone marrow transplantation could be planned better.

Dr. Wanders: The problem is that this variability is not only seen between families, but also within a single family, so you can have a child with ALD (adrenoleukodystrophy) presentation in the same family as another with AMN (adrenomyeloneuropathy). I know that the Baltimore group is looking for modifier genes, and there have been several reports on monozygotic twins showing widely discrepant clinical presentations. The idea now is that if you compare AMN with ALD, the basic difference is that in AMN the inflammatory reaction is missing. The crucial question is why this should be so. We don't know at the moment, but people are looking into it.

Dr. Sies: As well as peroxisomes, there are also microperoxisomes, especially in organs where there is steroid biosynthesis. Are there any diseases of the microperoxisomes?

Dr. Wanders: Microperoxisomes are identical to peroxisomes, but smaller. Both the enzyme content and everything that takes place in peroxisomes are the same as in microperoxisomes.

Dr. Baerlocher: Can you comment on the dietary treatment of adrenoleukodystrophy?

Dr. Wanders: "Lorenzo's oil" was supposed to be helpful, and all ALD patients were put on it. Unfortunately, the treatment was not double-blind or placebo-controlled. Now, after many years, it has been shown in Baltimore that there is no benefit from this treatment, but even so, it is still being used.

Dr. Baerlocher: I understand that the European Working Group on treatment with Lorenzo's oil has been dissolved.

Dr. Böhles: Dr. Martinez from Barcelona showed that C22: 6 was beneficial. She published a paper on this (1).

Dr. Wanders: That's another story obviously. In fact, patients with ALD don't have a deficit of C22: 6. Dr. Martinez claims that C22: 6 helps in Zellweger patients also, but I know many other people who disagree completely with this. Again, there have been no double-blind studies.

Dr. Batshaw: In Zellweger syndrome, have you any idea about the origin of the migrational defect, and how this primary defect could affect migration in early embryogenesis?

Dr. Wanders: We don't have any good clues about that. It appears to be related to one of the β oxidation intermediates which in some way or other inhibits neuronal migration.

Dr. Superti-Furga: This is speculative, but we could be dealing with the same mechanism as in the Smith–Lemli–Opitz syndrome—that is, interference mediated through a cholesterol deficit or membrane fluidity changes. Signal transduction in the so-called "sonic hedgehog" pathway is disturbed, and that is like a master switch for morphogenesis.

Dr. Wanders: That's a good possibility.

Dr. Batshaw: The problem with that hypothesis, however, is that sonic hedgehog is primarily involved in prosencephalic development, which occurs earlier than migration. There does not seem to be a problem with prosencephalic development in Zellweger syndrome.

REFERENCE

1. Martinez M. Docosahexaeonic acid therapy in patients with disorders of peroxisome biogenesis. *Lipids* 1996; 31:52–6.

Genetic Expression and Nutrition, edited by Claude Bachmann and Berthold Koletzko. Nestlé Nutrition Workshop Series, Pediatric Program, Vol. 50. Nestec Ltd., Vevey/Lippincott Williams & Wilkins, Philadelphia, © 2003.

Oxidative Stress: Antioxidants in Degenerative Neurologic and Ophthalmologic Diseases

Helmut Sies, Wilhelm Stahl, and Maria Cristina Polidori

Institut für Physiologische Chemie I, Heinrich-Heine Universität Düsseldorf, Germany

Reactive oxygen species, also called prooxidants, are produced physiologically during normal metabolic reactions in the aerobic organism. However, when they are produced in excessive amounts and are not sufficiently detoxified, the steady state balance between prooxidants and antioxidants may be disrupted. An imbalance in favor of the prooxidants and against the antioxidants then occurs. This has been called "oxidative stress" (1–3), and can lead to damage affecting all types of biologic molecules, including DNA, lipids, protein, and carbohydrates. Oxidative stress is thus involved in processes such as mutagenesis, carcinogenesis, cellular dysfunction, and tissue degeneration.

Although it is often suggested that antioxidants may play a critical role in human health, and that oxidative stress could be involved in the pathogenesis of various disorders, attempts to treat diseases with antioxidants have been less than satisfactory (4), and neurodegenerative diseases are no exception, However, there is increasing interest in the potential pathophysiologic role of oxidative stress and mitochondrial dysfunction in Alzheimer disease, Huntington disease, Parkinson disease, and amyotrophic lateral sclerosis. The fact that there is a multitude of oxidants as well as antioxidants with overlapping reactivities makes a biochemically rigorous assessment of the implications of oxidative stress difficult. Moreover, helpful as epidemiologic studies can be, sometimes there is a confounding of associations with cause–effect relations, leading to erroneous conclusions.

The aging process itself is associated with an overall increased oxidant load in the organism (5) and age is a major "risk factor" for neurodegenerative diseases. Accordingly, an involvement of oxidative stress has been suggested in accelerated aging processes such as progeria and Werner syndromes, which are also associated with neurodegenerative processes [for review, see Butterfield *et al.* (6)]. Studies conducted in patients, animal models, and human necropsy samples have shown that increased blood and tissue levels of biomarkers of oxidative damage to DNA, lipids, and proteins are associated with aging and neurodegeneration. Furthermore, in specific diseases, the gene mutations responsible for the impairment of antioxidant function are associated with severe neurodegeneration in humans (see below).

Another actively studied phenomenon in the field of oxidative damage and human disease is photooxidative stress experienced by light-exposed tissues. Reactive oxygen species, such as singlet molecular oxygen, are generated in reaction sequences involving the transfer of light energy through photosensitizers to oxygen (7,8). Photooxidative damage to proteins and lipids is thought to play a role in the pathologic events leading to degenerative eye diseases such as age-related macular degeneration (9) and cataract (10). As with neurodegenerative disorders, it has been suggested that antioxidants may prevent or delay the onset of these ophthalmic diseases.

In this chapter, we summarize relevant observations on the potential role of oxidants and antioxidants in both neurologic and ophthalmologic degenerative diseases.

NEURODEGENERATIVE DISEASES

Oxidative Stress and the Brain

There is much interest in the possibility that mitochondrial dysfunction and oxidative damage play a role in the pathogenesis of neurodegenerative diseases (11). Mitochondrial oxidative phosphorylation through the electron transport chain normally couples the reducing equivalents to oxygen with the simultaneous production of adenosine triphosphate (ATP). However, when supplies of oxygen are limited and oxidative phosphorylation slows or fails, the production of reactive oxygen species may increase. The result is a cumulative burden of oxidative damage to key cellular constituents, with widespread ramifications for cellular metabolism.

In addition, as ATP levels fall, a series of events ensues that exacerbates a condition of oxidative stress in the brain (12). These include partial neuronal depolarization, activation of voltage-dependent excitatory amino acid receptors, and opening of the calcium permeable channels linked to these receptors (13). The influx of calcium forms an important link between excitotoxicity and oxidative stress, because intracellular calcium can stimulate phospholipase with subsequent release of arachidonic acid, which may undergo free radical-induced lipid peroxidation and lead to further generation of free radicals. Arachidonic acid and oxygen radicals may, in turn, enhance the release of excitatory neurotransmitters and inhibit their uptake by inactivation of neuronal and glial transport processes, thus promoting a vicious cycle. Furthermore, calcium influx can activate peptidases such as calpain, which catalyzes the enzymatic conversion of xanthine dehydrogenase to xanthine oxidase, ultimately leading to superoxide anion generation.

In this cascade of damaging events, the simultaneous increase of lactic acid favors the liberation of cellular stores of iron, which promotes Fenton reactions to yield hydroxyl radicals. All these reactions lead finally to the so-called excitotoxic cell death (14), which constitutes a critical pathophysiologic mechanism not only in neurodegenerative brain changes (slow excitotoxicity) but also in acute brain injury (acute excitotoxicity).

The brain may be especially prone to oxidative damage not only because it consumes a large amount of the body's oxygen, but especially because the activity of antioxidant enzymes in brain tissue is relatively low (15).

Biomarkers of Oxidative Stress and Alzheimer Disease

Alzheimer disease is the most prominent cause of dementia in the elderly and is clinically characterized by memory dysfunction—loss of lexical access, spatial and temporal disorientation, and impairment of judgment. Histopathologically, Alzheimer disease is characterized by synaptic loss, nerve cell loss (mostly in the cerebral cortex, hippocampus, and amygdala), extracellular deposition of β-amyloid protein (forming senile plaques), and intracellular precipitation of hyperphosphorylated τ protein (forming neurofibrillary tangles). The exact biochemical mechanism of the pathogenesis of this disease is still unknown, but much attention is given to the role of the massive loss of the neurotransmitter acetylcholine (necessary for cognition and memory) and to the possible implication of oxidative stress in the development of Alzheimer disease.

There is impaired energy metabolism in Alzheimer disease (16). Consistently, cerebral blood flow and oxygen utilization have been shown to be decreased in this disorder in a temporoparietal pattern (17). Impaired mitochondrial function may lead to increased free radical–related damage affecting critical cellular key components. There is substantial evidence that mitochondrial function declines with age and is altered in Alzheimer disease (18). Mitochondrial DNA (mtDNA), which encodes subunits of the mitochondrial respiratory enzyme complex, is particularly susceptible to oxidative damage owing to its close proximity to the respiratory chain proteins, limited repair mechanisms, paucity of noncoding sequences, and absence of histones (19).

An increase in DNA, lipid, and protein oxidation products has been shown in blood and necropsy brain samples obtained from patients with Alzheimer disease in comparison with controls [for review, see Markesbery and Carney (20)]. A three-fold increase in 8-oxo-2′deoxyguanosine (8-oxo-dG) has been found in mtDNA isolated from cortical tissue of patients with Alzheimer disease in comparison with nuclear DNA (nDNA) (21). An increased 8-oxo-dG lymphocyte DNA content has also been found in Alzheimer disease donors compared with control subjects (22). This increase has been found recently to be inversely correlated with plasma concentrations of several antioxidant vitamins and micronutrients in patients with Alzheimer disease (23). Immunohistological studies on the antioxidant enzymes superoxide dismutase and catalase have shown that they are colocalized with a subset of neurofibrillary tangles and plaques in Alzheimer disease tissue (24). It has been suggested that the β-amyloid protein itself may cause oxidative damage to neurons [for review, see Butterfield *et al.* (25)].

Biomarkers of Oxidative Stress and Parkinson Disease

As a result of neurodegeneration occurring in the substantia nigra and the striatum and because of dopamine depletion, Parkinson disease is characterized clinically by bradykinesia, postural instability, gait difficulty, and tremor. Depigmentation, neuronal loss, and gliosis of the substantia nigra are typical brain abnormalities found in Parkinson disease. The mechanisms of cell death in this disorder have not yet been fully elucidated, but increased oxidative stress, abnormal mitochondrial function, and excitotoxicity are perhaps among the most important initiators or mediators of neuronal damage.

Oxidative damage is increased in Parkinson disease. The evidence for an involvement of free radicals comes from the observation that oxidation of dopamine yields potentially toxic semiquinones, and that the accelerated metabolism of dopamine by monoamine oxidase-B may induce an excessive formation of hydrogen peroxide, superoxide anions, and hydroxyl radicals (26).

Further evidence for the role of oxidative stress in Parkinson disease patients comes from studies on the selective toxicity against the substantia nigra of 1-methyl-4-phenyl-1,2,3,6-tetrahydropyridine (MPTP), which induces Parkinson-like symptoms in primates. MPTP acts through its metabolite MPP^+ to inhibit Complex I of the mitochondrial respiratory chain. Necropsy studies on Parkinson disease brains have shown that a disease-specific and drug-independent defect of mitochondrial Complex I accumulates in the substantia nigra of affected patients (27–29). It has also been suggested that MPP^+ acts by increasing the vulnerability of cells to oxidative stress (30).

The role of a Complex I dysfunction in Parkinson disease is supported by the finding that Complex I is inhibited *in vivo* and *in vitro* by neuroleptic drugs that have extrapyramidal side effects, but not by clozepine, an atypical antipsychotic agent that has minimal, if any, extrapyramidal toxicity (31). Mitochondrial Complex I defects have been also found in muscle (31) and platelets (32) in patients with Parkinson disease. Parkinson disease was found to be associated with increased lipid peroxidation in the brain (33) and with a marked increase in 8-oxo-dG in the caudatum and substantia nigra (11,34).

Biomarkers of Oxidative Stress and Huntington Disease

Huntington disease is a late-onset, progressive neurodegenerative disorder characterized by involuntary movements, cognitive loss, and personality changes. The disease is inherited as an autosomal dominant pattern.

Direct evidence for a defect in oxidative phosphorylation in Huntington disease was obtained applying proton nuclear magnetic resonance spectroscopy in 16 affected patients (35). A three-fold increase in lactate concentrations in the occipital cortex was observed in patients examined; increases were also found in the basal ganglia (35).

Several studies have shown DNA strand breaks in necropsy tissue from patients with Huntington disease, as detected by *in situ* end labeling of DNA.

Administration of 3-nitropropionic acid, a neurotoxic agent inhibiting succinate dehydrogenase, caused brain lesions accompanied by increased levels of 8-oxo-dG and increased staining for 8-oxo-dG in the basal ganglia (36). Increased 8-oxo-dG levels have been found in mtDNA from the parietal cortex of patients with Huntington disease (37).

Studies have also shown that the disease is associated with increased levels of malondialdehyde, 3-nitrotyrosine, and heme oxygenase in areas of degeneration (38). It has been proposed that glyceraldehyde-3-phosphate dehydrogenase, which binds specifically to proteins implicated in the pathogenesis of neurodegeneration and apoptosis, including Huntington, may affect neuronal energy production (39).

Biomarkers of Oxidative Stress in Other Neurodegenerative Diseases

Amyotrophic lateral sclerosis is a fatal neurodegenerative disease associated with point mutations and lowered activity of Cu–Zn superoxide dismutase in its familial form (40). Other reports have shown an increase in Mn superoxide dismutase protein and activity, in contrast with one study indicating potential gene defects of this enzyme in sporadic amyotrophic lateral sclerosis (41). Several mechanisms have been proposed for its pathophysiology (42), including a trigger exerted by the age-dependent decline in ascorbic acid availability in the presence of the specific superoxide dismutase defect (43).

It has been suggested that oxidative stress is involved in other neurodegenerative diseases, including genetic disorders of the central nervous system such as Friedreich ataxia and Down syndrome. Increased levels of 8-oxo-dG have been found in the urine from patients with Friedreich ataxia, the most common hereditary ataxia related to a gene encoding the protein frataxin (44). Down syndrome, which is caused by trisomy of chromosome 21, is characterized by pathologic brain features similar to those seen in Alzheimer disease. When looking at the consequences of chromosome 21–linked gene overexpression, it has been suggested that superoxide dismutase-1 and amyloid precursor protein genes may contribute to the phenotype of the syndrome. There are studies supporting the role for superoxide dismutase-1 and amyloid precursor protein genes in the pathogenesis of neural abnormalities of Down syndrome (45).

Is There a Role for Antioxidants in Neurodegenerative Diseases?

Higher ascorbic acid and β-carotene blood levels have been found to be associated with a better memory performance among subjects older than 65 years (46). On the other hand, lower levels of plasma vitamin C and E (47), as well as lower serum β-carotene and vitamin A (48), have been found in demented patients than in healthy controls. Plasma vitamin C levels are lower in Alzheimer disease patients than in controls, even in the presence of an adequate dietary vitamin C intake (49), and Alzheimer patients have significantly lower cerebrospinal fluid levels of vitamin E than controls (50). It is not known, however, whether Alzheimer disease is associated with a specific plasma antioxidant profile, as has been shown for other nonneurodegenerative diseases (51), because studies conducted so far assessed only a few antioxidants rather than a broad spectrum of water-soluble and lipophilic antioxidant vitamins and micronutrients.

Nonenzymatic antioxidants—particularly dietary antioxidants such as vitamin C, vitamin E, β-carotene, and polyphenols—have been shown to exert positive effects on cognitive performance in several studies (52–54). It has thus been suggested that a balanced diet containing large amounts of antioxidant-rich fruits and vegetables might be an efficient way of decreasing the incidence and the prevalence of dementia in elderly people (55). In the case of flavonoids, a significant inverse relation was found between flavonoid intake and the risk of incident dementia in a cohort of 1,367 elderly subjects followed for 5 years (56). No association between cognitive function and the intake of vitamins C and E was found in the Rotterdam study (57), whereas the cross-sectional observation of a potential protective effect of β carotene against cognitive impairment was made (57). Some of these results, however, may reflect confounding errors, among which calculations of the antioxidant intake with diet or changes in dietary habits may play an important role. Another study suggested that use of high doses of vitamin E and vitamin C supplements may reduce the risk of Alzheimer disease, but the proportion of subjects taking antioxidant supplements was too small to exclude the occurrence of chance (58).

When the effect of the cholinesterase inhibitor donepezil was compared with donepezil plus vitamin E in 19 patients with mild to moderate Alzheimer disease, no clear benefit of vitamin E was observed (59). In a large clinical trial, vitamin E (2,000 IU/d) and/or selegiline (10 mg/d) were given to patients with moderate Alzheimer disease for 2 years (60). Both compounds alone, but not the combination, proved to be effective in lowering the risk of reaching one of the primary endpoints—death, severe dementia, loss of ability to perform activities of daily living, or need for institutional care (60). One trial is currently ongoing on the preventive effect of vitamin E supplementation (2,000 IU/d for 3 years) against the development of dementia in patients with mild cognitive impairment (61). The American Psychiatric Association recommends the use of vitamin E in Alzheimer disease (62). Other clinical trials, including those carried out to assess the effect of extract of Gingko biloba and idebenone in Alzheimer disease, have not led to consistent results [for reviews, see Praticó and Delanty (63) and Rottkamp *et al.* (64)].

Epidemiologic studies suggest that a high vitamin E intake is associated with a lower risk of developing Parkinson disease [for review, see Ebadi *et al.* (65)]. Recently, however, it was found that plasma vitamin E levels were similar in Parkinson disease patients and in a control population (66). In a small study in patients with Parkinson disease taking 400 to 3,200 IU of vitamin E daily for an average of 7 years, the disease was significantly milder (better performance in the activities of daily living) in the supplemented group than in nonsupplemented subjects (67). In another study, patients with Parkinson disease receiving high doses of vitamin E and C did not require levodopa to treat their symptoms for 2.5 years longer than untreated patients (68). The DATATOP (deprenyl and tocopherol antioxidant therapy of parkinsonism) study, however, showed that 2,000 IU of vitamin E per day was not beneficial in slowing disease progression in patients with early, untreated Parkinson disease (69).

With respect to Huntington disease, MPTP-induced neurotoxic lesions have been shown to be attenuated by coenzyme Q_{10} administration in animal models [reviewed by Beal (70)]. Thus, trials are currently ongoing to assess the potential beneficial effects of coenzyme Q_{10} treatment in patients with Huntington disease. A recent multicenter, randomized clinical trial evaluating the effect of coenzyme Q_{10} or remacemide hydrochloride in patients with early Huntington disease has shown a beneficial trend of coenzyme Q_{10} treatment in slowing the decline in overall functional capacity, but neither treatment has produced significant slowing of functional decline (71).

Superoxide dismutase-1 transgenic mice supplemented with vitamin E have a delay in the onset of clinical symptoms resembling amyotrophic lateral sclerosis (72), and transgenic mouse models of familial amyotrophic lateral sclerosis supplemented orally with coenzyme Q_{10} show a significant increase in life span compared with untreated animals (73). N-acetylcysteine given to patients with amyotrophic lateral sclerosis showed only a trend toward benefit (74). In patients with Friedreich ataxia, raised urinary 8-oxo-dG levels have been found to decrease after treatment with idebenone (44).

Antioxidant deficiencies may be associated with severe neurologic deficits, as in the case of coenzyme Q_{10} deficiency (75). Copper is a cofactor of Cu–Zn superoxide dismutase, and copper deficiency is associated with severe neurologic disorders such as Menke disease and Wilson disease (76). Children with genetic abnormalities of the α tocopherol transfer protein suffer from severe neurologic dysfunction within 18 to 24 months of age, developing cerebellar ataxia, dysarthria, and abnormal tendon and Babinski reflexes (77). Adults may develop neurologic abnormalities after several years of chronic liver disease or fat malabsorption (78). In both cases, symptoms can be ameliorated by vitamin E supplementation (78).

OPHTHALMOLOGIC DISEASES

Age-Related Macular Degeneration

Degenerative processes in the retina and retinal pigment epithelium in the region of the macula lead to age-related macular degeneration, a major cause of visual loss and irreversible blindness among the elderly (79). The macula, the yellow spot of the eye, is the area of maximal visual acuity (9,80). The carotenoid pigments lutein and zeaxanthin are responsible for the yellow color of this tissue. Other dietary carotenoids, including β-carotene, which are present in blood, plasma, and most human tissues (81), are not found in the macula. Epidemiologic studies indicate that an increased dietary intake of the macular carotenoids is related to a lower risk of age-related macular degeneration. An increased risk is correlated with a lowered macular pigment density (82), whereas a lower risk is related with higher serum levels of the xanthophylls lutein and zeaxanthin (83). Further, a relation between a high dietary intake of these carotenoids and a lowered risk of age-related macular degeneration has been reported (84). It has also been shown that lutein and zeaxanthin levels in the macular region are lower in subjects with age-related macular degeneration than in controls (85).

Only limited information is available on the biochemical mechanisms involved in the development of this disease. It has been suggested that photooxidative damage plays an important role in the pathobiochemistry (9,80). In the presence of appropriate sensitizers, singlet molecular oxygen and further reactive oxygen species may be generated. Drusen, which are characteristic of age-related macular degeneration, are rich in lipids, and they may form when lipofuscin accumulates in the retinal epithelium. Lipofuscin is an indigestible, fluorescent product of lipid oxidation occurring in some of the oxidation products of retinol (86). Excessive blue light may increase its formation. *In vitro* studies have shown that lipofuscin acts as a photosensitizer (87); upon photoactivation of lipofuscin granules, singlet oxygen and other reactive oxygen species are generated.

Singlet oxygen formation in this reaction is most efficient at wavelengths around 420 nm (blue light). The absorbance maxima of lutein and zeaxanthin are around 450 nm. Both compounds may act as broadband filters, diminishing the intensity of blue light that reaches the macula. Like most carotenoids, they are also very efficient singlet oxygen quenchers and thus capable of scavenging the products of photooxidative reactions.

In homogeneous solutions, other carotenoids such as β-carotene also efficiently absorb blue light. However, *in vitro* data on membrane models indicate that the filter efficacy of lutein and zeaxanthin is superior to β-carotene and lycopene when the compounds are incorporated into lipid bilayers (88). It was suggested that this increased efficacy is related to the orientation of the incorporated molecules in the liposomal membrane and might be a reason why lutein and zeaxanthin, but not other carotenoids, are used as protective agents in the macula.

Cataract

Cataract is an eye disease that affects the lens and is characterized by opacities that impair visional function (89). Oxidative alterations of lens proteins such as crystallins, with subsequent aggregation and precipitation, appear to be a key step in the formation of cataract lesions. Reactive oxygen species may further inactivate repair systems that are responsible for the elimination of modified proteins and thus accelerate the generation of cataracts. There is increasing evidence that the ubiquitin–proteasome pathway plays a role in the removal of oxidatively damaged lens proteins (90).

In vitro data show that the proteasomal step of the ubiquitin-proteasome system is more susceptible to oxidative inactivation than the ubiquitination step. It could be that the attenuation of the ubiquitin–proteasome pathway activity is at least in part responsible for the cytotoxic accumulation of damaged lens proteins. It has further been proposed that the activities of ubiquitin-activating enzyme and ubiquitin-conjugating enzymes are regulated by cellular redox status (that is, the GSSG:GSH ratio) (91). Thus, changes in the antioxidant status that affect the redox state of a cell might have an impact on the ubiquitin–proteasome pathway by different mechanisms.

Epidemiologic studies provide evidence that an increased intake of a diet rich in antioxidants is associated with a lowered risk of cataract (92). For single antioxidants, the most promising data have been presented for vitamin C. It was shown that long-term supplementation with ascorbate is associated with a lower incidence (77%) of early lens opacities (93). Decreased levels of vitamin C in the lens are correlated with the severity of cataract.

Recent data provide additional evidence for a specific role of vitamin C (94). After adjustment for other nutrients, only vitamin C intake was significantly associated with the incidence of nuclear opacities. The prevalence of nuclear opacities was less in the highest vitamin C intake group than in the lowest. The study also showed a statistically significant trend of decreasing prevalence of nuclear opacities with increasing duration of vitamin C supplementation.

Vitamin C is the major dietary antioxidant in the lens and is detected in large amounts in this tissue, where concentrations of ascorbate are about ten-fold higher than in the blood. The endogenous antioxidant glutathione is also present in an unusually high concentration in the lens (94), exceeding the plasma level. In contrast, the levels of lipophilic antioxidants in the lens, such as tocopherol and the carotenoids, are much lower and less is known about their possible contribution to lens protection against oxidative damage.

ACKNOWLEDGEMENTS

M. C. Polidori is an EU Marie Curie Fellow for the "Quality of Life and Management of Living Resources Programme," project entitled "Nutritional health sustaining factors and determinants of healthy aging: oxidative stress-related biomarkers of successful aging and age-related diseases." H. S. is a Fellow of the National Foundation for Cancer Research (NFCR), Bethesda, Maryland, USA.

REFERENCES

1. Sies H. Oxidative stress: introductory remarks. In: Sies H, ed. *Oxidative stress*. London: Academic Press, 1985: 1–8.
2. Sies H. Biochemistry of oxidative stress. *Angewandte Chemie (International Edition in English)* 1986; 25: 1058–71.
3. Sies H. What is oxidative stress? In: Keaney JF, ed. *Oxidative stress and vascular disease*. Boston: Kluwer, 2000: 1–8.
4. Sies H, ed. Antioxidants in disease mechanism and therapy. In: *Advances in pharmacology*. Vol 38. San Diego: Academic Press, 1997: 1–691.
5. Harman D. Aging: a theory based on free radical and radiation chemistry. *J Gerontol* 1956; 11: 298–300.
6. Butterfield DA, Howard BJ, LaFontaine MA. Brain oxidative stress in animal models of accelerated aging and the age-related neurodegenerative disorders, Alzheimer disease and Huntington disease. *Curr Med Chem* 2001; 8: 815–28.
7. Halliwell B, Gutteridge JMC. *Free radicals in biology and medicine*, 3rd ed. Oxford: Oxford University Press, 1999.
8. Stahl W, Sies H. Protection against solar radiation–protective properties of antioxidants. In: Giacomoni PU, ed. *Sun protection in man*. Amsterdam: Elsevier, 2001: 561–72.
9. Landrum JT, Bone RA. Lutein, zeaxanthin, and the macular pigment. *Arch Biochem Biophys* 2001; 385: 28–40.
10. Taylor A, Nowell T. Oxidative stress and antioxidant function in relation to risk for cataract. *Adv Pharmacol* 1997; 38: 515–36.
11. Beal MF. Aging, energy, and oxidative stress in neurodegenerative diseases. *Ann Neurol* 1995; 38: 357–66.

12. Coyle JT, Puttfarcken P. Oxidative stress, glutamate, and neurodegenerative disorders. *Science* 1993; 262: 689–95.
13. Heininger K. A unifying hypothesis of Alzheimer's disease, IV: causation and sequence of events. *Rev Neurosci* 2000; 11: 213–328.
14. Sun AY, Chen Y-M. Oxidative stress and neurodegenerative disorders. *J Biomed Sci* 1998; 5: 401–14.
15. Floyd RA. Antioxidants, oxidative stress and degenerative neurological disorders. *Proc Soc Exp Biol Med* 1999; 222: 236–45.
16. Ojaimi J, Byrne E. Mitochondrial function and Alzheimer's disease. *Biol Signals Recept* 2001; 10: 254–62.
17. Duara R, Grady CL, Haxby JV, *et al.* Positron emission tomography in Alzheimer's disease. *Neurology* 1986; 36: 879–87.
18. Smith MA, Nunomura A, Zhu X, Takeda A, Perry G. Metabolic, metallic, and mitotic sources of oxidative stress in Alzheimer disease. *Antioxid Redox Signal* 2000; 2: 413–20.
19. Richter C, Park JW, Ames BN. Normal oxidative damage to mitochondrial and nuclear DNA is extensive. *Proc Natl Acad Sci USA* 1988; 85: 6465–7.
20. Markesbery WR, Carney JM, Oxidative alterations in Alzheimer's disease. *Brain Pathol* 1999; 9: 133–46.
21. Mecocci P, MacGarvey U, Beal MF. Oxidative damage to mitochondrial DNA is increased in Alzheimer's disease. *Ann Neurol* 1994; 36: 747–51.
22. Mecocci P, Polidori MC, Ingegni T, *et al.* Oxidative damage to DNA in lymphocytes from AD patients. *Neurology* 1998; 51: 1014–17.
23. Mecocci P, Polidori MC, Mattioli P, *et al.* Plasma antioxidants and oxidative damage in lymhocytes from normal aged people and Alzheimer disease patients [abstract]. *Neurobiol. Aging* 2000; 21[Suppl 1]: 273.
24. Pappolla MA, Omar RA, Kim KS, Robakis NK. Immunohistochemical evidence of antioxidant stress in Alzheimer's disease. *Am J Pathol* 1992; 140: 621–8.
25. Butterfield DA, Koppal T, Subramanian R, Yatin S. Vitamin E as an antioxidant/free radical scavenger against amyloid β-peptide-induced oxidative stress in neocortical synaptosomal membranes and hippocampal neurons in culture: insights into Alzheimer's disease. *Rev Neurosci* 1999; 10: 141–9.
26. Olanow CW. Oxidation reactions in Parkinson's disease. *Neurology* 1990; 40: 32–7.
27. Hattori N, Tanaka M, Ozawa T, Mizuno Y. Immunohistochemical studies on complexes I, II, III and IV of mitochondria in Parkinson's disease. *Ann Neurol* 1991; 30: 563–71.
28. Shapira AHV, Mann VM, Cooper JM. Mitochondrial function in Parkinson's disease. *Ann Neurol* 1992; 32: 116–24.
29. Burkhardt C, Kelly JP, Lim YH, Filley CM, Parker WD. Neuroleptic medications inhibit complex I of the electron transport chain. *Ann Neurol* 1993; 33: 512–17.
30. Lee H-S, Park C-W, Kim Y-S. MPP+ increases the vulnerability to oxidative stress rather than directly mediating oxidative damage in human neuroblastoma cells. *Exp Neurol* 2000; 165: 164–71.
31. Penn AMV, Roberts T, Hodder J, Allen PS, Zhu G, Martin WRW. Generalized mitochondrial dysfunction in Parkinson's disease detected by magnetic resonance spectroscopy of muscle. *Neurology* 1995; 45: 2097–9.
32. Blake CI, Spitz E, Leehey M, Hoffer BJ, Boyson SJ. Platelet mitochondrial respiratory chain function in Parkinson's disease. *Mov Disord* 1997; 12: 3–8.
33. Yoritaka A, Hattori N, Uchida K, Tanaka M, Stadtman ER, Mizuno Y. Immunohistochemical detection of 4-hydroxynonenal protein adducts in Parkinson's disease. *Proc Natl Acad Sci USA* 1996; 33: 2696–701.
34. Sanchez-Ramos JR, Overvik E, Ames BN. A marker of oxyradical-mediated DNA damage (8-hydroxy-2′deoxyguanosine) is increased in nigro-striatum of Parkinson's disease brain. *Neurodegeneration* 1994; 3: 197–204.
35. Jenkins BG, Koroshetz WJ, Beal MF, Rosen BR. Evidence for impairment of energy metabolism in vivo in Huntington's disease using localized ^{1}H NMR spectroscopy. *Neurology* 1993; 43: 2689–95.
36. Beal MF, Brouillet E, Jenkins BG, *et al.* Age-dependent striatal excitotoxic lesions produced by the endogenous mitochondrial inhibitor malonate. *J Neurochem* 1993; 61: 1147–50.
37. Polidori MC, Mecocci P, Browne SE, Senin U, Beal MF. Oxidative damage to mitochondrial DNA in Huntington's disease parietal cortex. *Neurosci Lett* 1999; 272: 53–6.
38. Browne SE, Ferrante RJ, Beal MF. Oxidative stress in Huntington's disease. *Brain Pathol* 1999; 9: 147–63.
39. Mazzola JL, Sirover MA. Reduction of glyceraldehyde-3-phosphate dehydrogenase activity in Alzheimer's disease and in Huntington's disease fibroblasts. *J Neurochem* 2001; 76: 442–9.

40. Rosen DR, Siddique T, Patterson D, *et al.* Mutations in Cu/Zn superoxide dismutase gene are associated with familial amyotrophic lateral sclerosis. *Nature* 1993; 362: 59–62.
41. MacMillan-Crow LA, Cruthirds DL. Manganese superoxide dismutase in disease. *Free Radic Res* 2001; 34: 325–36.
42. Cookson MR, Shaw PJ. Oxidative stress and motor neuron disease. *Brain Pathol* 1999; 9: 165–86.
43. Kok AB. Ascorbate availability and neurodegeration in amyotrophic lateral sclerosis. *Med Hypotheses* 1997; 48: 281–96.
44. Schulz JB, Dehmer T, Schöls L, *et al.* Oxidative stress in patients with Friedreich ataxia. *Neurology* 2000; 55: 1719–21.
45. Iannello RC, Crack PJ, de Haan JB, Kola I. Oxidative stress and neural dysfunction in Down syndrome. *J Neural Transm Suppl* 1999; 57: 257–67.
46. Perrig WJ, Perrig P, Stähelin HB. The relation between antioxidants and memory performance in the old and very old. *J Am Geriatr Soc* 1997; 45: 718–24.
47. Sinclair AJ, Bayer AJ, Johnston J, Warner C, Maxwell SRJ. Altered plasma antioxidant status in subjects with Alzheimer's disease and vascular dementia. *Int J Geriatr Psychiatry* 1998; 13: 840–5.
48. Jiménez-Jiménez FJ, Molina JA, de Bustos F, *et al.* Serum levels of β-carotene, α-carotene and vitamin A in patients with Alzheimer's disease. *Eur J Neurol* 1999; 6: 495–7.
49. Riviére S, Birlouez-Aragon I, Nourashémi F, Vellas B. Low plasma vitamin C in Alzheimer patients despite an adequate diet. *Int J Geriatr Psychiatry* 1998; 13: 749–54.
50. Jiménez-Jiménez FJ, de Bustos F, Molina JA, *et al.* Cerebrospinal fluid levels of alpha-tocopherol (vitamin E) in Alzheimer's disease. *J Neural Transm* 1997; 104: 703–10.
51. Polidori MC, Stahl W, Eichler O, Niestroj N, Sies H. Profiles of antioxidants in human plasma. *Free Radic Biol Med* 2001; 30: 456–62.
52. Behl C. Vitamin E and other antioxidants in neuroprotection. *Int J Vitam Nutr Res* 1999; 69: 213–19.
53. Christen Y. Oxidative stress and Alzheimer disease. *Am J Clin Nutr* 2000; 71: 621–9S.
54. Deschamps V, Barberger-Gateau P, Peuchant E, Orgogozo JM. Nutritional factors in cerebral aging and dementia: epidemiological arguments for a role of oxidative stress. *Neuroepidemiology* 2001; 20: 7–15.
55. Youdim KA, Joseph JA. A possible emerging role of phytochemicals in improving age-related neurological dysfunctions: a multiplicity of effects. *Free Radic Biol Med* 2001; 30: 583–94.
56. Commenges D, Scotet V, Renaud S, Jacqmin-Gadda H, Barberger-Gateau P, Dartigues JF. Intake of flavonoids and risk of dementia. *Eur J Epidemiol* 2000; 16: 357–63.
57. Warsama Jama J, Launer LJ, Witterman JCM, *et al.* Dietary antioxidants and cognitive function in a population-based sample of older persons: the Rotterdam Study. *Am J Epidemiol* 1996; 144: 275–80.
58. Morris MC, Beckett LA, Scherr PA, *et al.* Vitamin E and vitamin C supplement use and risk of incident Alzheimer disease. *Alzheimer Dis Ass Disord* 1998; 12: 121–6.
59. Sobow TM, Kloszewska I, Karlinska I. Addition of vitamin E to donepezil as a treatment for Alzheimer's disease. *Alzheimer Rep* 1999; 2: 143–6.
60. Sano M, Ernesto C, Thomas RG, *et al.* A controlled trial of selegiline, alpha-tocopherol, or both as treatment for Alzheimer's disease. *N Engl J Med* 1997; 336: 1216–22.
61. Grundman M. Vitamin E and Alzheimer disease: the basis for additional clinical trials. *Am J Clin Nutr* 2000; 71: 630–6S.
62. Anon. Practice guidelines covering management of Alzheimer's disease. *Am J Health System Pharm* 1997; 54: 1481–5.
63. Praticó D, Delanty N. Oxidative injury in diseases of the central nervous system: focus on Alzheimer's disease. *Am J Med* 2000; 109: 577–85.
64. Rottkamp CA, Nunomura A, Hirai K, Sayre LM, Perry G, Smith MA. Will antioxidants fulfill their expectations for the treatment of Alzheimer disease? *Mech Ageing Dev* 2000; 116: 169–79.
65. Ebadi M, Srnivasan SK, Baxi MD. Oxidative stress and antioxidant therapy in Parkinson's disease. *Prog Neurobiol* 1996; 48: 1–19.
66. Nicoletti G, Crescibene L, Scornaiechi M, *et al.* Plasma levels of vitamin E in Parkinson's disease. *Arch Gerontol Geriatr* 2001; 33: 7–12.
67. Factor SA, Weiner WJ. Retrospective evaluation of vitamin E therapy in Parkinson's disease. *Ann NY Acad Sci* 1989; 570: 441–2.
68. Fahn S. An open trial of high-dosage antioxidants in early Parkinson's disease [abstract]. *Am J Clin Nutr* 1991; 53: 380–2S.
69. Parkinson Study Group. Impact of deprenyl and tocopherol treatment on Parkinson's disease in DATATOP subjects not requiring levodopa. *Ann Neurol* 1996; 39: 29–36.
70. Beal MF. Coenzyme Q_{10} as a possible treatment for neurodegenerative diseases. *Free Radic Res* 2002; 36: 455–460.

71. The Huntington Study Group. A randomized, placebo-controlled trial of coenzyme Q_{10} and remacemide in Huntington's disease. *Neurology* 2001; 57: 397–404.
72. Gurney ME, Cutting FB, Zhai P, *et al.* Benefit of vitamin E, riluzole and gabapentin in a transgenic mouse model of familial amyotrophic lateral sclerosis. *Ann Neurol* 1996; 39: 147–57.
73. Matthews RT, Yang S, Browne SE, Baik M, Beal MF. Coenzyme Q10 administration increases brain mitochondrial concentrations and exerts neuroprotective effects. *Proc Natl Acad Sci USA* 1998; 95: 8892–7.
74. Louvel E, Hugon J, Doble A. Therapeutic advances in amyotrophic lateral sclerosis. *Trends Pharmacol Sci* 1997; 18: 196–203.
75. Musumeci O, Naini A, Slonim AE, *et al.* Familial cerebellar ataxia with muscle coenzyme Q10 deficiency. *Neurology* 2001; 56: 849–55.
76. Strausak D, Mercer JF, Dieter HH, Stremmel W, Multhaup G. Copper in disorders with neurological symptoms: Alzheimer's, Menkes, and Wilson diseases. *Brain Res Bull* 2001; 55: 175–85.
77. Ouahchi K, Arita M, Kayden H, *et al.* Ataxia with isolated vitamin E deficiency is caused by mutations in the α-tocopherol transfer protein. *Nat Genet* 1995; 9: 141–5.
78. Brigelius-Flohé R, Traber MG. Vitamin E: function and metabolism. *FASEB J* 1999; 13: 1145–55.
79. Vingerling JR, Klaver CW, Hofman A, de Jong PTVM. Epidemiology of age-related maculopathy. *Epidemiol Rev* 1995; 17: 347–60.
80. Schalch W, Dayhaw-Barker P, Barker FM. The carotenoids of the human retina. In: Taylor A, ed. *Nutritional and environmental influences on the eye*. Boca Raton: CRC Press, 1999: 215–50.
81. Stahl W, Schwarz W, Sundquist A, Sies H. Cis-trans Isomers of lycopene and β-carotene in human serum and tissue. *Arch Biochem Biophys* 1992; 294: 173–7.
82. Beatty S, Murray IJ, Henson DB, Carden D, Koh HH, Boulton ME. Macular pigment and risk for age-related macular degeneration in subjects from a northern European population. *Invest Ophthalmol Vis Sci* 2001; 42: 439–46.
83. EDCC (Eye Disease Case–Control) Study Group. Antioxidant status and neovascular age-related macular degeneration. *Arch Ophthalmol* 1993; 111: 104–9.
84. Seddon JM, Ajani UA, Sperduto RD, *et al.* Dietary carotenoids, vitamins A, C and E and advanced age-related macular degenration. *JAMA* 1994; 272: 1413–20.
85. Bone RA, Landrum JT, Mayne ST, Gomez CM, Tibor SE, Twaroska EE. Macular pigment in donor eyes with and without AMD: a case–control study. *Invest Ophthalmol Vis Sci* 2001; 42: 235–40.
86. Beatty S, Koh H, Phil M, Henson D, Boulton M. The role of oxidative stress in the pathogenesis of age-related macular degeneration. *Surv Ophthalmol* 2000; 45: 115–34.
87. Davies S, Elliott MH, Floor E, *et al.* Photocytotoxicity of lipofuscin in human retinal pigment epithelial cells *Free Radic Biol Med* 2001; 31: 256–65.
88. Junghans A, Sies H, Stahl W. Macular pigments lutein and zeaxanthin as blue light filters studied in liposomes. *Arch Biochem Biophys* 2001; 391: 160–4.
89. Taylor A, Nowell T. Oxidative stress and antioxidant function in relation to risk for cataract. *Adv Pharmacol* 1997; 38: 515–36.
90. Shang F, Nowell TR, Taylor A. Removal of oxidatively damaged proteins from lens cells by the ubiquitin-proteasome pathway. *Exp Eye Res* 2001; 73: 229–38.
91. Obin M, Shang F, Gong X, Handelman G, Blumberg J, Taylor A. Redox regulation of ubiquitin-conjugating enzymes: mechanistic insights using the thiol-specific oxidant diamide. *FASEB J* 1998; 12: 561–9.
92. Jacques PF, Chylack LT, Hankinson SE, *et al.* Long-term nutrient intake and early age-related nuclear lens opacities. *Arch Ophthalmol* 2001; 119: 1009–19.
93. Jacques PF, Taylor A, Hankinson SE, *et al.* Long-term vitamin C supplement use and prevalence of early age-related lens opacities. *Am J Clin Nutr* 1997; 66: 911–16.
94. Giblin FJ. Glutathione: a vital lens antioxidant. *J Ocul Pharmacol Ther* 2000; 16: 121–35.

DISCUSSION

Dr. Marquardt: About 10 years ago I heard one of the last talks by Linus Pauling, where he was propagating pharmacologic intakes of vitamin C, vitamin A, and vitamin E for different reasons. What is your opinion on this, and could an increased uptake of antioxidants in the general population be of benefit?

Dr. Sies: Linus Pauling's book, *How to Live Longer and Feel Better*, is still a valid exploration of this subject. He overinterpreted himself and was overinterpreted by others, but the basic message that these low-molecular-weight antioxidants are useful is certainly valid. The question is, at what dosage and under what circumstances? In general, Pauling was a pioneer in this field.

Dr. Scott: You stated categorically that antioxidants have been shown to be of benefit in Alzheimer disease. Is it your opinion that there have been well-conducted studies where that conclusion can be drawn? Do those studies have sufficient statistical power to answer that question?

Dr. Sies: In this area, the basic problem is that we are not looking at these compounds as pharmacologic agents but against a background of antioxidants already present in the tissues and whose activities may well overlap. If you remove one or increase one, everything else will also change, as will the pattern of gene expression of the antioxidant enzymes. So it's very difficult to do studies in this area. Having said that, there are indeed studies that have shown beneficial effects (1–4). If the numbers studied were larger, then the effect would certainly have been clearer, but I think the trend is convincing enough to show there is an effect.

Dr. Scott: I believe there may be associations, but they do not prove cause and effect. In your presentation, you made it sound like a proven positive result, but it is not.

Dr. Sies: I never said there was a causative effect—in other words that the lack of antioxidants causes Alzheimer disease. That is certainly something entirely different. There is a theory of pathogenesis involving lipid peroxidation in neural cell membranes (5,6), but you are right, this has to be proven in humans and at present that is very difficult to do. On the other hand, without antioxidants life would cease. There is no doubt about that.

Dr. Scott: Yes, but one must have properly conducted studies before making statements about benefits. It is possible to do such studies. If scientific journals don't insist on such evidence, then traditions of treatment emerge that are not evidence based and may not have a good scientific basis. Not only may they be untrue, but by clouding the picture, they may even hinder useful treatments from emerging.

Dr. Sies: I think there is no doubt that oxidative reactions can be stopped with antioxidants. What you are asking is whether the oxidative reactions are in any way related to the original pathogenesis of a disease. In Down syndrome, for example, there is evidence for excessive activity of superoxide dismutase, so the gene determining SOD activity may be very relevant (7). There are many studies suggesting effects of oxidative reactions in other diseases—Huntington disease is an example—but usually with small numbers, I accept.

Dr. Micheli: Is there any way to quantify oxidative stress? There are at least three diseases of the preterm baby where oxidative stress may play a role—retinopathy of prematurity, bronchopulmonary dysplasia, and perhaps necrotizing enterocolitis. It would be very helpful to be able to identify the babies at highest risk.

Dr. Sies: A first approximation is the profile of antioxidants in the plasma. If antioxidants are decreased, then this could reflect either a reduced dietary supply or an increased uptake of antioxidants because of reactions with prooxidants. There are some diseases where there are selective patterns of loss of antioxidants, for example cystic fibrosis. The problem is that in many of these diseases the harmful effects are occurring in a specific organ and have little to do with circulating plasma. To obtain information on the rate of generation of prooxidants and the rate of utilization of antioxidants at organ level is very difficult. So what one has to use are the biomarkers for oxidative stress that I showed you for some of the neurologic diseases, which comprise an accumulation of damaged products and are therefore a molecular measure of past oxidative stress. These include the lipofuscins, the carboxymethyl-lysines, malondialdehyde products, and other simple products of the degradation of lipids.

The problem is that, as with the antioxidants themselves, there is no single magic compound that one can measure that will confirm there has been oxidative stress. Oxidative stress occurs in the lipid phase, in the aqueous phase, at the nucleus, or at the mitochondria, so it is very diversified. In order to assess it, one needs different measures of antioxidants and damage products. These assays are relatively difficult because, fortunately for us, prooxidants are scarce. Some assays may be done on exhaled air—ethane and pentane, for example, are markers of lipid peroxidation. These are of course noninvasive, so could be of value in children. There have been clinical studies investigating these volatile products of lipid peroxidation (8).

Dr. Koletzko: Going back to your mention of the potential increase in oxidative stress in Down syndrome, there is already a thriving sales drive in Europe for very expensive megadose preparations of vitamins for people with Down syndrome, promoted by someone who claims to have been a coworker of Linus Pauling. Is there actually any controlled evidence that such an intervention in Down syndrome is justified?

Dr. Sies: There have been studies from Monash University in Melbourne on this topic (9). The problem with this whole field is that it gets taken over by people who do not really understand the subject, just riding the crest of the wave. The basic message of antioxidants is clear to my mind, but the dosage, timing, and applications are not. There is also another aspect, which is the matter of combinations. In fruits and vegetables, for example, there is not a single compound but a mixture of compounds. There are interesting synergistic effects between these compounds, so you need less of one compound if the correct amount of another is present. The best example is the synergism between vitamin E and vitamin C, because the vitamin E radical is repaired by vitamin C, which then becomes vitamin C radical and can be detoxified to nonradical end products. Therefore, the synergism is of great interest.

Dr. Bachmann: Do you know of any neurologic diseases where you think there might be primary involvement of prooxidants rather than secondary changes?

Dr. Sies: The Heidelberg group has formulated an oxidative stress theory of Alzheimer disease. I mentioned already the site for the cleavage of the APP (amyloid precursor protein), but they have also found copper binding sites on the APP, which generate a continual Fenton reaction type of oxidative damage (10). So the idea is that prooxidant generation is the major initial event. I think this is a plausible hypothesis, and the group is trying to find further support for the presence of copper-mediated neurologic disease. For example, no one knows the function of the prion protein, but several groups have now shown that it is a copper-binding protein (11–14); perhaps when it is falsely folded it can become an oxidant.

Dr. Bachmann: Have you looked at Wilson disease in that context?

Dr. Sies: Other people have of course looked at Wilson disease, and there is clear evidence of nitrotyrosine formation in that condition and in hemochromatosis. In all the metal diseases where there is inadequate binding and an increase in free metal, there is an oxidative component in a very prominent pathogenic position.

Dr. Kon: Is there any possibility of measuring metabolites of nitrotyrosine?

Dr. Sies: There are some studies measuring 8-oxo-dG in different parts of the brain. With immunocytochemistry, it is readily demonstrated that nitrotyrosine is formed at sites where you also have DNA damage. Therefore, these are very nice indicators that oxidative damage has taken place. In addition, when preventive treatment is given, there is a decrease in nitrotyrosine formation.

Dr. Scott: I want to return to the question of evidence-based medicine. There is a theory that raised homocysteine is a significant cause of Alzheimer disease, in that it might cause microvascular-type strokes. That is a plausible theory, and there is no doubt that people with

Alzheimer disease have elevated homocysteine. However, the groups that are putting that theory forward are doing a proper placebo-controlled trial, intervening with folic acid to see if it's true. That is the kind of burden of proof for Down syndrome or Alzheimer disease that we should be looking for from our colleagues. I do not think we should permit people who have studied a few cases of Down syndrome to say that it looks as though some intervention has worked. We need to insist that these studies are properly conducted and that they are not published unless that is the case. The public needs that reassurance from the scientific community. Many things are plausible and might happen, but proof must be provided.

Dr. Sies: It's a good thing to be scientifically critical, but one can easily turn this around the other way and say, as you said earlier today, that homocysteine might just be a marker of something else rather than a causative factor—but that would still be of interest. The papers I referred to on Alzheimer disease and so on were peer-reviewed and in good journals. I agree that what is translated into journalism and to the general public is the responsibility of the scientist, but I think these initial studies are excellent pilot studies, showing us the way to go in future research. Homocysteine is much more oxidizable than cysteine, and it could be that the key to an elevated homocysteine is an increased radical attack. In the end, we may find important links between our fields of homocysteine and antioxidants.

Dr. Scott: To resolve the issue of whether homocysteine is a marker or a risk factor, there are seven or eight trials in progress at the moment, so an evidence-based approach is being used which will, I think, resolve the issue.

Dr. Kunz: Many of the antioxidants that have been used in clinical studies are synthetic components, and it is known that their effects may be different from natural antioxidants. Could you comment on that?

Dr. Sies: Such antioxidant compounds are used because a patent can be applied. You cannot patent ascorbate or vitamin E in this day and age. The antioxidants selected for use in clinical studies are used as pharmacologic agents. Some of them are quite interesting, particularly N-acetylcysteine. It is clear that redox reactions occur in all cells and the redox balance can be regained with such antioxidants (15,16).

Dr. Daniel: With the experiences of the large β-carotene supplementation trials, nutritionists and health officials are increasingly concerned about the widespread use of supplements, particularly the enrichment of diets with vitamin A, vitamin C, and vitamin E. You mentioned that when metals are present in a free state they produce a prooxidant environment. I believe a study from Düsseldorf has shown that the incidence of iron overload in men is almost the same as the incidence of iron deficiency in women (17). Do we have to worry about men who have a particularly large amount of free iron?

Dr. Sies: Iron overload in men is relevant to the development of cardiovascular problems and should be taken seriously. Regarding β-carotene, there is certainly a plausible explanation for why high β-carotene levels, 4 to 5 μmol/l or more, might be detrimental, whereas in the normal range it has an important protective function (18). For example, it is essential in terms of protection against photooxidative reactions in the skin and the eye, and, I'm sure, in other organs. Your question was on levels, and how high one should go when carrying out a study. Should one use high doses to ensure an effect and risk the possibility of unforeseen potential side effects? These problems are important in designing a study.

REFERENCES

1. Grundman M. Vitamin E and Alzheimer disease : the basis for additional clinical trials. *Am J Clin Nutr* 2000; 71: 630–36.

2. Berr C, Balansard B, Arnaud J, Roussel AM, Alperovitch A. Cognitive decline is associated with systemic oxidative stress: the EVA study—Etude du Vieillissement Arteriel. *J Am Geriatr Soc* 2000; 48: 1285–91.
3. Schmidt R, Hayn M, Reinhart B, *et al.* Plasma antioxidants and cognitive performance in middle-aged and older adults: results of the Austrian Stroke Prevention Study. *J Am Geriatr Soc* 1998; 46: 1407–10.
4. Masaki KH, Losonczy KG, Izmirlian G, *et al.* Association of vitamin E and C supplement use with cognitive function and dementia in elderly men. *Neurology* 2000; 54: 1265–72.
5. Guo Q, Fu W, Holtsberg FW, Steiner SM, Mattson MP. Superoxide mediates the cell-death-enhancing action of presenilin-1 mutations. *J Neurosci Res* 1999; 56: 457–70.
6. Farooqui AA, Horrocks LA. Lipid peroxides in the free radical pathophysiology of brain diseases. *Cell Mol Neurobiol* 1998; 18: 599–608.
7. Turrens JF. Increased superoxide dismutase and Down's syndrome. *Med Hypotheses* 2001; 56: 617–9.
8. Knutson MD, Handelman GJ, Viteri FE. Methods for measuring ethane and pentane in expired air from rats and humans, *Free Radic Biol Med* 2000; 28: 514–9.
9. De Haan JB, Wolvetang EJ, Cristiano F, *et al.* Reactive oxygen species and their contribution to pathology in Down syndrome. *Adv Pharmacol* 1997; 38: 379–402.
10. Multhaup G, Ruppert T, Schlicksupp A, *et al.* Copper-binding amyloid precursor protein undergoes a site-specific fragmentation in the reduction of hydrogen peroxide, *Biochemistry* 1998; 37: 7224–30.
11. Requena JR, Groth D, Legname G, *et al.* Copper-catalyzed oxidation of the recombinant SHa(29-231) prion protein. *Proc Natl Acad Sci USA* 2001; 98: 7170–5.
12. Kramer ML, Kratzin HD, Schmidt B, *et al.* Prion protein binds copper within the physiological concentration range. *J Biol Chem* 2001; 276: 16711–9.
13. Brown DR, Clive C, Haswell SJ. Antioxidant activity related to copper binding of native prion protein. *J Neurochem* 2001; 76: 69–76.
14. Hornshaw MP, McDermott JR, Candy JM. Copper binding to the N-terminal tandem repeat regions of mammalian and avian prion protein. *Biochem Biophys Res Commun* 1995; 207: 621–9.
15. Malorni W, Rivabene R, Matarrese P. The antioxidant N-Acetyl-cysteine protects cultured epothelial cells from Menadione-induced cytopathology. *Chem Biol Interact* 1995; 96: 113–23.
16. Dröge W. Free radicals in the physiological control of cell function, *Physiol Rev* 2002; 82: 47–95.
17. Niederau C. Iron overload and atherosclerosis. *Hepatology* 2000; 32: 672–4.
18. Pryor WA, Stahl W, Rock CL. Beta carotene: from biochemistry to clinical trials, *Nutr Rev* 2000; 58: 39–53.

Genetic Expression and Nutrition, edited by Claude Bachmann and Berthold Koletzko. Nestlé Nutrition Workshop Series, Pediatric Program, Vol. 50. Nestec Ltd., Vevey/Lippincott Williams & Wilkins, Philadelphia, © 2003.

Carbohydrate Therapy of Congenital Disorders of Glycosylation

Thorsten Marquardt

Klinik und Poliklinik für Kinderheilkunde, Münster, Germany

Inherited disorders caused by defects in the assembly or processing of glycoconjugates are called congenital disorders of glycosylation (CDG). Until recently, these diseases were called carbohydrate-deficient glycoprotein syndromes (CDG syndromes, CDGS), but rapidly improving knowledge in recent years has led to the need for a more general term (1). All disorders known so far have been identified by their effects on the N-glycosylation of proteins, but several of them are also supposed to affect other glycosylation processes such as O-glycosylation.

N-glycans are assembled on dolichol-pyrophosphate in a stepwise manner until a final oligosaccharide consisting of 14 monosaccharides has been synthesized. This oligosaccharide is then cotranslationally transferred to newly synthesized glycoproteins. The current nomenclature classifies all CDG that are caused by defects in the assembly of N-glycans on dolichol-pyrophosphate as CDG-I (or type I), and all defects in the processing of the N-glycans *after* their transfer to the protein as CDG-II (or type II). Once the molecular defect has been identified, a small letter is added to the name. Currently, six different disorders based on defects in the oligosaccharide assembly have been identified (CDG-Ia to CDG-If), as well as three disorders based on processing defects (CDG-IIa to CDG-IIc). In at least 30% of all patients with CDG, the molecular defect is different from that of known CDG types. These patients are classified as CDG-x, and most current research is focused on identifying the biochemical basis of the glycosylation defects in these patients.

PRINCIPLES OF N-GLYCOSYLATION

To understand the hypoglycosylation pattern of glycoconjugates in patients with CDG as well as possible therapeutic approaches in some of them, it is necessary to understand the processes leading to the correct glycosylation of proteins.

The two major types of glycosylation of a protein are N- and O-glycosylation. N-glycosylation occurs cotranslationally—that is, during the biosynthesis of a protein—in the lumen of the rough endoplasmic reticulum (RER), whereas O-glycosylation occurs further on in the secretory pathway in the Golgi apparatus. As all known

CDGs are largely characterized by their effects on N-glycosylation, I will describe the processes involved in more detail (Fig. 1).

The oligosaccharide assembly starts by transfer of N-acetylglucosamine (GlcNAc) from UDP-GlcNAc to dolichol-phosphate. Dolichol-phosphate resides in the membrane of the RER and the phosphate is located on the cytoplasmic side of the membrane. A different enzyme then transfers a second GlcNAc, leading to $GlcNAc_2$-P-P-dol. Step by step, five mannoses are then added from GDP-mannose by different enzymes. The resulting structure, $Man_5GlcNAc_2$-P-P-dol, flips from the cytoplasmic side of the membrane to the luminal side—that is, to the inside of the RER—where all further elongation processes take place.

The RER membrane is impermeable to sugar nucleotides, so further elongation with mannoses does not occur using GDP-mannose. Instead, GDP-mannose reacts on the outside of the RER with dolichol-phosphate to form Man-P-dolichol, which then itself flips to the luminal side of the membrane to serve as a donor for further elongation of the oligosaccharide precursor with mannoses. Four more mannoses are added in order to generate $Man_9GlcNAc_2$-P-P-dol. This oligosaccharide has three branches, one of which is then further elongated by the stepwise addition of three glucose residues from the Glc-P-Dol, which was generated from UDP-Glc on the cytoplasmic side and which then also flipped to the inside of the RER.

The final 14-monosaccharide structure is then transferred *en bloc* to the asparagine residues in special amino acid contexts just after passage of this part of the nascent chain through the translocation channel. The multimeric enzyme complex responsible for this transfer is called oligosaccharyltransferase (OST). The timeframe for the oligosaccharide transfer is very short, because transfer can only take place just after the glycosylation consensus site has come through the translocation channel and is still in spatial proximity to the OST complex. OST transfers full-length oligosaccharides with a high affinity, but the k_m for truncated oligosaccharides differs considerably so that they are transferred far less efficiently. Enzymatic defects in the assembly line of N-glycans thus lead to a uniform hypoglycosylation pattern of proteins.

Owing to inefficient transfer of truncated oligosaccharides, assembly line defects normally result in missing N-glycans at some glycosylation consensus sites, resulting in proteins that have fewer oligosaccharides attached than would normally be the case. Although dolichol-linked oligosaccharides are often truncated in CDG-I, protein-derived oligosaccharides in general have a normal size but are fewer in number.

Experimental inhibition of N-glycosylation, either by deletion of glycosylation consensus sites or by inhibition of precursor synthesis with tunicamycin or other drugs, has clearly shown that a correct transfer of N-glycans is necessary to increase the solubility of protein subdomains during protein folding. Loss of a critical number of glycans leads to insolubility and the formation of irreversible covalent protein aggregates that are subsequently degraded (2,3).

Immediately after the oligosaccharide transfer—that is, when protein translation is still ongoing—an extensive modification of the oligosaccharide takes place which begins with the trimming of the three terminal glucoses by glucosidases I and II.

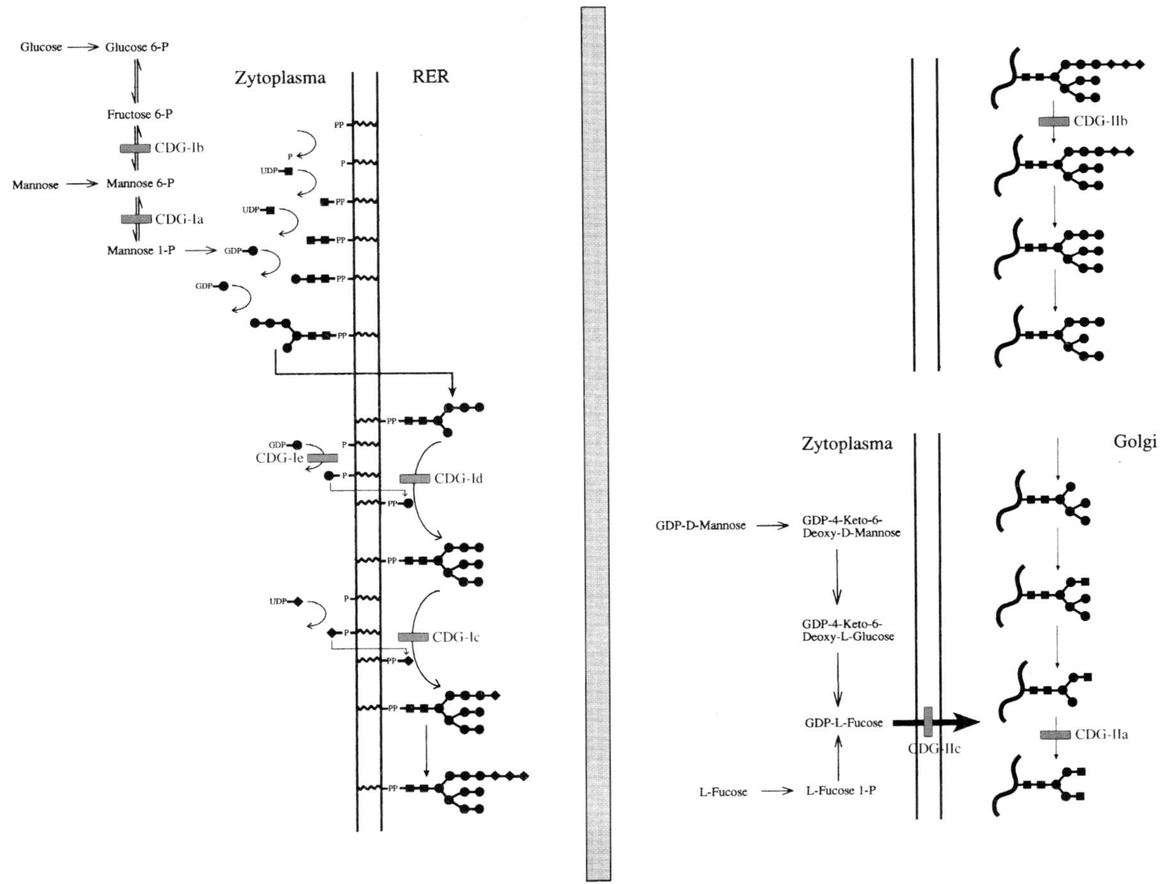

FIG. 1. Biosynthesis of GDP-mannose and assembly of the dolichol-linked oligosaccharide.

Monoglucosylated glycoproteins are the substrate for binding to an endogenous endoplasmic reticular lectin called calnexin, a process that is essential for the so-called quality control inside the RER [a system responsible for the retention and degradation of incorrectly folded glycoproteins (4,5)]. Subsequently, the oligosaccharide is transported to the Golgi apparatus and further trimmed until a $Man_5GlcNAc_2$ species is reached, which is then elongated with GlcNAc on one branch. Trimming and elongation events take place, leading to the final structure $Sia_2Gal_2GlcNAc_2Man_3GlcNAc_2$.

Although the functions of most of the processing steps are still not understood, knock-outs of some of the genes involved in the process have resulted in early death in mice in several cases, illustrating the fundamental importance of each of these complicated modification steps. In addition to the biantennary structure described, many different carbohydrate structures with higher branching or different monosaccharide compositions are generated in the Golgi apparatus, leading to a substantial diversity of carbohydrate structures. Based on this complexity, it has to be assumed that many more inherited disorders of glycosylation will be discovered in the not so distant future.

CARBOHYDRATE TREATMENT

With the elucidation of the molecular defects of CDG, it has become possible to develop specific treatments for children with CDG that use alternative metabolic pathways to circumvent the blocked pathways caused by enzyme defects. Presently, CDG-Ib can be treated satisfactorily in all patients and CDG-IIc in some. In the most common form of the disease, CDG-Ia, the hypoglycosylation can easily be corrected in cell culture experiments but the same approach has so far been unsuccessful in children. Therapeutic strategies for the different CDG types will now be discussed in more detail.

CDG-Ia

Even before the molecular defect of CDG-Ia had been discovered, it was known that the addition of more than 250 μmol/l of D-mannose to tissue culture medium containing CDG-Ia fibroblasts would correct the hypoglycosylation phenotype in these cells (6). In contrast to the secreted mostly liver-derived glycoproteins found in the plasma of CDG-Ia patients, all consensus sites in CDG-Ia fibroblasts are glycosylated; however, instead of full-length oligosaccharides, truncated forms like $Man_5GlcNAc_2$ are the predominant type. In the presence of mannose, the truncation is corrected and full-length oligosaccharides are transferred (6,7). This correction effect occurs immediately after the addition of mannose to the medium and disappears just as rapidly once mannose is removed. Complete correction is achieved with concentrations of 1 mmol/l or more, but a concentration of 250 μmol/l is sufficient for the attachment of full-length oligosaccharides to most consensus sites.

CDG-Ia is caused by an intracellular decrease in the GDP-mannose pool needed for the first mannose elongation steps of the oligosaccharide precursor, as well as for the synthesis of Man-P-dolichol. It is caused by a defect of phosphomannomutase 2, an enzyme that converts Man-6-P to Man-1-P in the cytoplasm (Fig. 1). The biochemical basis for the correction induced by mannose in cell culture experiments is still unclear. Exogenous mannose is converted to mannose-6-phosphate by hexokinase, which then has to be further processed by the defective enzyme. In eukaroytes, no mannokinase for the direct conversion of mannose to mannose-1-phosphate has been found. Nevertheless, it has been confirmed by different groups that D-mannose does indeed correct the hypoglycosylation phenotype of CDG-Ia fibroblasts, and that the GDP-mannose pool in these cells is nearly completely restored (8).

It was speculated early on that D-mannose could be used for the treatment of CDG-Ia patients. Initially, only a few reports from the 1960s and 1980s were available, which addressed the question of resorption of mannose from the gut and its short-term toxicity (9–11). There appeared to be no data addressing the issue of long-term toxicity or the presence of D-mannose in naturally occurring foods. Resorption kinetics were established at the beginning of mannose treatment trials by enthusiastic researchers who took mannose themselves and determined its concentrations in subsequent plasma samples (12). Methods for the correct determination of D-mannose had to be established (13). All analytical methods are complicated by the small concentrations of mannose compared to glucose, which is about 100 times more abundant. The enzymatic assay for D-mannose relies on the complete depletion of glucose in the sample of interest, whereas the more reliable determination by gas chromatography/mass spectrometry is complicated by the very close migration of the mannose and glucose peaks.

Mannose concentrations in serum samples from healthy individuals are about 50 μmol/l. The concentration of free D-mannose in serum samples of children with CDG-Ia is considerably lower (10–30 μmol/l). Oral mannose can be given without major side effects at up to 200 mg/kg body weight per single dose up to five times a day, if the dose is increased slowly. If this dose is exceeded, most people will get osmotic diarrhea from incomplete mannose absorption.

Serum mannose concentrations rise after oral ingestion of mannose, and peak concentrations are reached within 60 to 90 minutes after ingestion if there has been no simultaneous food intake. In healthy controls, peak concentrations often are above 300 μmol/l. In contrast, D-mannose absorption is impaired in children with CDG-Ia. Peak serum concentrations in these children are nearly always below 200 μmol/l. These levels do not rise significantly, even if the children are treated for several years with D-mannose. In severely affected children, peak concentrations can even be significantly lower and may not exceed the concentrations of healthy controls who have not ingested mannose (Fig. 2).

When mannose is given five times a day, peak concentrations increase during the day, as serum mannose concentrations are still above baseline level when the next dose is given. Nevertheless, children with CDG-Ia who are given the highest tolerated oral mannose doses have serum mannose concentrations during the day that are nearly always below the minimum concentration needed for phenotype correction in fibroblast cultures. In addition, the substrate requirements for N-glycan biosynthesis

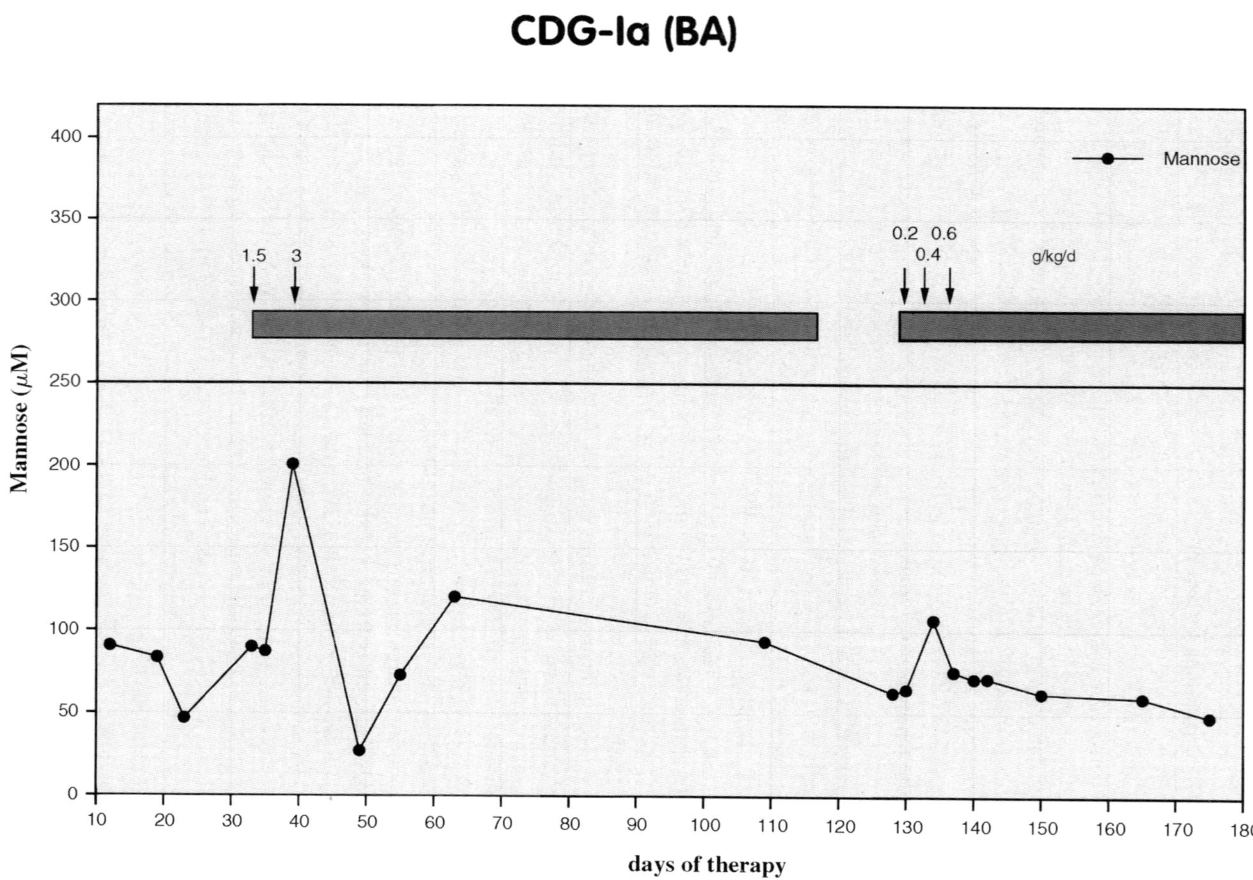

FIG. 2. Serum mannose concentrations in a patient severely affected with congenital disorder of glycosylation–Ia (CDG-Ia) during treatment. Mannose was given by continuous infusion using a gastric tube.

in fibroblasts may be considerably lower than in liver cells and other cell types, where phenotype correction might require much higher mannose concentrations.

Children with CDG-Ia who have been treated with oral mannose for up to 3 years did not show any improvement in the hypoglycosylation of plasma glycoproteins, as seen on two-dimensional gel electrophoresis, or in Western blots of β-trace cerebrospinal fluid (CSF) glycoprotein isolated from serum samples (Hüer C, Marquardt T, manuscript in preparation). Although many parents and other caretakers have reported improvements in the psychomotor performance of the children, this could not be proven statistically owing to the small numbers of cases and the impossibility of doing placebo-controlled studies.

The major objective of further mannose treatment studies in CDG-Ia has to be the achievement of increased serum mannose concentrations by parenteral administration. A short-term study involving intravenous mannose infusion has been done without any effect on transferrin hypoglycosylation (14,15), but experience with mannose treatment in CDG-Ib indicates that the timeframe of observation might have been much too short. Experimental approaches using lipids attached to mannose-1-phosphate are currently being investigated in cell culture experiments, but it is doubtful whether this will ever result in a therapeutic approach for patients with CDG-Ia.

In some patients, self-correction of the hypoglycosylation phenotype has been observed. In the first patient with CDG-Ia diagnosed in our department, a follow up over several years showed nearly complete disappearance of the hypoglycosylation of transferrin and other plasma glycoproteins, which is still present to date (Fig. 3). Phosphomannomutase activity is still well below 5% of normal. The reason for this phenomenon is unknown but it shows that a normal isoelectric focusing of serum transferrin—the test normally performed for the diagnosis of CDG—does not always exclude the presence of this disorder.

CDG-Ib

CDG-Ib is caused by a defect in the endogenous production of mannose-6-phosphate caused by defective phosphomannose isomerase, leading to inefficient conversion of fructose-6-phosphate to mannose-6-phosphate (Fig. 1). The conversion of exogenous mannose by hexokinase and its subsequent metabolism is unaffected. For this reason, an increase in the exogenous mannose supply can normalize the decreased GDP-mannose pool in CDG-Ib fibroblasts and completely reverses the hypoglycosylation of plasma proteins, as well as the symptoms of the disease in patients with CDG-Ib (16).

CDG-Ib is characterized by hepatic fibrosis, protein-losing enteropathy, thrombosis, and episodes of hypoglycemia. Several patients are known who have died during childhood from complications of the disease, but there are also cases of apparently healthy adults with the disorder. When CDG-Ib patients are treated with oral D-mannose, mannose is very efficiently absorbed and peak serum concentrations can exceed 500 μmol/l. The hypoglycosylation of serum glycoproteins almost completely disappears after a few months of treatment. Although the biologic half-life of serum transferrin is

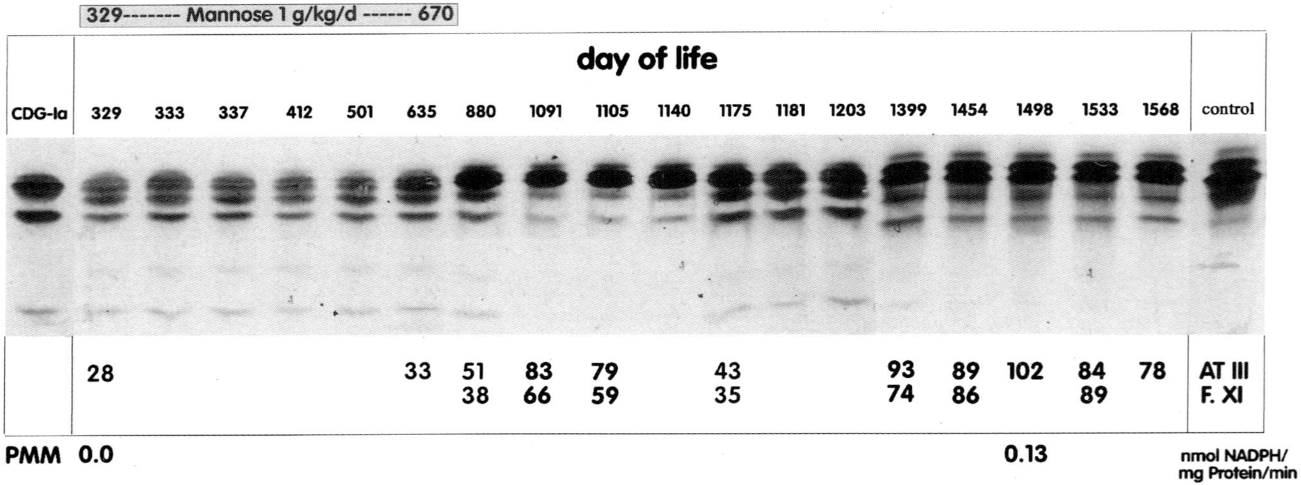

Patient : RL

PMM 2 mutations : F119L / R141H

FIG. 3. IEF pattern during long-term follow up of a patient with congenital disorder of glycosylation–Ia (CDG-Ia).

only approximately 1 week, it takes several months for the correction to occur. The exact reason for this phenomenon is not known, but it might indicate that the treatment effects in CDG-Ia also require several months before becoming apparent.

CDG-Ic, CDG-Id, CDG-Ie

There is no current treatment option for these forms of CDG.

CDG-If

CDG-If has recently been recognized as a new entity, and the papers describing this disorder have been submitted but are still under review (Kranz C *et al* and Shenk B *et al.* unpublished data). It is a disorder with severe psychomotor retardation, impaired vision, and, occasionally, ichthyosislike skin manifestations. It is caused by alterations in the Lec35/MPDU1 protein, which is responsible for the bioavailability of mannose-P-dolichol and glucose-P-dolichol. Its exact function is unknown, but the mutations in its gene, as seen in CDG-If patients, lead to a reduced bioavailability of Man-P-Dol and Glc-P-Dol for elongation of the oligosaccharide precursor; thus $Man_5GlcNAc_2$ and $Man_9GlcNAc_2$ are the predominant oligosaccharide species attached to dolichol.

In streptolysin-O-permeabilized CDG-If fibroblasts, initial experiments have shown that increased amounts of UDP-glucose could partially correct the glycosylation of the oligosaccharide precursor (Kranz C, Marquardt T, unpublished data; Fig. 4). Whether this effect can also be seen with high glucose concentrations in unpermeabilized cells is currently under investigation. No attempts at treating patients with CDG-If have been initiated so far.

CDG-IIa, CDG-IIb

There is no current treatment option for these forms of CDG.

CDG-IIc

CDG-IIc is a very rare disorder that is also known as leukocyte adhesion deficiency syndrome type II (LAD II). Three living patients are known in Israel and one in Germany (17). The disorder is characterized by a profound reduction of fucose in different glycosidic linkages in all glycoconjugates investigated. Constant leukocytosis is present, because neutrophils are unable to interact with selectins at the endothelial wall because of the absence of fucosylated ligands on their surface. The molecular basis of the condition is a mutation of the GDP-fucose transporter gene leading to a reduced import of GDP-fucose into the Golgi apparatus, where it is normally used for fucosylation reactions by different fucosyltransferases (18,19).

In cell culture experiments with CDG-IIc fibroblasts from various different patients, the addition of L-fucose to the culture medium could restore fucosylation.

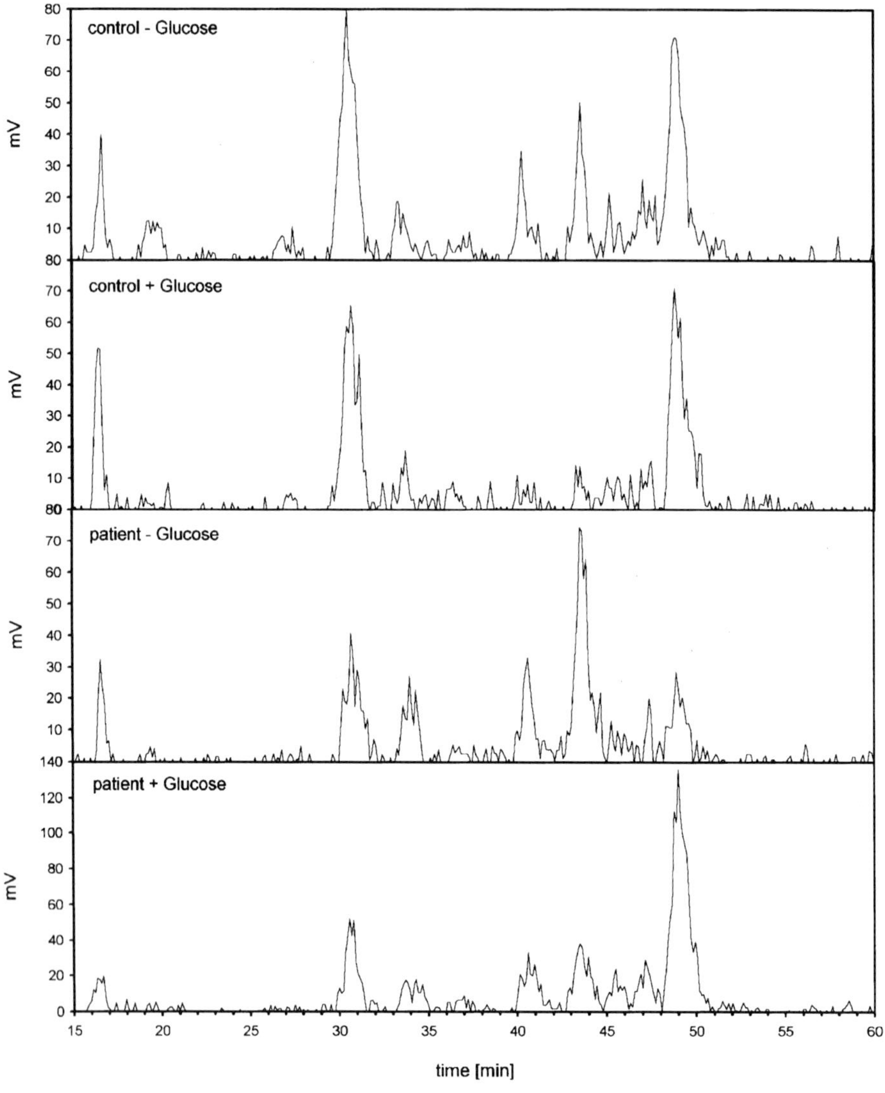

FIG. 4. Partial phenotype correction in SLO-permeabilized congenital disorders of glycosylation (CDG)-If fibroblasts. Cells were permeabilized with streptolysin O on ice and subsequently labeled with ^{3}H-GDP-mannose in the absence (−glucose) or presence (+glucose) of UDP-glucose. Lipid-linked oligosaccharides were hydrolyzed from dolichol and separated by high-performance liquid chromatography. Control, upper two panels; CDG-If, lower two panels. In CDG-If cells, UDP-glucose induces a shift from Man_9 $GlcNAc_2$ to $Glc_3Man_9GlcNAc_2$. (From Kranz C, Marquardt T, unpublished data.)

When this was also found to be true for fibroblasts from the German CDG-IIc patient, oral L-fucose supplementation was begun in this patient. Major precautions were taken, as few data on toxicity were available and autoimmune hemolysis was anticipated if the ABO blood groups were restored by the treatment. The researchers established fucose resorption kinetics were established after oral ingestion. Fucose doses as low as 125 mg/kg body weight daily were sufficient to reduce the elevated neutrophil counts in the patient to normal, and higher doses restored E- and P-selectin binding (20). After 3 years of L-fucose treatment, leukocyte counts and infection frequency in the patient are normal and no side effects have been observed. Weight gain and body growth are still insufficient and the patient has severe psychomotor retardation.

CONCLUSIONS

Oral monosaccharide supplementation appears to be an effective treatment for some types of congenital disorders of glycosylation. These treatments literally provide a "sugar pill" for these otherwise life-threatening inherited diseases. Long-term side effects are unknown, although patients with CDG-Ia, CDG-Ib, and CDG-IIc have been treated for years without obvious side effects. It is to be hoped that it will be possible to use alternative metabolic pathways in other CDG types in the near future, so that more patients will benefit from a specific treatment for these disorders.

REFERENCES

1. Participants of the first international workshop on CDGS. 1999. Carbohydrate-deficient glycoprotein syndromes become congenital disorders of glycosylation: an updated nomenclature for CDG. *Glycoconj J* 1999; 16: 669–71.
2. Marquardt T, Helenius A. Misfolding and aggregation of newly synthesized proteins in the endoplasmic reticulum. *J Cell Biol* 1992; 117: 505–13.
3. Helenius A, Hammond C, Marquardt T, Braakman I. How glycoproteins acquire their three-dimensional structure in the endoplasmic reticulum. *Nature* 1992; 356: 260–2.
4. Hebert DN, Foellmer B, Helenius A. Glucose trimming and reglucosyation determine glycoprotein association with calnexin in the endoplasmic reticulum. *Cell* 1995; 81: 425–33.
5. Hammond C, Helenius A. Quality control in the secretory pathway. *Curr Opin Cell Biol* 1995; 7: 523–9.
6. Panneerselvam K, Freeze HH. Mannose corrects altered N-glycosylation in carbohydrate-deficient glycoprotein syndrome fibroblasts. *J Clin Invest* 1996; 97: 1478–87.
7. Körner C, Lehle L, von Figura K. Carbohydrate-deficient glycoprotein syndrome type 1: correction of the glycosylation defect by deprivation of glucose or supplementation of mannose. *Glycoconj J* 1998; 15: 499–505.
8. Rush JS, Panneerselvam K, Waechter CJ, Freeze HH. Mannose supplementation corrects GDP-mannose deficiency in cultured fibroblasts from some patients with congenital disorders of glycosylation (CDG). *Glycobiology* 2000; 10: 829–35.
9. Wood FC, Cahill GF. Mannose utilization in man. *J Clin Invest* 1963; 42: 1300–12.
10. de la Fuente M, Penas P, Sols A. Mechanism of mannose toxicity. *Biochem Biophys Res Commun* 1986; 140: 51–5.
11. de la Fuente M, Hernanz A. Enzymes of mannose metabolism in murine and human lymphocytic leukaemia. *Br J Cancer* 1988; 58: 567–9.
12. Alton G, Kjaergaard S, Etchinson JR, Skovby F, Freeze HH. Oral ingestion of mannose elevates blood mannose levels: a first step toward potential therapy for carbohydrate-deficient glycoprotein syndrome type I. *Biochem Mol Med* 1997; 60: 127–33.
13. Etchison JR, Freeze HH. Enzymatic assay of D mannose in serum. *Clin Chem* 1997; 43: 533–8.

14. Mayatepek E, Schröder M, Kohlmüller D, Bieger WP, Nützenadel W. Continuous mannose infusion in carbohydrate-deficient glycoprotein syndrome type I. *Acta Paediatr* 1997; 86: 1138–40.
15. Mayatepek E, Kohlmüller D. Mannose supplementation in carbohydrate-deficient glycoprotein syndrome type I and phosphomannomutase deficiency. *Eur J Pediatr* 1998; 157: 605–6.
16. Niehues R, Hasilik M, Alton G, *et al.* Carbohydrate-deficient glycoprotein syndrome type Ib. Phosphomannose isomerase deficiency and mannose therapy. *J Clin. Invest* 1998; 101: 1414–20.
17. Marquardt T, Brune T, Lühn K, *et al.* A new patient with leukocyte adhesion deficiency (LAD) II syndrome, a generalized defect in fucose metabolism. *J Pediatr* 1999;134:681–8.
18. Lübke T, Marquardt T, von Figura K, Körner C. A new type of carbohydrate-deficient glycoprotein syndrome due to a decreased import of GDP-fucose into the Golgi. *J Biol Chem* 1999; 274: 25986–9.
19. Lübke T, Marquardt T, Etzioni A, Hartmann E, von Figura K, Körner C. Complementation cloning identifies CDG-IIc (LAD II), a new type of congenital disorders of glycosylation, as a GDP-fucose transporter deficiency. *Nat Genet* 2001; 28: 73–6.
20. Marquardt T, Lühn K, Srikrishna G, Freeze HH, Harms E, Vestweber D. Correction of leukocyte adhesion deficiency type II with oral fucose. *Blood* 1999; 94: 3976–85.

DISCUSSION

Dr. Belli: It seems that isoelectric focusing of transferrin can be falsely negative in some forms of CDG. Is that true, and if so, when should we undertake determination of the mutation?

Dr. Marquardt: The problem is that isoelectric focusing of serum transferrin is the best technique we have, but I agree it's not that good. First, it fails to identify some of the CDG syndromes at all, such as CDG-IIb and CDG-IIc. These are missed because they have no abnormal transferrin. Then there are hypoglycosylation disorders like CDG-Ia, where there are affected patients with a normal transferrin test, and nobody knows why that should be. I suggest that if you have a strong suspicion that a patient has CDG-Ia—which is relatively easy because these patients have internal strabismus and psychomotor retardation—but there is a normal transferrin test, then you should do a phosphomannomutase assay and measure the enzyme directly. This can easily be done from a blood sample because the enzyme is present in leukocytes. For other disorders, it is more difficult, and there is certainly a need to develop new tests, for example based on lectin binding. Right now, though, transferrin is the best we can offer. You can look at the mutations as well, which is easy in the CDG-Ia and in the others too, but it is easier to do the enzymatic assay.

Dr. Böhles: Chronic excessive alcohol intake appears to cause a defect of glycosylation. What is the mechanism?

Dr. Marquardt: This was first described in 1980, but it took a few years to find out that there was a glycosylation problem. If you perform isoelectric focusing on blood from alcoholics, you see hypoglycosylation of serum transferrin. There appears to be inhibition of the enzymes involved in mannose metabolism, probably caused by acetaldehyde (1,2). If you drink more than about half a bottle of wine a day, you get hypoglycosylation. The half-life of transferrin is about 1 week, so it takes a while to get rid of this phenotype.

Dr. Bachmann: You have been looking at serum samples. Is that a good place to look? It seems to me that the liver might be better. When you look at patterns in the liver—I'm thinking particularly of CDG-Ib—you see retention of glycosylated forms. It surprises me that you usually appear to find normally glycosylated proteins. Do you think that the disease results from retention of these abnormal forms? Why should quality control in the cell be so poor that it doesn't get rid of the abnormal forms? Moreover, why don't the normally glycosylated proteins that are present still fulfill their function?

Dr. Marquardt: Most of the CDG-Is are simply caused by a lack of substrate, for example the GDP-mannose pool is reduced and that diminishes the population of fully assembled glycans. That's the reason why some glycosylation sites are still occupied by fully assembled glycans, whereas others are not because oligosaccharyltransferase has no glycan to transfer. The glycans are very important not only for the function of some proteins but because they increase the solubility of protein domains during protein folding in the endoplasmic reticulum, thereby preventing aggregation. If you experimentally inhibit glycosylation, you have covalent aggregation and the proteins never reach their destinations. For example, if you look at endoplasmic reticulum in patients with CDG-Ia, you see dilatation and precipitates. There are certainly some proteins that don't make it at all to their final destination.

In terms of quality control, obviously some proteins escape and can get out. What is intriguing to me is why CDG-IIb causes such severe illness. There has only been one patient described so far and this child died at 2 1/2 months of age after a very severe illness. In this condition there is no activity of glucosidase I. As a result, glucosidase II cannot act either. Monoglucosylated carbohydrates bind to calnexin and calreticulin, and this is a very important interaction for quality control, because these are endogenous chaperones in the endoplasmic reticulum; as long as a protein is bound to those chaperones, it can still fold and acquire its correct structure. However, in CDG-IIb, there will be impaired quality control because monoglucosylated species cannot be formed. I believe that may be the reason why these children are so very sick.

Where should we look for the abnormal proteins? Serum is not a good sample I agree, because most of the major proteins are of course liver-derived. I am sure you also need to look at other organ systems, and maybe you should look at cellular proteins as well. You also need to look with different methods, but these have to be developed first!

Dr. Kunz: Your presentation was very interesting because for the first time it has become apparent that the lack of a simple monosaccharide can lead to very severe consequences. From a nutritional point of view—and referring to fucosylation defects—where do we get fucose from? There is no fucose in the diet as far as I know.

Dr. Marquardt: The fact that you can influence CDG-IIc with fucose is very fortunate. On the other hand, I am of course worried about this kind of treatment. As you say, there is no free fucose in the diet, and if you look at body fluids in healthy people—such as CSF or serum samples—they do not contain any free fucose. Nevertheless, we are raising the fucose levels in these patients to several hundred micromoles per liter. Therefore, this treatment is certainly not physiologic and I am concerned about possible long-term side effects. To obtain these sugars is another problem. There is only one company in the world that can supply fucose, which is manufactured from seaweed. It is extremely expensive, and of course, you have to have quality control to ensure that there are no toxic substances present.

Dr. Wanders: With regard to CDG-Ia, there's a strange problem in that your original paper in the *Journal of Clinical Investigation* reported correction by mannose of the glycosylation defect, but you do not see this in patients. Why is there this difference between *in vitro* and *in vivo* experiments? Second, can you explain the difference between CDG-Ia and Ib, because there are tremendous differences in clinical presentation but they follow one another in the pathway up to GDP-mannose.

Dr. Marquardt: There are many things we do not understand. For example, why does mannose work in the cell culture at all? Mannose is phosphorylated to mannose-6-phosphate and then it encounters the block. People have though that there might be a mannokinase that makes mannose-1-phosphate directly, but no one has found it. Therefore, it is a mystery why mannose works in cell culture, but there is no doubt that it does and this has been confirmed by several groups. There are other strange effects. If you starve the cells of glucose, you get a correction as well—it takes longer, but you do get a correction, and nobody understands that either. So why don't we see similar effects in patients? First, I'm sure we don't get enough mannose into the patient. Fibroblasts do not make as many glycoproteins as liver cells, and the focus of our follow-up is liver proteins. Therefore, you may need higher concentrations, if it works at all in these patients. That could be the simple reason why we do not see any effects clinically. There was a very elegant and courageous study by Mayatepek in Heidelberg a few years ago (3,4) where he infused mannose by a central venous line in a child with CDG-Ia for 3 weeks. He obtained mannose levels that were consistently around 2 mmol/l, which of course is very high, and he didn't see any effect. I wonder whether there simply is no effect, or whether the study was not carried on for long enough. In our patients with CDG-Ib, it takes quite a long time for an effect to occur, maybe months. We do not know why that is, but based on that observation, I imagine that Mayatepek's trial was too short.

CDG-Ia and CDG-Ib are only one step apart, nevertheless the phenotype is completely different. There is no scientific basis for why the phenotypes are so different.

Dr. Superti-Furga: Concerning your fucose transporter–deficient patient, you were able to correct the leukocytosis but you mentioned that the patient had a very severe brain phenotype. Based on that patient and the patient with CDG-Ia, to what extent is the disorder actually a developmental defect that you are not going to correct even with the best possible treatment?

Dr. Marquardt: All these disorders start during intrauterine life. For example, in CDG-Ia there is cerebellar hypoplasia that is present from birth. There must already be a lot of damage before birth, which can only partially be corrected.

Dr. Superti-Furga: Do you have any necropsy data on the type and extent of the damage?

Dr. Marquardt: There are necropsy data from a few CDG-Ia children showing abnormal migration of certain cell types in the brain and cerebellar hypoplasia (5,6). For some reason, there is no hypoglycosylation of serum proteins during intrauterine life, which is another thing we do not understand.

Dr. Endres: Are there any unpleasant side effects of mannose therapy—for instance, increased renal water loss?

Dr. Marquardt: Not as far as we know. The only side effect we have seen is osmotic diarrhea if you give too much mannose. In the patient with CDG-Ib, there is increased glycosylated hemoglobin with high mannose doses. This may simply reflect mannosylation of hemoglobin because mannose is very reactive. Otherwise, we have not found any side effects.

Dr. Braegger: Why did your patient with CDG-Ib have severe intestinal bleeding? What is the pathogenic mechanism?

Dr. Marquardt: In many of the CDGs, there is hypoglycosylation of coagulation factors. If you do functional assays, for example for antithrombin III or factor XI, you will see that their activities are severely reduced. Most of the children are asymptomatic in this respect. Thrombosis may also occur because the system that dissolves coagulum is affected as well.

Dr. Braegger: Would this be a general coagulation disorder, not purely intestinal?

Dr. Marquardt: Yes. I think there is a delicate balance between coagulation and noncoagulation that can very easily be disturbed in one way or another.

REFERENCES

1. Stibler H. Carbohydrate-deficient transferrin in serum: A new marker of potentially harmful alcohol consumption reviewed. *Clin Chem* 1991; 37: 2029–37.
2. Stibler H, Borg S. Postmortem findings in two patients with the carbohydrate-deficient glycoprotein syndrome. *Acta Paediatr Scand Suppl* 1991; 275: 55–62.
3. Mayatepek E, Kohlmüller D. Mannose supplementation in carbohydrate-deficient glycoprotein syndrome type I and phosphomannomutase deficiency. *Eur J Pediatr* 1998; 157: 605–6.
4. Mayatepek E, Schröder M, Kohlmüller D, *et al.* Continuous mannose infusion in carbohydrate-deficient glycoproteine syndrome type I. *Acta Paediatr* 1997; 86: 1138–40.
5. Strømme P, Maehlen J, Strøm EH. Postmortem findings in two patients with the carbohydrate-deficient glycoprotein syndrome. *Acta Paediatr Scand Suppl* 1991; 275: 55–62.
6. Jaeken J, Hagberg B, Strømme P. Clinical presentation and natural course of the carbohydrate glycoprotein syndrome. *Acta Paediatr Scand Suppl* 1991; 375: 6–13.

Genetic Expression and Nutrition, edited by Claude Bachmann and Berthold Koletzko. Nestlé Nutrition Workshop Series, Pediatric Program, Vol. 50. Nestec Ltd., Vevey/Lippincott Williams & Wilkins, Philadelphia, © 2003.

Genetic Variability of Human Milk Oligosaccharides: Are There Biologic Consequences?

Clemens Kunz, Lars Bode, and Silvia Rudloff

Institute of Nutrition, University of Giessen, Giessen, Germany

CONCENTRATIONS OF OLIGOSACCHARIDES IN HUMAN MILK

Human milk is a rich source of complex oligosaccharides synthesized within the mammary gland. The concentration of this fraction, however, varies widely with lactational stage, decreasing from large amounts of up to 50 g/l or more in colostrum to an average of 5 to 10 g/l in mature milk (1). The interindividual variability can partly be explained by the genetic variance of the donors leading to different blood group specific components as well as to a characteristic set of milk oligosaccharides.

Table 1 summarizes the results on oligosaccharides in term and preterm milk in relation to their structure (1,2). Major components are mono- and difucosyllactose, lacto-N-tetraose (mainly type 1) and their mono- and difucosylated derivatives, as well as fucosylated lacto-N-hexaoses and lacto-N-octaoses. Among the sialylated components we isolated sialyllactose, the three isomeric forms of sialyl-lacto-N-tetraose (LST a, b, and c), and disialyl-lacto-N-tetraose. Because of the fucosylation of these oligosaccharides there are many more components in milk that need further characterization. There appear to be no qualitative or quantitative differences between term milk and preterm milk. Compositional changes in oligosaccharides occur during lactation, particularly during the early stages (1).

The major components of the complex oligosaccharides are lacto-N-tetraose (0.5–1.5 g/l) and their monofucosylated derivatives, lacto-N-fucopentaose I and II (up to 1.7 g/l) (Table 2). Lacto-N-tetraose and their monofucosylated derivatives comprise up to 50% to 70% of the total complex carbohydrates. Of the sialylated components, the content of sialyllactose (NeuAcα2-6Lac and NeuAcα2-3Lac) is about 1.0 g/l, followed by isomers of monosialylated lacto-N-tetraose and disialylated lacto-N-tetraose. In bovine milk, only small amounts of oligosaccharides are detectable, with sialyllactose being the major component (Table 2).

TABLE 1. *Major oligosaccharides in preterm and term human milk*

Trivial name	Abbreviation	Structure
Lactose	Lac	Galβ1-4Glc
2N-Fucosyl-lactose[a]	2N-Fuc-Lac	Fucα1-2Galβ1-4Glc
3-Fucosyl-lactose	3-Fuc-Lac	Galβ1-4Glc 3 □ 1 Fucα
Lacto-N-tetraose (type 1)	LNT	Galβ1-3GlcNAcβ1-3Galβ1-4Glc
Lacto-N-neo-tetraose (type 2)	neo-LNT	Galβ1-4GlcNAcβ1-3Galβ1-4Glc
Lacto-N-fucopentaose I[a]	LNFP I	Fucα1-2Galβ1-3GlcNAcβ1-3Galβ1-4Glc
Lacto-N-fucopentaose II[b]	LNFP II	Galβ1-3GlcNAcβ1-3Galcβ1-4Glc 4 □ 1 Fucα
Lacto-N-fucopentaose III	LNFP III	Galβ1-4GlcNAcβ1-3Galβ1-4Glc 3 □ 1 Fucα
Lacto-N-fucopentaose V	LNFP V	Galβ1-3GlcNAcβ1-3Galβ1-4Glc 3 □ 1 Fucα
Lacto-N-difuco-hexaose I[c]	LNDFH I	Fucα1-2Galβ1-3GlcNAcβ1-3Galβ1-4Glc 4 □ 1 Fucα
Lacto-N-difuco-hexaose II	LNDFH II	Galβ1-3GlcNAcβ1-3Galβ1-4Glc 4 3 □ □ 1 1 Fucα Fucα
Sialyα2-3lactose	NeuAcα2-3Lac	NeuAcα2-3Galβ1-4Glc
Sialyα2-6lactose	NeuAcα2-6Lac	NeuAcα2-6Galβ1-4Glc
Sialyl-lacto-N-tetraose a (LST a)	NeuAc-LNT	NeuAcα2-3Galβ1-3GlcNAcαβ1-3Galβ1-4Glc
Sialyl-lacto-N-tetraose b (LST b)	NeuAc-LNT	Galβ1-3GlcNAcβ1-3Galβ1-4Glc 6 □ 2 NeuAcα
Sialyl-lacto-N-tetraose c (LST c)	NeuAc-LNT	NeuAcα2-6Galβ1-4GlcNAcβ1-3Galβ1-4Glc
Disialyl-lacto-N-tetraose	$NeuAc_2LNT$	NeuAcα2-3Galβ1-3GlcNAcβ1-3Galβ1-4Glc 6 □ 2 NeuAcα

[a] Milk of women with secretor status.
[b] Le^a-active components.
[c] Le^b-active components.

TABLE 2. *Oligosaccharides in human milk and cows milk*

	Human milk[a,b] (g/l)	Cow's milk[c,d] (g/l)
Lactose	55–70	40–50
Oligosaccharides		
Lacto-N-tetraose	0.5–1.5	Traces
Lacto-N-fucopentaose I	1.2–1.7	—
Lacto-N-fucopentaose II	0.3–1.0	—
Lacto-N-fucopentaose III	0.01–0.2	—
Lacto-N-difucohexaose I	0.1–0.2	—
NeuAc(α2-6)lactose	0.3–0.5	—
NeuAc(α2-3) lactose	0.1–0.3	0.03–0.06
NeuAc-lacto-N-tetraose a	0.03–0.2	Traces
NeuAc-lacto-N-tetraose c	0.1–0.6	Traces
$NeuAc_2$-lacto-N-tetraose	0.2–0.6	Traces
Oligosaccharides (total)	5.0–10.0	Traces

[a] Montreuil J, Mullet S. Etude des variations des constituents glucidiques du lait de femme au cours de la lactation. *Bull Soc Chim Biol* 1960; 42: 365.
[b] Kunz C, Rudloff S. Biological functions of oligosaccharides in human milk. *Acta Paediatr* 1993; 82: 903.
[c] Kobata A. Isolation of oligosaccharides from human milk. *Methods Enzymol* 1972; 28: 262.
[d] Parkkinen J, Finne FJ. Isolation of sialyl oligosaccharides and sialyl oligosaccharide phosphates from bovine colostrum and human urine. *Methods Enzymol* 1987; 138: 289.

THE INFLUENCE OF THE BLOOD GROUP (ABO/LEWIS) AND SECRETOR STATUS OF THE DONOR ON MILK COMPOSITION

Biological functions of milk oligosaccharides are closely related to their configuration. The monomers of milk oligosaccharides are D-glucose (Glc), D-galactose (Gal), N-acetylglucosamine (GlcNAc), L-fucose (Fuc), and sialic acid (N-acetylneuraminic acid, NeuAc). With few exceptions, all oligosaccharides isolated so far carry lactose at their reducing end (Table 1). An elongation is achieved by an enzymatic attachment of GlcNAc residues linked in β1-3 or in β1-6 linkage to a Gal residue, followed by further addition of Gal in β1-3 or β1-4 linkage. Thus, many core structures can be formed. Further variations occur because of the attachment of lactosamine (Galβ1-3GlcNAc$\beta_{n=1-15}$), Fuc, or NeuAc residues at different positions of the core region and of the core elongation chain (3,4). The complexity of the oligosaccharides can be grasped by looking at the variability resulting from the attachment of just one Fuc moiety to lacto-N-tetraose (LNT), one of the major components in human milk, which can then yield four different fucosylated components: lacto-N-fucopentaose I, II, III, and V (Table 1). With regard to the Lewis blood group and secretor status, it is important to realize that the type of the Fuc linkage is dependent on at least three different fucosyltransferases in a genetically determined process (5,6). Oligosaccharides characteristic of the different Lewis blood groups are shown in Table 3. With regard to potential functions in inflammatory processes, Lewis x and its sialylated derivatives are currently of particular interest (see below). Here, the terminal Gal is linked β1–4 to the subterminal GlcNAc, leading to type 2 epitopes; in type 1 epitopes the linkage is Galβ1-3GlcNAc. The biosynthetic pathway of ABO and Lewis blood group determinants is shown in Fig. 1.

TABLE 3. *Epitopes of Lewis a and b active components in glycolipids, glycoproteins, and oligosaccharides*

	Type 1 epitopes		Type 2 epitopes
H type 1	Galβ1-3GlcNAcβ1-3Gal- 2 | Fucα1	H type 2	Galβ1-4GlcNAcβ1-3Gal- 2 | Fucα1
Le a	Galβ1-3GlcNAcβ1-3Gal- 4 | Fucα1	Le x	Galβ1-4GlcNAcβ1-3Gal- 3 | Fucα1
Le b	Galβ1-3GlcNAcβ1-3Gal- 2 4 | | Fucα1 Fucα1	Le y	Galβ1-4GlcNAcβ1-3Gal- 2 3 | | Fucα1 Fucα1

The Effect of Secretor Status

An α1-2-fucosyltransferase is expressed in about 77% of all North Europeans who are classified as secretors. Thus, oligosaccharides in milk from North European women are characterized by the presence of 1-2-fucosylated components, for example 2′-fucosyllactose (Fucα1-2Galβ1-4Glc), lacto-N-fucopentaose I (Fucα1-2Galβ1-3GlcNAc1-3Galβ1-4Glc), or more complex oligosaccharides, all possessing Fucα1-2Galβ1-3GlcNAc residues (Table 1).

Neutral Oligosaccharides and Lewis Blood Group Specificity

In Lewis (a^+b^-) individuals, another fucosyltransferase attaches Fuc residues in α1–4 linkages to a subterminal GlcNAc residue of type 1 chains. Thus in "Lewis

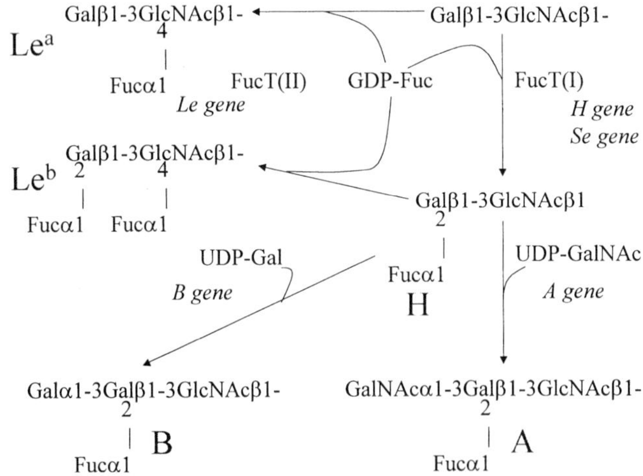

FIG. 1. Biosynthesis of blood group Lewis active oligosaccharides.

(a^+b^-), nonsecretor milk" the major fucosylated oligosaccharide is lacto-N-fucopentaose II (Galβ1-3[Fucα1-4]GlcNAcβ1-3Galβ1-4Glc; Table 1). This characteristic component is found in about 20% of the population.

In Lewis (a^-b^+) donors, both fucosyltransferases—the secretor gene and the Lewis gene–dependent form—are expressed. Here, one of the major milk oligosaccharides is lacto-N-difucohexaose I (Fucα1-2Galβ1-3[Fucα1-4]GlcNAcβ1-3Galβ1-4Glc; Table 1).

In about 5% of the population who belong to blood group Lewis (a^-b^-), a third type of milk oligosaccharide is found carrying Fuc α1–3 linked to GlcNAc in type 2 chains. The major oligosaccharide in milk of these donors is lacto-N-fucopentaose III (Galβ1-4[Fucα1-3]GlcNAcβ1-3Galβ1-4Glc; Table 1).

Acidic Oligosaccharides and Blood Group Specificity

In addition to fucosyltransferases, several sialyltransferases add NeuAc in different linkages (α2–3 or α2–6) at different positions to oligosaccharide core structures (3). The various actions of fucosyltransferases and sialyltransferases have led to the identification of more than 150 different components in human milk characterized so far.

The fact that different glycosyltransferases are active in the human lactating mammary gland, obviously has an impact on both the total amount of oligosaccharides and their pattern (Fig. 2). Many neutral oligosaccharides can be detected in Lewis (a^+b^-) and Lewis (a^-b^+) milk, whereas there are fewer components in Lewis (a^-b^-/secretor milk) (Fig. 2, lower right panel). In particular, only a few fucosylated components, which elute in the first 15 minutes of the chromatographic separation, are detectable in milk of Lewis (a^-b^-) individuals. However, the number of acidic oligosaccharides in the same donors does not seem to vary much (Fig. 2; retention time > 15 minutes). With very few exceptions, NeuAcα2-6Lac is the predominant sialyllactose present (Fig. 3).

Whether these obvious differences between individuals in the specific components of the oligosaccharides and their total amounts have any influence on specific biologic functions is not yet known.

POTENTIAL INFLUENCE OF A SPECIFIC OLIGOSACCHARIDE PATTERN ON BIOLOGIC PROCESSES

Carbohydrates as Soluble Receptors for Pathogenic Microorganisms

There is increasing evidence that oligosaccharides in human milk have a direct inhibitory effect on certain virulent abilities of pathogenic microorganisms. The decisive pathophysiologic factor for infectious diseases such as diarrhea seems to be the ability of microorganisms to adhere to the mucosal surface and their subsequent spreading, colonization, and invasion (for example, for *Escherichia coli*, *Helicobacter jejuni*, *Shigella* strains, *Vibrio cholerae*, and *Salmonella* species) in the gut (7,8). Bacterial adhesion is often a receptor-mediated interaction between structures on the bacterial surface and complementary ligands on the mucosal surface of the host (Fig. 4).

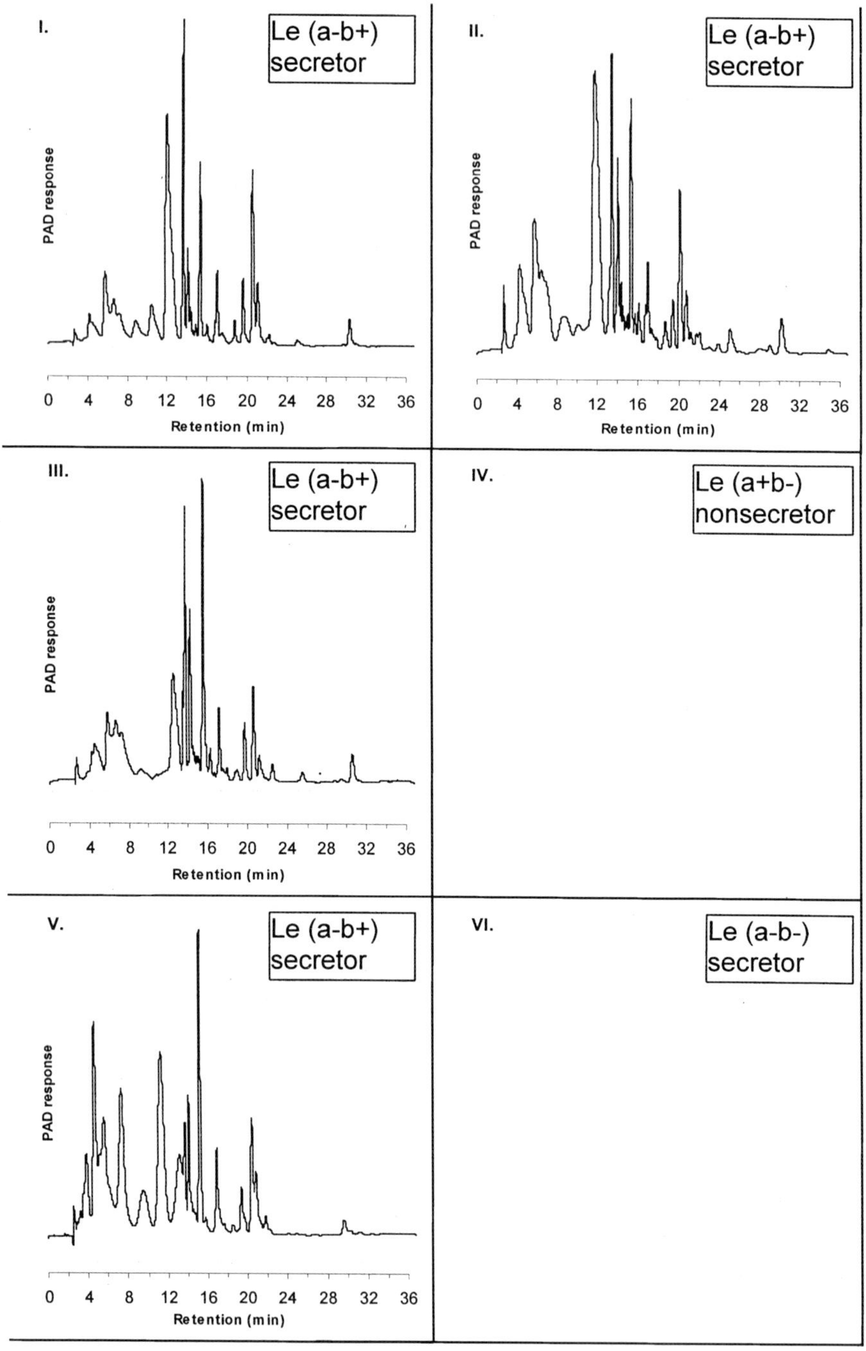

FIG. 2. High pH anion exchange chromatography with pulsed amperometric detection of the milk carbohydrate fraction after Sephadex G25 gel filtration from six individuals with different blood group specificities. Neutral oligosaccharides elute within the first 15 minutes, acidic oligosaccharides thereafter.

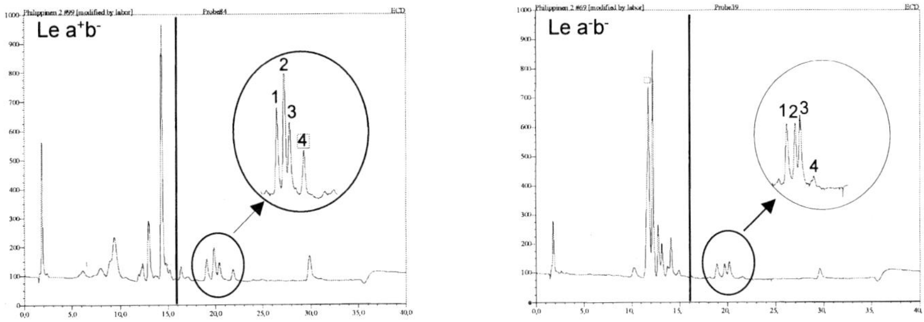

1 NeuAc α2-6 Gal β1-4 GlcNAc β1-3 Gal β1-4 Glc

2 NeuAc α2-6 Gal β1-4 Glc

3 NeuAc α2-3 Gal β1-4 Glc

4 Gal β1-3 GlcNAc β1-3 Gal β1-4 Glc
6
|
2
NeuAc α

FIG. 3. High pH anion exchange chromatography with pulsed amperometric detection of the milk carbohydrate fraction after Sephadex G25 gel filtration from two individuals with a different ratio of sialyllactoses.

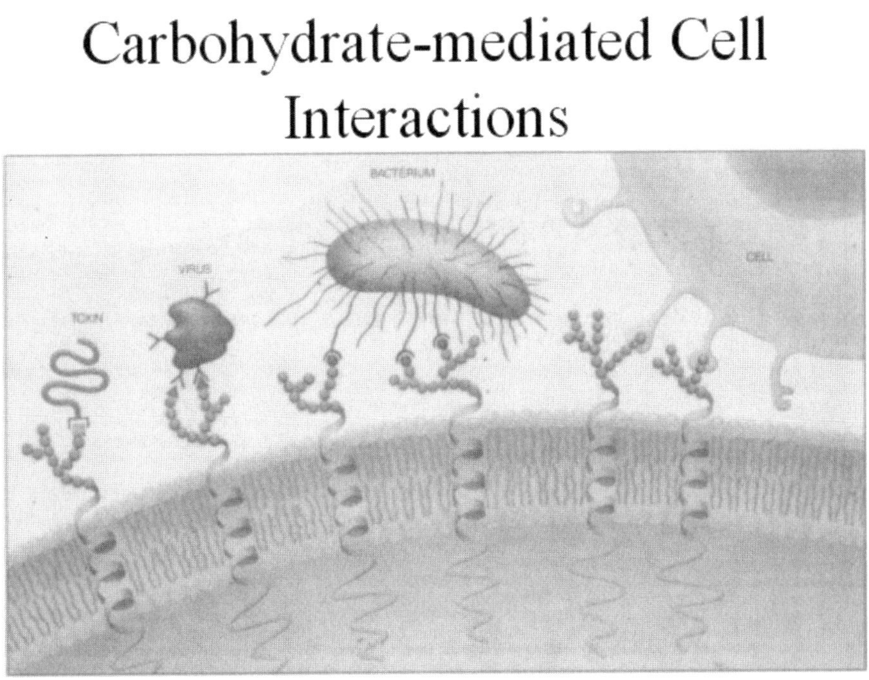

FIG. 4. Interactions of pathogenic microorganisms or their toxins with epithelial cell surface carbohydrates. Different monosaccharides are shown as black circles and triangles or grey squares.

The variable ligand specificity—for example, of *E. coli* strains—could explain the different intestinal colonization of breastfed versus formula-fed neonates (9). The large quantity and variety of oligosaccharides in human milk might prevent the intestinal attachment of microorganisms by acting as soluble analogs competing with epithelial receptors for bacterial binding (10).

To illustrate the similarities between oligosaccharides in human milk and the receptors for pathogenic microorganisms, some receptor specificities are listed in Table 4. This only shows those ligands that occur naturally in term and preterm milk (Table 1). Two reviews (8,11) have dealt in more detail with the interactions between microbial adhesins and host cell surfaces.

There are many examples for *in vitro* studies on the specific host–pathogen interactions that might be influenced by oligosaccharides. Recently, carbohydrates as antiadhesive drugs have also been tested in animals—for example, in rabbits and infant rats (12). In these animals, experimental pneumonia caused by *Streptococcus pneumoniae* was markedly reduced by intranasal or intratracheal administration of either free oligosaccharides or neoglycoproteins similar or identical to those naturally occurring in human milk.

In another recent study, *Helicobacter pylori*–positive rhesus monkeys were treated with 3′-sialyllactose alone or in combination with either one of the commonly used antiulcer drugs, bismuth subsalicylate (a gastric-coating agent) or omeprazole (a proton-pump inhibitor) (13). Of the six monkeys that were given the milk oligosaccharide only, two were permanently cured and a third was transiently cleared, while three of the animals remained persistently colonized. According to the authors, antiadhesive treatment is safe and can cure or reduce *H. pylori* colonization in some rhesus monkeys.

TABLE 4. *Oligosaccharides as receptors for microbes*

Receptors	Microbes
Mannose-containing glycoproteins	*Escherichia coli* (type 1 fimbriae)
Fucosylated oligosaccharides	*E. coli* (heat-stable enterotoxin)
Fucosylated tetra- and pentasaccharides	*E. coli*
Sialyl(α2-3)lactose and glycoproteins	*E. coli* (S fimbriae)
Sialyl(α2-3)galactosides in mucins	*E. coli* (S fimbriae)
Neutral oligosaccharides (LNT, neo LNT)	*Streptococcus pneumoniae*
Gal(β1-4)GlcNAc or Gal(β1-3)GlcNAc	*Pseudomonas aeruginosa*
Fucα1-2Gal epitopes	*Candida albicans*
Sialyl-lactose	*Helicobacter pylori*
Sialyl-lactose	*Streptococcus sanguis*
Sialyl-lactose and sialylated glycoproteins	*H. pylori*
Sialylated glycoproteins (α2-3 linked)	*Mycoplasma pneumoniae*
Sialylated poly-N-acetyllactosamine	*M. pneumoniae*
Sialylated (α2-3)poly-N-acetylactosaminoglycans	*Streptococcus suis*
Sialyl(α2-6)lactose	Influenza virus A
Sialyl(α2-3)lactose	Influenza virus B
9-O-Ac of NeuA(α2-3)R	Influenza virus C

For references see: Rudloff S, Kunz C. Protein and nonprotein nitrogen components in human milk, bovine milk, and infant formula: quantitative and qualitative aspects in infant nutrition. *J Pediatr Gastroenterol Nutr* 1997; 24: 328.

Disialyl-Fucosyl-Lacto-N-Hexaose

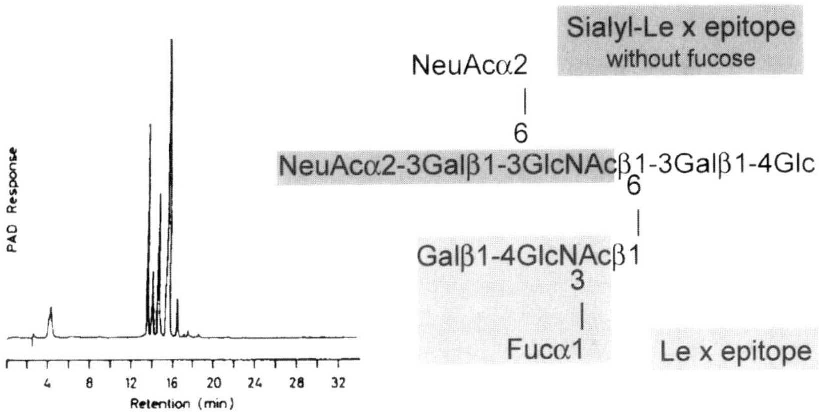

FIG. 5. High pH anion exchange chromatography with pulsed amperometric detection of an acidic milk fraction with disialyl-fucosyl-lacto-N-hexaose as major component. On the right panel, the structure is shown revealing multiple Lewis epitopes within the molecule.

Clinical experience with oligosaccharides as antiadhesive drugs is still very limited. In a recent study, the efficacy of 3′-sialyllacto-N-neotetraose was investigated as prophylaxis for acute otitis media (14). In this randomized, double-blind, placebo-controlled study, 507 healthy children were assigned either to the acidic milk oligosaccharide or to placebo as intranasal sprays twice daily for 3 months. This study failed to reduce the incidence of nasopharyngeal colonization with *S. pneumoniae* or *Haemophilus influenzae*, or the occurrence of acute otitis media. However, it is possible that during natural infection, bacteria can express multiple lectins with diverse specificities, and the inhibition of these may require a cocktail of different oligosaccharides (8).

The advantage of human milk oligosaccharides may lay in the large variety of different components. In addition, there may be several epitopes for microorganisms on just one oligosaccharide, leading to a polyvalent bioactive component. This is demonstrated in Fig. 5, which shows several epitopes for sialyl-Lewis x or its precursor. Such oligosaccharides could interfere with the attachment of microorganisms to the epithelial cell surface and also with leukocyte–endothelial cell interactions. This would require that at least some human milk oligosaccharides be absorbed as intact moieties without being digested. The ability to resist digestion has been shown recently by *in vitro* methods (15,16). More important, it has been observed in term and preterm infants that some human milk oligosaccharides are not only absorbed as intact moieties but are also excreted unchanged in the infant's urine (17,18). Thus, there is potential for human milk oligosaccharides to have systemic effects in these infants.

Leukocyte–Endothelial Cell Interactions

The quantity of oligosaccharides in the infant's circulation is comparable to the dose of carbohydrate drugs currently under investigation and may thus be sufficient to affect immune reactions.

The initial phase of leukocyte extravasation is mediated by cell adhesion molecules of the selectin family—for example on the surface of endothelial cells and their carbohydrate ligands on leukocytes. These interactions slow down the leukocytes, which then roll on the vessel wall and adhere to endothelial cells (Fig. 6). The physiologic ligand seems to require characteristic oligosaccharide epitopes like sialyl-Lewis x or sialyl-Lewis a (Fig. 7) (19–21).

The acidic (NeuAc-containing) fraction of human milk oligosaccharides contains various structures that affect these epitopes or their precursors. Thus, they may serve as soluble ligand analogs leading to competitive inhibition of leukocyte adhesion. This hypothesis is currently being investigated by our group, using a cell culture system under flow conditions (22). The addition of acidic human milk oligosaccharides to the cell media reduced leukocyte adhesion to the TNF-activated endothelial cell layer by about 50% (Fig. 8, See Color Plate 1). This effect was even more pronounced than that of the physiologic ligand sialyl-Lewis x used as a soluble additive.

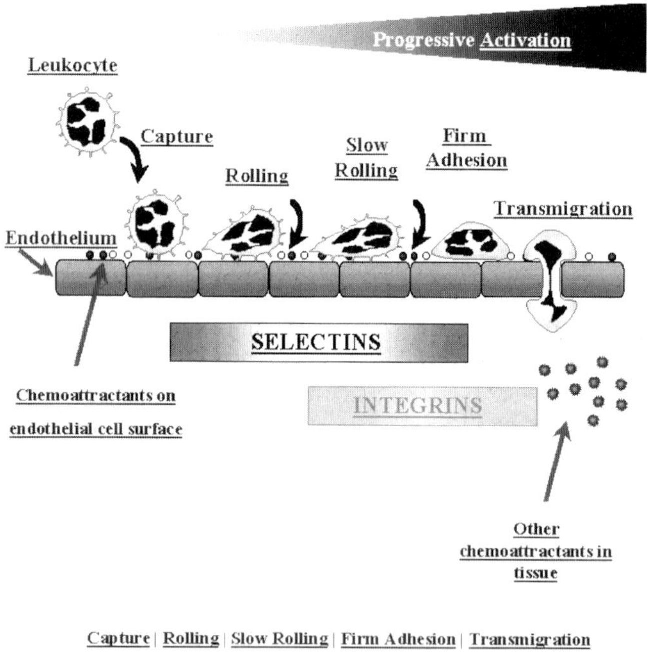

FIG. 6. Leukocyte–endothelial cell interactions are characterized by a cascade of events in which selectins play an important role. They mediate the first cell–cell contact and initiate firmer adhesion. (Available at http://www.med.virginia.edu/medicine/basic–sci/biomed/).

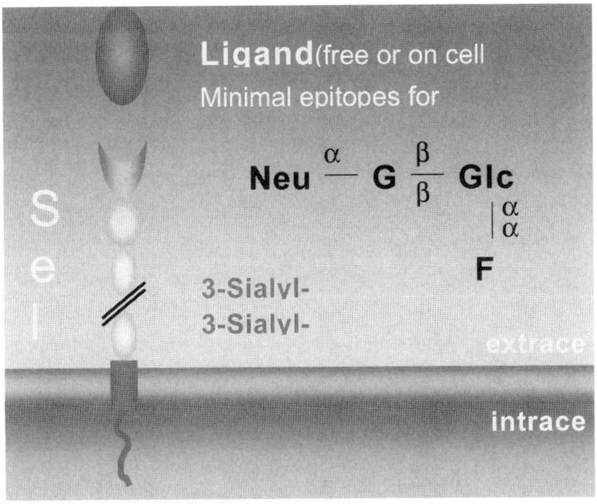

FIG. 7. Illustration of the selectin–ligand specificity with sialyl-Lewis x and sialyl-Lewis a as physiologic ligand epitopes.

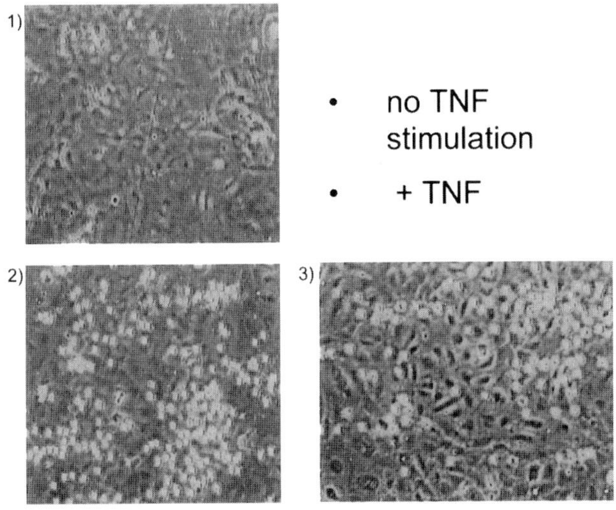

FIG. 8. Microscopic images of the leukocyte adhesion to endothelial cells, also shown in the presence of acidic human milk oligosaccharides (panel 3).

These effects of acidic human milk oligosaccharides could be due to the natural clustering of epitopes on high-molecular-weight oligosaccharides (Fig. 5). Our observations indicate that human milk oligosaccharides may function as immune regulatory components in infants. However, the capacity of specific oligosaccharide patterns in milk donors to be antiinfective or antiinflammatory is still a matter of speculation. The genetic layout of certain blood groups does align with the presence and quantity of specific (fucosylated) milk proteins, but these may not necessarily affect their potential immunologic properties.

The carboxyl group of NeuAc seems to be essential for the function of oligosaccharides as selectin ligand analogs. NeuAc in sialyl-Lewis x is 2–3 linked, whereas 2–6 linkages are abundant in human milk oligosaccharides. There are a very few exceptions among the milk donors (interestingly all of these have so far been found in the Philippines), where the ratio of 3′- to 6′-sialyllactose is reversed (Fig. 3). It will be necessary to investigate whether this might affect the antiadhesive properties of the oligosaccharides in inflammatory processes.

ACKNOWLEDGMENTS

Our work on human milk glycoconjugates is supported by the German Research Foundation (grant nos. DFG Ku 781/2-1/-2; 781/8-1, and Ru 529/4-1/-3).

REFERENCES

1. Kunz C, Rudloff S, Schad W, Braun D. Lactose-derived oligosaccharides in the milk of elephants: comparison with human milk. *Br J Nutr* 1999; 82: 391–9.
2. Kunz C, Rudloff S, Hintermann A, Pohlentz G, Egge H. High-pH anion exchange chromatography with pulsed amperometric detection and molar response factor of human milk oligosaccharides. *J Chromatogr B Biomed Appl* 1996; 685: 211–21.
3. Egge H. The diversity of oligosaccharides in human milk. In: Renner B, Sawatzki G, eds. *New perspectives in infant nutrition.* Stuttgart: Thieme, 1993: 12–26.
4. Stahl B, Thurl S, Zeng J, et al. Oligosaccharides from human milk as revealed by matrix-assisted laser desorption/ionization mass spectrometry. *Anal Biochem* 1994; 223: 218–26.
5. Morgan WTJ, Watkins WM. Unravelling the biochemical basis of blood group ABO and Lewis antigenic specificity. *Glycoconj J* 2000; 17: 501–30.
6. Lloyd KO. The chemistry and immunochemistry of blood group A, B, H, and Lewis antigens: past, present and future. *Glycoconj J* 2000; 17: 531–41.
7. Kunz C, Rudloff S. Biological functions of oligosaccharides in human milk. *Acta Paediatr* 1993; 82: 903–12.
8. Sharon N, Ofek I. Safe as mother's milk: carbohydrates as future anti-adhesion drugs for bacterial diseases. *Glycoconj J* 2000; 17: 659–64.
9. Adlerberth I. The establishment of a normal intestinal microflora in the newborn infant. In: Hanson LA, Yolken RB, eds. *Probiotics, other nutritional factors, and intestinal flora (42nd Nestlé Nutrition Workshop, Beijing).* New York: Raven Press, 1997: 63–78.
10. Kunz C. Microbial receptor analogs in human milk. In: Hanson LA, Yolken RB, eds. *Probiotics, other nutritional factors, and intestinal flora (42nd Nestlé Nutrition Workshop, Beijing).* New York: Raven Press, 1997: 157–74.
11. Karlsson KA. Microbial recognition of target-cell glycoconjugates. *Curr Opin Struct Biol* 1995; 5: 622–35.
12. Idnpn-Heikkil I, Simon PM, Zopf D, et al. Oligosaccharides interfere with the establishment andprogression of experimental pneumococcal pneumonia. *J Infect Dis* 1997; 176: 704–12.
13. Mysore JV, Wigginton T, Simon PM, Zopf D, Heman-Ackah LM, Dubois A. Treatment of Helicobacter pylori infection in Rhesus monkeys using a novel antiadhesion compound. *Gastroenterology* 1999; 117: 1316–26.

14. Ukkonen P, Varis K, Jernfors M, et al. Treatment of acute otitis media with an antiadhesive oligosaccharide: a randomised, double-blind, placebo-controlled trial. *Lancet* 2000; 356: 1398–402.
15. Engfer MB, Stahl B, Finke B, Sawatzki G, Daniel H. Human milk oligosaccharides are resistant to enzymatic hydrolysis in the upper gastrointestinal tract. *Am J Clin Nutr* 2000; 71: 1589–96.
16. Gnoth M J, Kunz C, Kinne-Saffran E, Rudloff S. Human milk oligosaccharides are minimally digested in vitro. *J Nutr* 2000; 130: 3014–20.
17. Rudloff S, Pohlentz G, Diekmann L, Egge H, Kunz C. Urinary excretion of lactose and oligosaccharides in preterm infants fed human milk or infant formula. *Acta Paediatr* 1996; 85: 598–603.
18. Kunz C, Rudloff S, Baier W, Klein N, Strobel S. Oligosaccharides in human milk. Structural, functional and metabolic aspects. *Annu Rev Nutr* 2000; 20: 699–722.
19. McEver RP. Selectins. *Curr Opin Immunol* 1994; 6: 75–84.
20. Varki A. Selectin ligands. *Proc Natl Acad Sci USA* 1994; 91: 7390–7.
21. Lasky L A. Selectin–carbohydrate interactions and the initiation of the inflammatory response. *Annu Rev Biochem* 1995; 64: 22. Bode L, Muhly-Reinholz M, Seeger W, Rudloff S. Effects of human milk oligosaccharides (HMO) on the adhesion of monocytes to endothelial cells under flow. *Eur J Biochem* 2001; 268: 221. 113–39.
22. Bode L, Muhly-Reinholz M, Seeger W, Rudloff S. Effects of human milk oligosaccharides (HMO) on the adhesion of monocytes to endothelial cells under flow. *Eur J Biochem* 2001; 268: 221.

DISCUSSION

Dr. Steinmann: What are the relative proportions of these oligosaccharides with respect to lactose?

Dr. Kunz: The lactose concentration in human milk is around 55 to 70 g/l, and the total concentration of oligosaccharides in our analyses was about 5 to 10 g/l. Others report somewhat higher concentrations, maybe up 15 g/l. In any case, these oligosaccharides are major components of human milk. The concentration may be as high as the protein concentration. Little was known about these components before about 10 years ago because it's rather difficult from an analytical point of view to separate large oligosaccharides with many isomeric components.

Dr. Steinmann: Are they recognized by lactase?

Dr. Kunz: We have done *in vitro* studies and found very little effect of lactase on the oligosaccharides. They appear not to be digested, which does not necessarily mean that they are not absorbed. In fact, about 2 to 3 mg/d are excreted in the urine of the infant, which means that substantially larger amounts are absorbed and circulate in the blood. In addition, we know that about 100 to 200 mg of total oligosaccharides per suckling perfuse the whole gastrointestinal tract—that's a very large amount.

Dr. Marquardt: So there is a large variety of lactose-based oligosaccharides in human milk, which are unique because they are only produced by the mammary gland, and in the newborn, at least some of these oligosaccharides are taken up as complete structures without digestion. What are the implications of this for the immune response? For example, do they induce antibody production against bacterial-like antigens that gives protection against possible infection? Or, do they have some relevance to ABO blood group antibodies?

Dr. Kunz: We don't know. The problem with the ABO and particularly the Lewis blood group systems is that we don't understand their function. With respect to Lewis antigens, some components might interfere with the leukocyte–endothelial cell interactions that I discussed, but there are no clinical studies at all and only a few *in vitro* studies trying to show such an effect. We need urgently to obtain larger quantities of very specific components so that we can do preclinical and clinical studies. Unfortunately they are extremely expensive. However, there are companies that are very interested in this topic and are trying to produce these components.

Dr. Marquardt: It would be interesting to look at the immune response, because there are so many unique compounds getting into the bloodstream of newborn infants. There must be some kind of immune response to these components.

Dr. Kunz: Yes, I am sure there is, but it is difficult to investigate. You cannot use whole human milk because it contains a wide variety of different components that might have the same effects.

Dr. Superti-Furga: The Lewis blood groups have a long history. They have been thought of as modifying factors in cystic fibrosis and even in fucosidosis. It seems to me that as they have been known about for so many years, there should be data by now on whether the Lewis blood group and secretor status of the mother have any clinical influence on the child.

Dr. Kunz: The answer to that is not straightforward because one would need to analyze milk from hundreds of mothers, and the analyses are not that easy. Another problem—and this is known for many cell types—is that there is a change in glycosylation pattern during development, for example during tumor processes and so on. So how do we know whether we are comparing the right things?

Dr. Superti-Furga: I meant just knowing whether the mother is Lewis A, Lewis B, or Lewis-negative, and secretor-positive or -negative.

Dr. Kunz: This doesn't help much. The status of the mother's red blood cell antigens may not reflect the milk oligosaccharides.

Dr. Superti-Furga: I assume that these complex structures must be synthesized on a protein or some other sort of core. Are they free in the milk, or are they still bound to their synthesizing core?

Dr. Kunz: No, they are not protein-bound. We don't know how this happens, but there is a rather well-investigated glycosyltransferase, which usually adds galactose to N-acetyl-glucosamine, and then you get the N-linked glycans that Dr. Marquardt has just discussed. However, during lactation α-lactalbumin, which is only produced in the lactating mammary gland, acts as a specifier protein that switches the affinity from GLC-NAC to glucose, and then you get lactose; if you have lactose, you can then get elongation by lactosamine. That's all we know about the biosynthesis.

Dr. Takada: Is there any difference in oligosaccharides compounds between early and late stages of lactation?

Dr. Kunz: Yes, like almost all other milk components there is a change during lactation, with very high concentrations immediately after birth, declining rapidly to a fairly constant but much lower level that persists to 3 to 5 months of lactation. We have recently analyzed some samples from Denmark at a very late stage (1.5 to 2 years of lactation), and there are still some oligosaccharides, but the concentration is very low. In bovine milk colostrum, the concentration of sialylated oligosaccharides can be extremely high (>1g/l).

Dr. Endres: You described four studies involving different infections in animals. Why was 3-sialyllactose used for *H. pylori* infection in rhesus monkeys, and how did the investigators decide on that particular oligosaccharide, when we know now that there are about 500 different complex oligosaccharides in human milk? What would be your future approach for determining which oligosaccharide would make a good prebiotic or antiadhesive agent for the prevention of infections or allergy or whatever?

Dr. Kunz: No one knows what kinds of oligosaccharide to use at the moment. We might need a combination of certain oligosaccharides, depending on what one wants to prevent. The reason sialyllactose and sialyl-lacto-N-tetraose were chosen is that in previous *in vitro* experiments by the same team, they found these components to be the most effective. They are also relatively simple to obtain, either from human milk or by synthesis by various enzymatic means or by gene modification.

Dr. Endres: Are you more in favor of the specificity of single oligosaccharides, or do you believe that it is a combination of several oligosaccharides that is effective, when bound to ligands or whatever?

Dr. Kunz: In relation to preventive aspects, I would prefer to use a combination of certain oligosaccharides, but until we have actual data, this will remain speculative.

Dr. Braegger: Is there any evidence that breastfed infants are less often colonized by *H. pylori* than formula-fed infants in later life?

Dr. Kunz: I do not know the answer to that.

Dr. S. Koletzko: You said that during early lactation there is a dramatic decrease in oligosaccharides in breast milk, reaching very low levels later on. Are there any indications from *in vitro* studies that giving the wrong oligosaccharides or the wrong dose at a certain time of life could have any adverse affects?

Dr. Kunz: I'm fairly sure that there are no data on that, but I might have missed some publications. The major question here is about the analyses. Colostrum is a very interesting fluid containing many peptides, proteins, and glycopeptides. Some of these components may have similar effects as oligosaccharides. To answer your question, it will be necessary to isolate these different components, which is extremely difficult. I'm not aware of any adverse effects in *in vitro* studies.

Dr. B. Koletzko: A recent study in the *New England Journal of Medicine* (1) reported repeated anaphylactic reaction to oligofructosylsaccharose, interesting enough both to the inulin type, with up to 60 fructosyl units, and to short-chain oligofructose with, I think, up to six fructosyl units. Do we have any reason to believe that there are specific allergenic aspects to the fructosyl oligosaccharides or do we assume that such allergic anaphylactic reactions could occur to a variety of different compounds?

Dr. Kunz: It is important to differentiate between the fructosylated oligosaccharides, which do not have much in common with the oligosaccharides in human milk, and the fucosylated oligosaccharides. In prebiotics, we are talking about the fructosylated compounds, which have a completely different structure. There is no evidence for allergy to human milk oligosaccharides, and it would be unexpected. However, reactions to fructosylated components are common. Even galactose linkage in inulin or oligofructuse is completely different from the fucosylated compound in human milk. It's always α1–2, α1–4, α1–6 linkage. In human milk, we always have β1–4, β1–3, or β1–6 linkages, which is important with regard to function.

Dr. Böhles: It was reported in the 1970s that an extract of carrots prevented adhesion of *E. coli* (2). Do you know of any other natural products that have an effect on specific bacteria?

Dr. Kunz: I believe there must be many components in food that might have similar antiadhesive effects to human milk oligosaccharides. The advantage of human milk is that you have this whole variety as a cocktail.

Dr. Superti-Furga: I wonder whether these compounds might play a role in the induction of tolerance to antigens. For antigens that are secreted, colostrum is the ideal first contact. Do you know whether these compounds are good antigens? Has anyone ever tried to put them in a lymphocyte culture to see what happens?

Dr. Kunz: I don't know. I'm not aware of any studies looking at these aspects. There are some indications that signaling events might be influenced by LNT or its fucosylated components. You might get stimulation of B cell proliferation, interleukin production, and down-regulation of T helper cells and so on. But, these are all *in vitro* data.

Dr. Bier: Unless someone puts their cards on the table and carries out experiments with some or any of these substances to tell us whether there is any activity in humans, it seems to me we're going to be at the 60th Nestlé Symposium with better analytic methods and a further

500 compounds whose role in the feeding of the human infant is still completely unknown! I'd like to see us move this on somehow.

Dr. Kunz: I agree absolutely. But we *are* now moving ahead because there are companies in the United States that are able to produce large amounts of these compounds.

Dr. Yamashiro: You didn't mention oligosaccharides in the brain. I would guess that the oligosaccharides of breast milk are a very important component source for brain oligosaccharides. What do you think about that possibility?

Dr. Kunz: That's a very interesting question. In a study that was done about 25 years ago, the effect of radioactively labeled N-acetylneuraminic acid was compared with radiolabeled sialyllactose, and surprisingly it was found that the incorporation of the label into brain glycolipids was much greater from the sialyllactose than from the free sialic acid. One could easily explore this further in an animal study, maybe with measures that are more specific.

Dr. Yamashiro: The reason I raise the question is that if you give sphingomyelin orally, as has been done in animal studies, it reaches the brain. Perhaps milk oligosaccharides can also reach the brain.

Dr. Steinmann: Could you tell us the difference between the oligosaccharides of cow's milk *versus* human milk?

Dr. Kunz: In bovine milk, there are almost no neutral oligosaccharides—that is, no fucosylated oligosaccharides. There are some acidic components like sialyllactose, which occurs in two different isomers, but the concentration of these is about 10 times lower than in human milk, though in bovine colostrum there is a very large amount of sialyllactose. This would be a good source for obtaining large amounts of this compound.

Dr. Yamashiro: We have added oligosaccharides to infant formulas for over 10 years in Japan and no side effects have been reported.

REFERENCES

1. Gay-Crosier F, Schreiber G, Hauser C. Anaphylasis from inulin in vegetables and processed food. *N Engl J Med* 2000; 342: 1372.
2. Guggenbichler JP. Adherence of enterobacteria in infantile diarrhea and its prevention. *Infection* 1983; 11: 239–42.

Genetic Expression and Nutrition, edited by Claude Bachmann and Berthold Koletzko. Nestlé Nutrition Workshop Series, Pediatric Program, Vol. 50. Nestec Ltd., Vevey/Lippincott Williams & Wilkins, Philadelphia, © 2003.

From Arginine to Creatine: Regional and Cellular Gene Expression of Enzymes and Transporters Linking Nitrogen and Energy Metabolism in Brain

Claude Bachmann, Hugues Henry, and Olivier Braissant

Centre Hospitalier Universitaire Vaudois (CHUV), Laboratoire Central de Chimie Clinique, Lausanne, Switzerland

Arginine, a semiessential amino acid, is provided to the organism by protein intake and newly synthesized from citrulline, in the kidney. In the liver and intestine it is the final useful product of the urea cycle. Arginine is a substrate for different metabolic pathways (Fig. 1), including the synthesis of nitric oxide, ornithine, creatine, and creatine phosphate, which, in the presence of the creatine kinase isozymes, function as an energy buffer and for channeling energy-rich phosphates between the mitochondrion and the cytosol. Metabolic schemes do not usually identify the organs in which enzymes and transporters are expressed. Some recent reviews have given an overview of our knowledge on this (1,2), and metabolite exchange between organs has been investigated.

The important role of the kidney in synthesizing arginine from citrulline, which is roughly equal to intestinal citrulline formation, has been shown experimentally and is supported by the clinical finding of increased plasma citrulline in chronic renal insufficiency. Interestingly, the renal cortex (the proximal convoluted tubule) is also the major site of guanidinoacetate formation, the rate-limiting step in creatine synthesis; this is catalyzed by arginine:glycine amidino transferase (AGAT), which is pretranslationally repressed in the kidney by creatine, the endproduct of the pathway.

Inherited disorders of the urea cycle enzymes (except for arginase deficiency) and of transporters of urea cycle metabolites lead to hyperammonemia and low plasma arginine. The latter is also found in malnutrition. The mechanisms leading to irreversible damage of the central nervous system (CNS) in these inherited disorders are not understood. Our research involves investigating whether the metabolism of arginine and creatine are affected in hyperammonemia.

The CNS is a well-isolated compartment. All the creatine-synthesizing enzymes have been shown to be present in the brain, in contrast to muscle. Thus it is appropriate to ask whether, or to what extent, the CNS depends on an external supply of arginine or creatine. The transport of metabolites between blood and brain tissue, and

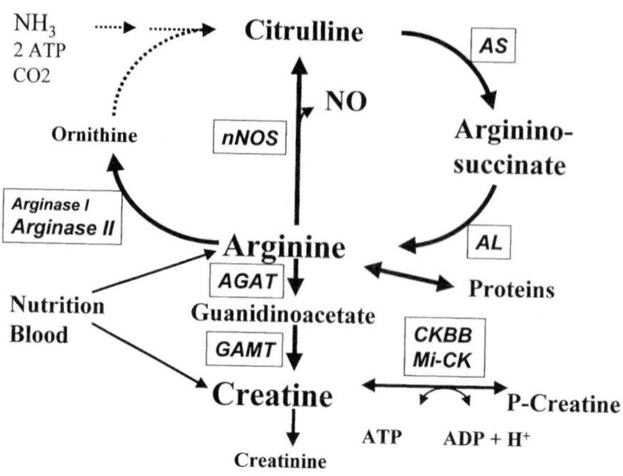

FIG. 1. Synthesis of L-arginine and creatine, including the nitric oxide (NO) cycle. The dotted arrows indicate the enzymatic steps that are not expressed in brain but solely in liver and gut (urea cycle). AGAT, arginine:glycine aminidino transferase; AL, argininosuccinate lyase; AS, argininosuccinate synthetase; CKBB, creatine kinase, brain isoenzyme; GAMT, guanidinoacetate:S-adenosylmethionine methyl transferase; MiCK, mitochondrial creatine kinase; NO, nitric oxide; nNOS, nitric oxide synthetase, brain isoenzyme.

within the brain between the different types of cell, may be specific to the brain because some transporter proteins differ from those of the plasma membranes of other organs.

The genes for most enzymes and transporters involved in the different pathways for arginine and creatine synthesis and utilization have now been described, mainly in man and rodents (3–6). Genetic defects of the creatine transporter 1 (CRT1) and of all the enzymes involved in *de novo* synthesis of creatine have been described (7–20). They lead to a decrease or absence of creatine, as detected by *in vivo* magnetic resonance spectroscopy. In some but not all of these disorders, supplementation with high doses of creatine over months allows brain creatine to increase but not to normalize.

Because genes represent cellular potential only, a further step in understanding the physiologic mechanisms must be made. We are currently studying the expression, at the messenger RNA (mRNA) level, of enzymes and transporters concerned with arginine and creatine. We have used normal adult rat brains to establish how these genes are expressed in brain regions and how they are distributed among the cell types. To analyze the effects of hyperammonemia on the CNS, we used aggregate cultures of embryonic rat brain as an experimental model.

The semiquantitative data presented below and in more detail in published reports (21–24) should make it possible to interpret quantitative data in regional brain homogenates more adequately.

ARGININE AND CREATINE PATHWAYS IN THE NORMAL ADULT RAT BRAIN

First, we sought a detailed basis of knowledge on the regional localization and cell-specific expression in the CNS of the genes involved in the transport, synthesis, and

TABLE 1. *Argininosuccinate synthetase, argininosuccinate lyase, nitric oxide synthase and arginase II mRNA expression patterns in normal adult rat brain [for details see (21)]*

	nNOS	AS	AL	AII
Neurons: variable expression	+	+	+	+
depending on neuron identity	−	+	+	+
	−	−	+	+
Astrocytes	−	−	+	+
	−	−	−	+
Oligodendrocytes	−	−	−	+

AS, arginosuccinate synthetase; AL, arginosuccinate lyase, nNOS, nitric oxide synthase; AII, arginase II; mRNA, messenger RNA.

utilization of arginine and creatine. Nonradioisotopic *in situ* hybridization experiments were performed on the brain of adult normal rats, based on the method of Braissant and Wahli (25). The mRNAs analyzed were those of arginine transporters and of the enzymes involved in arginine utilization [the nitric oxide (NO) cycle] and degradation: the cationic amino acid transporters (CATs) CAT1, CAT2(B), and CAT3, brain nitric oxide synthase (nNOS), argininosuccinate synthetase, argininosuccinate lyase, and arginase II. We also analyzed the mRNA of the genes involved in creatine synthesis and transport, namely L-arginine:glycine amidinotransferase (AGAT), guanidinoacetate:S-adenosylmethionine methyl transferase (GAMT), and CRT1 (see Fig. 1 for the description of the genes and pathways involved). The characterization of the CNS cells was done by coupling *in situ* hybridization with immunohistochemistry against microtubule associated protein 2 (MAP2, neurons), glial fibrillary acidic protein (GFAP, astrocytes), and myelin basic protein (MBP, oligodendrocytes).

EXPRESSION OF GENES OF ARGININE AND CREATINE METABOLISM

Tables 1 and 2 summarize the mRNA expression of the enzymes and transporters involved in arginine and creatine metabolism in the different cell types of the brain (21–23). nNOS, argininosuccinate synthetase, argininosuccinate lyase, and arginase II presented an interesting differential pattern of neuronal expression. nNOS was restricted to scattered neurons, which without exception also expressed argininosuccinate synthetase, argininosuccinate lyase, and arginase II. In those neurons from which

TABLE 2. *Cationic amino acid transporters CAT1, CAT2(B), and CAT3, L-arginine: glycine amidinotransferase (AGAT), guanidonoacetate: S-adenosylmethionine methyl transferase (GAMT), and creatine transporter 1 (CRT1) mRNA expression in normal adult rat brain [(for details see (21–23)]*

	CAT1	CAT2(B)	CAT3	CRT1	AGAT	GAMT
Neurons	+	+	+	+	+	+
Astrocytes	+	−	−	−	+	+
Oligodendrocytes	+	+	−	+	+	+

mRNA, messenger RNA.

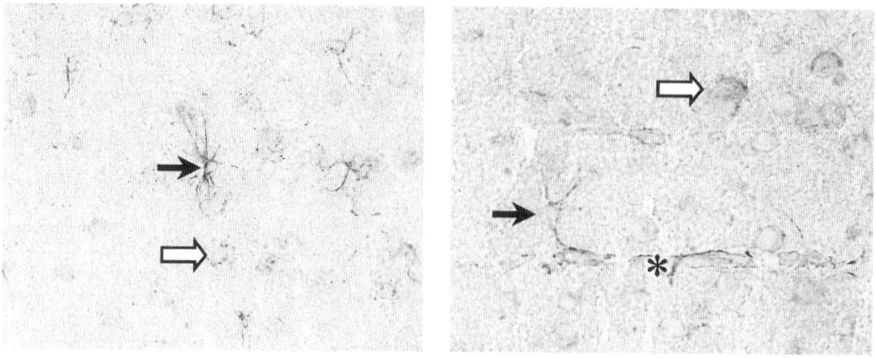

FIG. 2. Expression of creatine transporter 1 (CRT1) messenger RNA (mRNA) (*in situ* hybridization, blue) and glial fibrillary acidic protein (GFAP) as astrocyte marker (immunohistochemistry, red) in the adult rat brain cortex. Left panel: CRT1 mRNA is not expressed in astrocytes. Right panel: CRT1 mRNA is not expressed in astrocytes contacting capillary endothelia. *Solid arrows*, astrocytes; *open arrows*, neurons; *asterisk*, capillary.

nNOS was absent, some expressed argininosuccinate synthetase, argininosuccinate lyase, and arginase II, some expressed argininosuccinate lyase and arginase II, and others only expressed argininosuccinate lyase. nNOS and argininosuccinate synthetase could never be detected in glial cells. Argininosuccinate lyase could be detected in some but not all of the astrocytes, but was absent from oligodendrocytes. Arginase II was expressed in all glial cells. Thus the intracellular recycling of citrulline to arginine seems restricted to some neurons that are not further characterized at present. For those cells not expressing the whole citrulline–NO cycle, intercellular trafficking of citrulline and argininosuccinate or arginine must be assumed unless expression of the genes is induced in these cells under specific physiologic or pathologic conditions by signaling pathways to be detected in the future. The cationic amino acid transporter CAT1 was expressed in neurons, astrocytes, and oligodendrocytes and was found in all brain regions. In contrast, the brain-specific CAT3 was solely expressed in neurons. Interestingly, CAT2(B) was found expressed in neurons and oligodendrocytes throughout the brain but could never be detected in astrocytes. This was paralleled by the mRNA expression of the creatine transporter, CRT1, which also was present in neurons and oligodendrocytes throughout the brain but was absent from astrocytes, including those contacting brain capillaries (the blood–brain barrier) (Fig. 2; See Color Plate 2). Transcripts of AGAT and GAMT were found in the whole brain in all three types of cell described.

WHAT COULD THE EXPRESSION PATTERNS IN DIFFERENT CELL TYPES OF ADULT RAT BRAIN MEAN?

Based on mRNA expression of the arginine transporters, it appears that CAT1 is the prime candidate for basic arginine trafficking between blood and brain and between brain cells. The neuronal restriction of CAT3 and its expression in nNOS-expressing neurons leads us to speculate that it might play a role in providing substrate for the NOS pathway. On the other hand, the correlation between CAT2(B) and CRT1 expression suggests that CAT2(B) could be involved in the arginine trafficking between oligodendrocytes and neurons, especially in relation to creatine synthesis.

As AGAT and GAMT mRNAs were found in all brain regions in all three types of cell, all cells of the CNS seem to be equipped for *de novo* synthesis of creatine. The important finding of the absence of detectable CRT1 mRNA in the astrocytes of adult rat brain contrasts with previous reports of its presence in astroglial cultures (26) as well as in rat brain glial cells located in the white matter (corpus callosum) (27,28), which might be interpreted as oligodendrocytes, as in our study (23). We are aware that the absence of detectable mRNA in rat brain astrocytes does not exclude the presence of polypeptides or functional proteins. Minimal amounts of RNA might not be detectable, or the gene expression might be altered—for example, by diurnal, hormonal, or nutritional factors.

The absence of CRT1 mRNA in the rat astrocytes contacting brain capillaries suggests that under physiologic conditions, creatine import from the blood to the brain is probably not of major importance in the brain of adult rats. Additional indirect evidence supporting this view is the limited therapeutic efficacy of creatine supplementation in human inherited GAMT deficiency, where even very high doses of creatine given over months do not allow normalization of brain creatine content, as assessed by magnetic resonance spectroscopy (17). Such supplementation leads to a 10- to 25-fold increase in plasma creatine concentrations compared with the normal level; thus creatine is more likely to penetrate into the brain by diffusion than by active transport. The absence of creatine signals in the brain of patients who have a genetic defect of the CRT1 gene (13) does not contradict our interpretation either. If our findings can be confirmed at the level of the functional protein, the pathology found in inherited defects of creatine metabolism would have to be interpreted on the basis of the disturbance of intercellular trafficking of creatine between astrocytes, oligodendrocytes, and neurons. It is possible that other means of supplementing creatine to the brain should be considered in this instance—for example the use of creatine derivatives with better diffusion across the blood–brain barrier and plasma membranes, which can be cleaved in the brain. The correlation of the degree of expression of enzymes and transporters involved in arginine recycling (NO cycle) or creatine synthesis in different brain regions suggests that selective regulatory mechanisms might govern the different pathways. The comparison of the regional and cell-type–dependent patterns of expression under physiologic and pathologic conditions in the pathways of arginine—which is at a crossing point between endproducts of waste nitrogen excretion, energy buffering and channeling, and NO synthesis—might suggest hypotheses about the regulation and coordination of these important pathways.

ARGININE AND CREATINE PATHWAYS IN AGGREGATING CELL CULTURES OF FETAL RAT BRAIN EXPOSED TO AMMONIUM CHLORIDE

Animal models of subchronic hyperammonemia are not ideal because hyperammonemic animals markedly decrease their nutritional intake, including water, and

TABLE 3. *Composition of culture medium*

	Final concentration		Final concentration
NaCl	110 mM	Folic acid	4 mg/l
KCl	5.36 mM	Choline chloride	4 mg/l
$NaH_2PO_4.H_2O$	1.01 mM	Nicotinamide	4 mg/l
$MgSO_4.7H_2O$	0.90 mM	Ca panthothenate	4 mg/l
$CaCl_2$	2.40 mM	Pyridoxal HCl	4 mg/l
D-Glucose	25 mM	Thiamine.HCl	4 mg/l
		Riboflavin	0.4 mg/l
Arginine HCl	0.40 mM	i-Inositol	7.2 mg/l
Cystine	0.20 mM		
Glutamine	4.0 mM	Insulin	0.8 μM
Glycine	0.40 mM	Tri-iodothyronine	30 nM
Histidine $HCl.H_2O$	0.20 mM	Hydrocortisone-21-phosphate	20 nM
Isoleucine	0.80 mM	Transferrin	1 μg/ml
Leucine	0.80 mM	Biotin	4 μM
Lysine HCl	0.80 mM	Vitamin B_{12}	1 μM
Methionine	0.20 mM	Linoleate	10 μM
Phenylalanine	0.40 mM	Lipoic acid	1 μM
Serine	0.40 mM	L-carnitine	10 μM
Threonine	0.80 mM	Trace elements	See Ref. 30
Tryptophan	0.08 mM	Gentamicin sulfate	25 μg/ml
Tyrosine	0.40 mM	Bicarbonate	44 mM
Valine	0.80 mM	Hepes	20 mM

show signs of acute-phase reaction; in the animal portocaval shunt model, the metabolic situation is further complicated by involution of the bypassed liver associated with hyperinsulinism (29). These problems make it difficult to interpret the specific effects of hyperammonemia or distinguish them from secondary changes. To control as many variables as possible, we used the model of aggregating cell cultures of fetal rat brain that was developed and extensively studied by Honegger and Monnet-Tschudi (30). The use of totally artificial media devoid of fetal calf serum (Table 3) for the aggregate cultures limits the number of variables. The fact that the culture conditions allow neurons, oligodendrocytes, and astrocytes to grow simultaneously in an organized structure preserves intercellular interactions, including metabolite exchange. The model is thus probably more apt to reflect metabolism of the developing or mature CNS than single-cell–type cultures, and certainly closer to physiologic expression than tumor-derived or eternalized cell lines.

Our studies were thus done in aggregating cell cultures exposed to ammonia between 5 and 13 days—that is, in the phase of brain cell development and maturation before the mature "adult" level (30 days of culture). We chose this time period because patients with urea cycle defects develop irreversible brain damage mainly if hyperammonemia of more than 3 days' duration occurs within the first 2 years of life. Patients who experience such crises only after this major phase of brain maturation has occurred have a milder clinical course, unless there is persistent gross hyperammonemia leading to brain edema. This situation corresponds with the nNOS-independent increase in cerebral blood flow (but not in CSF glutamine) that occurs in portocaval-shunted rats (31).

We showed earlier that in the aggregate model, the expression of argininosuccinate synthetase and argininosuccinate lyase was induced in astrocytes by ammonium chloride exposure (1–5 mM), whereas arginase II, CAT1, and CAT3 tended to be down-regulated (24,32). We have also observed that the uptake of arginine from the medium into the aggregate cells is reduced by 30% to 32% ($p < 0.001$) in the presence of ammonia compared with controls, especially in the period between 8 and 13 days of culture: the intracellular arginine on day 13 is increased by 36% ($p < 0.005$) (unpublished results). Thus, the inhibition of utilization and export of arginine from various different cells is more pronounced than its uptake. One should be aware that most of the arginine in the medium is used for protein synthesis (PCA-precipitable material).

Because arginine and glycine are the initial substrates of the *de novo* synthesis of creatine, we studied the creatine synthesis pathway under conditions of increased ammonia concentrations in the aggregate culture medium, which does not contain any creatine. We show, as preliminary data, that the treatment of the aggregates with ammonium chloride leads to a reduced creatine concentration in the aggregate cells [Fig. 3 and (33)]. If creatine (1 mM) is added to the medium, its intracellular concentration doubles; in the presence of ammonia this increase is, however, less pronounced but nevertheless greater than in controls. Guanidinoacetic acid (1 mM)—the precursor of creatine and the substrate of GAMT—reduces the intracellular creatine concentration even more than ammonium chloride, and is no substitute for creatine (Fig. 3). Competition with creatine for export by CRT1 in some cells might explain the effect of guanidinoacetic acid. This hypothesis assumes that the CRT1 transporter is not en-

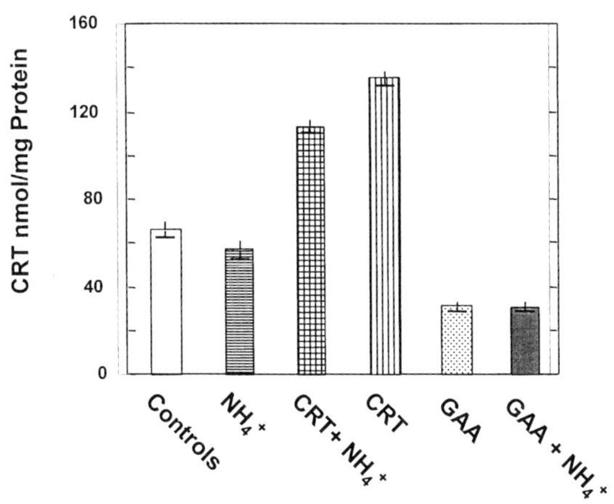

FIG. 3. Intracellular creatine (nmol/mg protein; enzymatic creatine assay) in fetal rat brain aggregate cultures harvested at 13 days, cultured from days 5 to 13 either in medium alone (controls) or with added ammonium (NH_4^+; 5 mM) and/or creatine (CRT; 1 mM), and also after adding guanidino acetic acid (GAA, 1 mM) with or without ammonium. Error bars = standard deviation, $n = 3$. The addition of creatine to the medium restores the reduction of intracellular creatine effected by ammonium, but to a lesser extent than in aggregates cultured with creatine alone. Guanidino acetate strongly reduces intracellular creatine; additional ammonium does not further reduce intracellular creatine.

tirely unidirectional (uptake). To our knowledge, proof of unidirectional transport has not been provided; it has only been deduced from the concentration gradient between plasma and muscle, a tissue lacking the machinery for *de novo* creatine synthesis. No additional decrease in creatine in the aggregates is achieved by combining guanidinoacetic acid with ammonia.

The morphology of the brain cell aggregates is also altered by treatment with ammonium chloride, as illustrated by Bielschowsky silver impregnation (Fig. 4; See Color Plate 3). A massive decrease in the number of axons, which normally grow mostly in the periphery of the aggregates, is observed. Our preliminary finding of the decrease in intracellular creatine in aggregates exposed to NH_4Cl led us to treat the aggregates with creatine plus NH_4Cl: this double treatment prevented the deleterious effect of ammonia on axonal development to a large extent (Fig. 4).

Axonal development is affected by increased ammonia, possibly because of reduced synthesis or intercellular transport of creatine. Our data indicate that hyper-

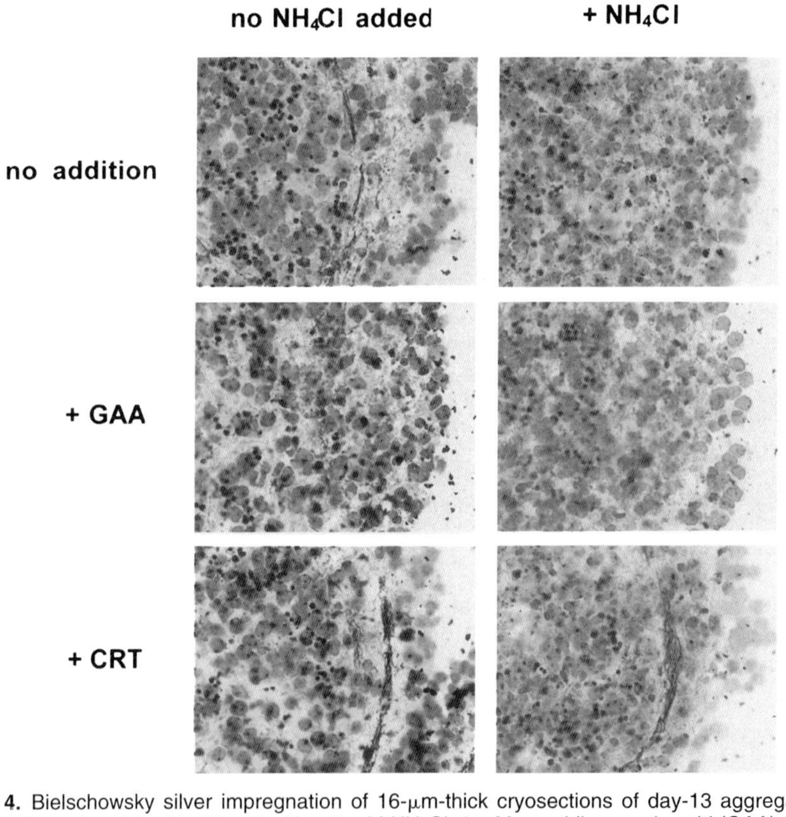

FIG. 4. Bielschowsky silver impregnation of 16-μm-thick cryosections of day-13 aggregates treated from days 5 to day 13 with either 5 mM NH_4Cl, 1 mM guanidino acetic acid (GAA), or 1 mM creatine (CRT). NH_4Cl and guanidinoacetate alter the development of axons at the periphery of the brain cell aggregates. Cotreatment with creatine prevents the axonal loss due to NH_4Cl. Creatine alone enhances the axonal development at the aggregate periphery.

ammonemia is likely to affect neuronal growth during brain development and maturation through a decrease in intracellular creatine. The localization of the creatine-synthesizing enzymes AGAT and GAMT in the aggregate model, as well as of its transporter CRT1, is necessary for interpretation. Preliminary data indicate that AGAT and GAMT are expressed in all cell types in the aggregates, as in the adult rat brain. CRT1 is present in neurons and oligodendrocytes, but is only faintly expressed in astrocytes. A cell-specific modulation of expression of these three genes seems to occur under exposure to NH_4Cl, including induction of GAMT in oligodendrocytes, induction of CRT1 in astrocytes, and repression of CRT1 in oligodendrocytes and of AGAT in neurons (unpublished data). These preliminary results support the hypothesis that metabolite exchange (for example, creatine–arginine) between the different types of brain cells plays an important role, and thus that experiments *in vitro* limited to cultures with only a single predominant cell type should be interpreted with great care. Our data on AGAT, GAMT, and CRT1 expression might explain the ammonia-induced axonal loss that is thought to occur because of decreased creatine concentration, which is at least partially restored by adding creatine to the culture medium.

CONCLUSIONS AND OUTLOOK

Much is still to be learned about the mechanisms of regulation of the AGAT, GAMT, and CRT1 genes under hyperammonemic conditions, the effect of creatine deprivation on neurons, and the intercellular interactions concerning creatine in the developing and mature brain. Results in this domain could help determine whether supplementation of hyperammonemic patients with creatine could be an efficient way of preventing or reducing neuronal damage.

De novo synthesis of creatine from arginine in the brain might be physiologically more important than the import of creatine from the blood. Such a mechanism could protect the brain from major changes in creatine concentration provoked by external (that is, nutritional) variability and excess load.

In view of the suspected predominance of local synthesis of creatine in the brain and the increased demand for arginine by the developing brain (for protein synthesis), systemic arginine deficiency might be more deleterious to axonal development in the first 2 years of life in humans than later on. L-arginine, as a semiessential amino acid, should thus be supplied in sufficient amounts, or supplemented in conditions of protein restriction or when there are genetic defects that lead to arginine deficiency, to avoid substrate limitation for creatine (and creatine phosphate) synthesis. Creatine, along with CK-BB and the mitochondrial creatine kinase, plays a crucial role in energy buffering and transport (34).

ACKNOWLEDGMENTS

We are grateful to P. Honegger, Physiological Institute, University of Lausanne, for the culture of the brain aggregates. This work is supported in part by the Swiss National Research Foundation, grant no. 3100-063892.

REFERENCES

1. Wiesinger H. Arginine metabolism and the synthesis of nitric oxide in the nervous system. *Prog Neurobiol* 2001; 64: 365–91.
2. Wyss M, Kaddurah-Daouk R. Creatine and creatinine metabolism. *Physiol Rev* 2000; 80: 1107–213.
3. Henry H, Speer O, Braissant O, Eilers B, Bachmann C, Wallimann T. Expression of splice variants of the creatine transporter in rat brain, muscle and kidney. In: *Sixth International Conference on Guanidino Compounds in Biology and Medicine, August 30–September 3, 2001*, Cincinnati, USA.
4. Isbrandt D, von Figura K. Cloning and sequence analysis of human guanidinoacetate N-methyltransferase cDNA. *Biochim Biophys Acta* 1995; 1264: 265–7.
5. Jenne DE, Olsen AS, Zimmer M. The human guanidinoacetate methyltransferase (GAMT) gene maps to a syntenic region on 19p13.3, homologous to band C of mouse chromosome 10, but GAMT is not mutated in jittery mice. *Biochem Biophys Res Commun* 1997; 238: 723–7.
6. Sandoval N, Bauer D, Brenner V, *et al.* The genomic organization of a human creatine transporter (CRTR) gene located in Xq28. *Genomics* 1996; 35: 383–5.
7. Carducci C, Leuzzi V, Prudente S, Mercuri L, Antonozzi I. Two new severe mutations causing guanidinoacetate methyltransferase deficiency. *Mol Genet Metab* 2000; 71: 633–8.
8. Cecil KM, Salomons GS, Ball WS, *et al.* Irreversible brain creatine deficiency with elevated serum and urine creatine: a creatine transporter defect? *Ann Neurol* 2001; 49: 401–4.
9. Ganesan V, Johnson A, Connelly A, Eckhardt S, Surtees RA. Guanidinoacetate methyltransferase deficiency: new clinical features. *Pediatr Neurol* 1997; 17: 155–7.
10. Ilas J, Muhl A, Stöckler-Ipsiroglu S. Guanidinoacetate methyltransferase (GAMT) deficiency: non-invasive enzymatic diagnosis of a newly recognized inborn error of metabolism. *Clin Chim Acta* 2000; 290: 179–88.
11. Item C, Stromberger C, Alessandri MG, *et al.* Arginine:glycine amidinotransferase (AGAT) deficiency: the third inborn error of creatine metabolism. *J Inherit Metab Dis* 2001; 24 (suppl 1): 118.
12. Leuzzi V, Bianchi MC, Tosetti M, *et al.* Brain creatine depletion: guanidinoacetate methyltransferase deficiency (improving with creatine supplementation). *Neurology* 2000; 55: 1407–9.
13. Salomons GS, van Dooren SJ, Verhoeven NM, *et al.* X-linked creatine transporter defect: the second family. *J Inherit Metab Dis* 2001; 24[Suppl 1]: 119.
14. Salomons GS, van Dooren SJ, Verhoeven NM, *et al.* X-linked creatine-transporter gene (SLC6A8) defect: a new creatine-deficiency syndrome. *Am J Hum Genet* 2001; 68: 1497–500.
15. Schulze A, Hess T, Wevers R, *et al.* Creatine deficiency syndrome caused by guanidinoacetate methyltransferase deficiency: diagnostic tools for a new inborn error of metabolism. *J Pediatr* 1997; 131: 626–31.
16. Schulze A, Mayatepek E, Bachert P, *et al.* Therapeutic trial of arginine restriction in creatine deficiency syndrome. *Eur J Pediatr* 1998; 157: 606–7.
17. Stöckler S, Hanefeld F. Guanidinoacetate methyltransferase deficiency: a newly recognized inborn error of creatine biosynthesis. *Wien Klin Wochenschr* 1997; 109: 86–8.
18. Stöckler S, Hanefeld F, Frahm J. Creatine replacement therapy in guanidinoacetate methyltransferase deficiency, a novel inborn error of metabolism. *Lancet* 1996; 348: 789–90.
19. Stöckler S, Holzbach U, Hanefeld F, *et al.* Creatine deficiency in the brain: a new, treatable inborn error of metabolism. *Pediatr Res* 1994; 36: 409–13.
20. Stöckler S, Isbrandt D, Hanefeld F, Schmidt B, von Figura K. Guanidinoacetate methyltransferase deficiency: the first inborn error of creatine metabolism in man. *Am J Hum Genet* 1996; 58: 914–22.
21. Braissant O, Gotoh T, Loup M, Mori M, Bachmann C. L-arginine uptake, the citrulline-NO cycle and arginase II in the rat brain: an in situ hybridization study. *Brain Res Mol Brain Res*1999; 70: 231–41.
22. Braissant O, Gotoh T, Loup M, Mori M, Bachmann C. Differential expression of the cationic amino acid transporter 2(B) in the adult rat brain. *Brain Res Mol Brain Res* 2001; 91: 189–95.
23. Braissant O, Henry H, Loup M, Eilers B, Bachmann C. Endogenous synthesis and transport of creatine in the rat brain: an in situ hybridization study. *Brain Res Mol Brain Res* 2001; 86: 193–201.
24. Braissant O, Honegger P, Loup M, Iwase K, Takiguchi M, Bachmann C. Hyperammonemia: regulation of argininosuccinate synthetase and argininosuccinate lyase genes in aggregating cell cultures of fetal rat brain. *Neurosci Lett* 1999; 266: 89–92.
25. Braissant O, Wahli W. Differential expression of peroxisome proliferator-activated receptor-alpha, -beta, and -gamma during rat embryonic development. *Endocrinology* 1998; 139: 2748–54.
26. Möller A, Hamprecht B. Creatine transport in cultured cells of rat and mouse brain. *J Neurochem* 1989; 52: 544–50.

27. Happe HK, Murrin LC. In situ hybridization analysis of CHOT1, a creatine transporter, in the rat central nervous system. *J Comp Neurol* 1995; 351: 94–103.
28. Saltarelli MD, Bauman AL, Moore KR, Bradley CC, Blakely RD. Expression of the rat brain creatine transporter in situ and in transfected HeLa cells. *Dev Neurosci* 1996; 18: 524–34.
29. Bachmann C. Ornithine carbamoyl transferase deficiency: findings, models and problems. *J Inherit Metab Dis* 1992; 15: 578–91.
30. Honegger P, Monnet-Tschudi F. Aggregating neural cell culture. In: Fedoroff S, Richardson A, eds. *Protocols for neural cell culture*. Totowa, NJ: Humana Press, 1997: 25–49.
31. Larsen FS, Gottstein J, Blei AT. Cerebral hyperemia and nitric oxide synthase in rats with ammonia-induced brain edema. *J Hepatol* 2001; 34: 548–54.
32. Braissant O, Henry H, Loup M, Eilers B, Honegger P, Bachmann C. Hyperammonemia, arginine and the citrulline-nitric oxide cycle in aggregating cell cultures of fetal rat brain. In: *NO-2000*, March 10, 2000, Paris.
33. Braissant O, Henry H, Honegger P, *et al.* Hyperammonemia alters the gene expression for creatine transport and metabolism in aggregating cell cultures of fetal rat brain. In: *Sixth International Conference on Guanidino Compounds in Biology and Medicine, August 30–September 3, 2001, Cincinnati, USA.*
34. Hornemann T, Rutishauser D, Wallimann T. Why is creatine kinase a dimer? Evidence for cooperativity between the two subunits. *Biochim Biophys Acta* 2000; 1480: 365–73.

DISCUSSION

Dr. Fowler: Could you tell us how the concentrations of ammonia that you used relate to the physiological situation, and leading on from that, did you find the same effects with 1 mM as with 5 mM? The former is more physiologic.

Dr. Bachmann: It's not exactly the same with 1 mM as with 5 mM, although at 2.5 and 5 effects are significant. As to the physiologic concentration in the tissues, data vary quite a bit in published reports, but I think that you easily might reach 3 or 4 mM in cells, which are more acid than plasma, if there is hyperammonemia in the blood.

Dr. Fowler: Can we extrapolate from mRNA expression to enzyme activity?

Dr. Bachmann: No, we don't know much about the turnover of the proteins, or what their half-life is, or whether their breakdown is affected, although we know of course that a lot of arginine is used in protein synthesis in the first 2 years. So we don't have all the answers; thus I tried not to overinterpret the data. We are, however, one step further on in that we now have some very specific markers; if you want to do immunohistochemistry with antibodies the interpretation can be difficult because the antibodies are often not very specific. We need very specific antibodies to look at protein, and even if we identify protein, we don't have information on its function. How do you measure function in an oligodendrocyte?

Dr. Wanders: How do you control the level of ammonia in your culture medium?

Dr. Bachmann: You can find the details in our published report (1).

Dr. Wanders: What do you think is the effect of creatine? Does it work by creating phosphate? Is it an energetic phenomenon?

Dr. Bachmann: Olivier Braissant has speculated that it might have something to do with energy needed in the axonal growth button.

Dr. Wanders: Why can't you replace creatine by its precursor?

Dr. Bachmann: That's an interesting question because it fits our concept that the oligodendrocyte mothers the axons—that is, the synthesis of creatine from arginine occurs to a major degree in the oligodendrocytes. Then a creatine transporter or transporter isoform must be present in the oligodendrocyte to export or transport the creatine to the axon of the neuron. That is our interpretation. Then under hypoammonemic conditions, in a way we do not understand, this creatine transporter would be down-regulated in the oligodendrocytes. The fact that it is not down-regulated generally probably shows that this effect of ammonia possesses some specificity.

Dr. Pencharz: I want to clarify the issue of arginine synthesis, at least in the neonate. In both the infant human and the infant pig, arginine is synthesized in the gut—we've shown that very clearly. Outside of the neonatal period, it is synthesized in the kidney, as Brosnan has shown, but if you feed an arginine-free enteral nutrition via the gut, you will end up with hyperammonemia. Proline will partially ameliorate that, but glutamate will not.

Dr. Bachmann: I said that an amount of arginine equal to that of citrulline synthesized in the gut was formed in the kidney. I wanted to emphasize all of the sources of arginine synthesis other than the liver, especially as the kidney also plays an important role in the peripheral synthesis of creatine, or at least the first step in its synthesis.

Dr. Daniel: With respect to the dramatic effect of ammonium chloride on axonal growth, how do you know whether you are talking about an ammonia effect or a pH effect? Ammonium chloride is an excellent acid-loading compound, so there were probably extremely acidic conditions. And, could creatine also play a role in restoring the pH?

Dr. Bachmann: We measured and controlled the pH, and we did not see an effect that could be attributed to pH on its own.

Dr. Daniel: But did you measure intracellular pH?

Dr. Bachmann: No, we did not. It is a good question but I have difficulty in imagining why the effect of low pH should be strongly expressed in oligodendrocytes and much less in astrocytes.

Dr. Daniel: I cannot give an answer to that, but it is an important thing to measure.

Dr. Daniel: You raised the question of whether the citrulline transporters can also efflux the amino acids. Those transporters are electrogenic but most probably they are not carriers that efflux amino acids. Many newly discovered amino acid transporters are expressed in all kind of different cell types. These are exchangers, which means they can take up amino acids for exchange with intracellular amino acids, and they can even take up cationic amino acids and then pump out neutral ones in the same step (2,3).

Dr. Bachmann: Is there one for creatine too?

Dr. Daniel: Unfortunately not, but with respect to arginine flux and the transport of the other neutral amino acids, these are candidates.

Dr. Wanders: A good control would be to include methylamine. It's not metabolizable but it has the same properties as the ammonium ion. You should do such an experiment because then you will really be able to find out whether the effect has to do with metabolism of ammonia or whether it has to do with its pH properties.

Dr. Bachmann: I agree. That should be done.

Dr. Batshaw: The urea cycle disorders are primarily gray matter disorders except for argininemia, which behaves like a white matter disorder. So you would expect to find damaged neurons, and that's what you mostly see. Based on your findings, do you really think that creatine plays a significant role from a clinical perspective? What you are seeing is an effect on oligodendrocytes that is not clinically apparent in affected children.

Dr. Bachmann: It is too early to say. I am looking for explanations for mental retardation as a long-term effect. We know about the reversible symptoms, which you yourself showed, mediated by serotonin. We know that serotonin-induced dilatation of blood vessels leads to edema whether the glutamine is increased or not, and the terminal symptoms are related to brain edema. Then we have the glutamate effects on the NMDA receptor neurons, although we do not really know what happens to them. However, I feel we do not know enough about what happens to neuronal development in the longer term.

Dr. Superti-Furga: I notice that even in untreated cultures creatine has a stimulating effect on neurons. I remember that Wallimann presented data from his myotube culture model showing that creatine has a differentiating effect on these cells. Have you considered that possibility? In addition, do you know whether your cell culture expresses creatine kinase?

Dr. Bachmann: In relation to CK, we did investigate mitochondrial CK and found at least one of the three forms expressed. For the first part of your question, you cannot compare muscle with brain because muscle does not have the creatine synthesizing capacity that the brain has.

Dr. Superti-Furga: But the differentiating effects might still be there.

Dr. Endres: Your title mentions energy metabolism. What are the implications of interference with energy metabolism in the physiologic state? Can you deduce anything from diseases of the urea cycle—for example, where there is an arginine deficiency as in argininosuccinate lyase deficiency?

Dr. Bachmann: There have been contradictory results on total adenosine triphosphate (ATP) measurements under hyperammonemic conditions—some have found it diminished and others have not. It's not always easy to do those measurements. One could speculate that if creatine phosphate and creatine diminish, you might be much more vulnerable to changes in cellular ATP, but it would be pure speculation to answer your question.

Dr. Micheli: You said that it was important to avoid arginine deficiency under conditions of hyperammonemia. We are often faced with borderline hyperammonemia (around 150 μmol/l) in very preterm babies who are on mixed enteral and parenteral feeding. How would you define arginine deficiency in those preterm babies?

Dr. Bachmann: I would monitor serum arginine in these babies. I would try to maintain an arginine concentration above 70 μmol/l, and not more than around 120 to 150. That would be a pragmatic approach. Dr. Batshaw described asymptomatic hyperammonemia due to arginine deficiency (4).

Dr. Batshaw: We have found that 1 mmol/kg/d of arginine will normalize the serum arginine and stabilize the ammonia levels. We also did a follow-up study on those patients for a few years to see if it made any difference developmentally, but we could not identify any developmental differences between treated and untreated groups, so it is unclear whether it has any clinical significance.

Dr. Braegger: I have another clinical question. Patients with liver cirrhosis and portal hypertension tend to develop hepatic encephalopathy, which is associated with high ammonia concentrations in the serum. These patients have to be kept on a protein-restricted diet. Do you see any role for creatine supplementation in these patients?

Dr. Bachmann: I do not think you can affect the central nervous system very much by feeding creatine.

REFERENCES

1. Braissant O, Honegger P, Loup M, *et al.* Hyperammonemia: regulation of argininosuccinate synthetase and argininosuccinate lyase genes in aggregating cell cultures of fetal rat brain. *Neurosci Lett* 1999; 266: 89–92.
2. Meier C, Ristic Z, Klauser S, Verrey F. Activation of system L heterodimeric amino acid exchangers by intracellular substrates. *EMBO J* 2002; 21: 580–9.
3. Chillaron J, Roca R, Valencia A, Zorzano A, Palacin M. Heteromeric amino acid transporters: biochemistry, genetics and physiology. *Am J Physiol Renal Physiol* 2001; 281: 995–1018.
4. Batshaw ML, Wachtel RC, Cohen L, *et al.* Neurologic outcome in premature infants with transient asymptomatic hyperammonemia. *J Pediat* 1986; 108: 271–5.

Genetic Expression and Nutrition, edited by Claude Bachmann and Berthold Koletzko. Nestlé Nutrition Workshop Series, Pediatric Program, Vol. 50. Nestec Ltd., Vevey/Lippincott Williams & Wilkins, Philadelphia, © 2003.

Current Understanding of the Molecular and Cellular Basis of Celiac Disease

Ludvig M. Sollid

Institute of Immunology, Rikshospitalet, University of Oslo, Oslo, Norway

Celiac disease has received increased attention in recent years. It has become evident that this is a common disorder in Western societies, affecting around 1:130 to 1:300 individuals (1). Celiac disease is an acquired disorder. It may be diagnosed in early childhood with classical symptoms such as diarrhea and malabsorption, but it may also be diagnosed later in life. In adults there is a broader spectrum of symptoms (1). Some have no subjective symptoms, some have the classical symptoms, but often the disease is manifested by anemia, osteoporosis, fatigue, depression, or infertility. The vague character of these symptoms does not suggest a gut disease, and persons who are diagnosed later in life often report protracted disease histories.

Celiac disease develops because of intolerance to ingested wheat gluten (consisting of the subcomponents gliadin and glutenin) or related proteins from rye and barley. There is chronic inflammation in the small intestine, with resultant flattening of the mucosa. The disease is effectively treated by a gluten-exclusion diet. Poor dietary compliance and undiagnosed disease are associated with long-term complications that include osteoporosis and an increased risk of intestinal lymphoma. Patients with celiac disease on a gluten-containing diet have increased levels of serum antibodies to various antigens including gluten and the autoantigen tissue transglutaminase (tTG) (2,3). The presence of antibodies to gluten and tTG is strictly dependent on dietary exposure to gluten.

Celiac disease belongs to the group of chronic inflammatory diseases of multifactorial etiology in which both genetic and environmental components are involved. Among these disorders, celiac disease is a particularly good model. In this chapter, I will outline briefly some recent advances in our understanding of this disorder.

GENETIC FACTORS IN CELIAC DISEASE

A high prevalence (10%) among first-degree relatives of patients with celiac disease indicates that susceptibility to the disease is strongly influenced by inherited (genetic) factors (4). Familial clustering is stronger in celiac disease than in most other chronic inflammatory diseases of multifactorial etiology (5,6). The strong genetic influence in celiac disease is further supported by a high concordance rate of 70% to

100% in monozygotic twins (7). Both HLA and non-HLA genes contribute to the genetic predisposition, and assuming a multiplicative model of disease genetics, it has been estimated that the overall importance of non-HLA genes is greater than that of HLA genes (5,6). These figures should be interpreted with caution, however, as increased sharing of environmental factors by the siblings would tend to overestimate the role of the non-HLA genes.

HLA Genes

Most patients with celiac disease carry the DRB1*0301-DQA1*0501-DQB1*0201 haplotype (the DR3, DQ2 haplotype) or are DRB1*11/12-DQA1*0505-DQB1-*0301/DRB1*07-DQA1*0201-DQB1*0202 heterozygotes (carrying the DR5-DQ7/DR7-DQ2 haplotypes) (8–12). The DQα chains encoded by DQA1*0501 and DQA1*0505 differ by one residue in the leader peptide, and the DQβ chains encoded by DQB1*0201 and DQB1*0202 differ by one residue in the membrane proximal domain. These substitutions are highly unlikely to have any functional consequences. Patients with celiac disease with the above-mentioned DR–DQ combinations share the same functional DQ molecule on the cell surface, encoded by genes carried in the *cis* position (for example, DQA1*05 and DQB1*02 carried on the same haplotype) or the *trans* position (for example, DQA1*05 carried on a different haplotype to DQB1*02) (Fig. 1) (13).

Recombination (crossing over; rearrangement of genes on the same chromosome) is an important mechanism for the generation of HLA haplotypes (14). Accumulating evidence suggests that the DR3-DQ2, DR7-DQ2, and DR5-DQ7 haplotypes have a close evolutionary relationship. Fragments of DNA flanking the DQA1 gene of the

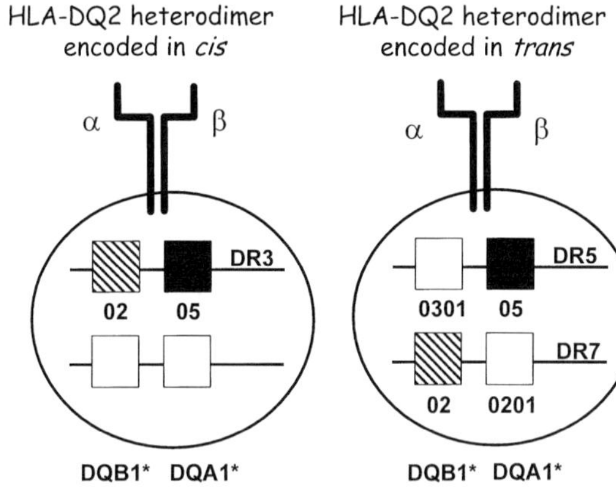

FIG. 1. Patients with celiac disease who are DR3 or DR5/DR7 heterozygous express the same HLA-DQ2 molecule, HLA-DQ(α1*05, β1*02). The DQA1*05 and DQB1*02 genes are located in *cis* (on the same chromosome) in DR3 individuals, whereas they are located in *trans* (on opposite chromosomes) in DR5/DR7 heterozygous individuals. Figure adapted from Sollid and Thorsby (17), with permission.

DR3-DQ2 haplotype have been identified on the DR5-DQ7 haplotype, and fragments of DNA flanking the DQB1 gene of the DR3-DQ2 haplotype have been identified on the DR7-DQ2 haplotype (15,16). The genetic information in the DQ subregion of the DR3-DQ2 haplotype is thus reestablished in DR5-DQ7/DR7-DQ2 heterozygotes, although the sequence information is split between two chromosomes. Susceptibility to celiac disease therefore probably depends on an interaction between at least two genes on the DR3-DQ2 haplotype that are reunited in DR5-DQ7/DR7-DQ2 heterozygous individuals. The DQA1 and DQB1 genes are the primordial candidates because their products interact to form an HLA class II heterodimer and they are situated close to a putative recombination site.

Almost all the patients with celiac disease who are DQA1*05 and DQB1*02-negative bear the DRB1*04, DQA1*03, DQB1*0302 haplotype (that is, the DR4-DQ8 haplotype) and it is likely that these patients have an HLA association that is different from those who are DQ2-positive. Although it is less clear what is the primary disease susceptibility determinant of the DR4-DQ8 haplotype, most data favor DQ8 (17).

Overall, the existing data suggest that the susceptibility for developing celiac disease is primarily associated with two conventional DQ molecules DQ(α1*05,β1*02) (=DQ2) and to a lesser extent DQ(α1*03,β1*0302) (=DQ8). DQ molecules bind peptides and present these to $CD4^+$ T helper cells carrying the $\alpha\beta$ T-cell receptor (TCR). The genetic evidence thus points toward a central role $CD4^+$ T cells in controlling the development of celiac disease.

CTLA4/CD28/ICOS Genes

Genome-wide linkage studies in celiac disease have indicated numerous susceptibility regions with weak genetic effects, and the indications are strongest for susceptibility genes located at 5qter and 11qter (18,19). However, no disease associations have been established for any genes of these regions. The only gene for which there are relatively consistent reports on disease association is the CTLA4 gene on chromosome 2q33. This has a single nucleotide polymorphism (A/G) at position 49 in exon 1, and the A allele has been found to be associated with celiac disease in both French and Swedish/Norwegian populations (20,21). In the Swedish/Norwegian material, there was also suggestive linkage for several markers, and in a Finnish study linkage was found for the marker D2S116, located immediately adjacent to CTLA4 (21). A study of Italian and Tunisian patients failed to confirm the association with the +49 polymorphism or linkage in the CTLA4 region (22). CTLA4 is involved in down-regulating T-cell responses, and the A allele of the +49 dimorphism is associated with increased CTLA-4 expression and enhanced control of T cell proliferation (23,24). Thus, the +49 dimorphism is a prime suspect for the observed genetic effect, although other polymorphisms in linkage disequilibrium cannot be ruled out. In this respect, it is worth noting that the CD28 and ICOS genes, whose gene products are central players in T- and B-cell activation, are located very close to the CTLA4 gene.

ENVIRONMENTAL FACTORS

Gluten is obviously a critical environmental factor in celiac disease. Whether other environmental factors are also involved is still an open question. Gut infections may be involved. Adenovirus 12 has been in the forefront among the candidate microorganisms. This virus was originally proposed as a candidate because of partial linear homology over 12 amino acids in a virus protein and an α gliadin (25). Based on our current understanding of the pathogenesis of celiac disease, the rationale for the candidacy of this virus is weak, and there are few epidemiologic data supporting a role for this virus (26).

CHARACTERISTICS OF THE CELIAC LESION

The lesion in celiac disease is localized in the proximal part of the small intestine. The full-blown lesion is characterized by villous atrophy, crypt cell hyperplasia, and leukocytic cell infiltration of the epithelium and the lamina propria. Intraepithelial lymphocytes (IELs) of both the $TCR\alpha\beta^+$ $CD8^+CD4^-$ and the $TCR\gamma\delta+$ $CD8^-CD4^-$ phenotypes are expanded in celiac disease. In contrast to the $TCR\alpha\beta^+$ $CD8^+$ IELs, which return to normal when gluten is removed from the diet, the $TCR\gamma\delta^+$ IELs appear to remain at an increased level (27). Interestingly, most $TCR\gamma\delta^+$ IELs express the $V\delta1$ TCR variable region (28,29). Spies and coworkers (30) have shown that $\gamma\delta$ T cells expressing this variable region recognize MICA and MICB molecules—molecules that are expressed by intestinal epithelial cells (31). Activated human IELs are able to produce various cytokines including interferon-γ (IFN-γ), interleukin 2 (IL-2), IL-8, and tumor necrosis factor-α (TNF-α), and are known to have a lytic potential (32).

In the lamina propria, there is a marked infiltration of $TCR\alpha\beta^+$ T cells in the active lesion. These T cells are mostly $CD4^+$ and carry a memory phenotype ($CD45RO^+$) (33). Lamina propria T cells actively produce cytokines, of which IFN-γ seems to be dominant (34,35). This observation is consistent with the finding that gluten-specific T cells isolated from celiac lesions produce large amounts of IFN-γ (36).

Just beneath the epithelium in the normal mucosa there are many macrophage/dendriticlike cells that stain positive for CD68 (37). It is conceivable that these cells are involved in sampling of luminal antigens. The expression of HLA class II, ICAM-1, and CD25 molecules is increased in these macrophage/dendritic-like cells, suggesting that they are activated in the disease state (38–40).

In the normal small intestine, there is a balance between extracellular matrix formation by stromal cells and extracellular matrix degradation mediated by matrix metalloproteinases (MMPs). It has been suggested that increased extracellular matrix degradation plays a role in the villous atrophy of celiac disease. This is supported by the finding of a decreased ratio of cells expressing collagen I and tissue inhibitor of metalloproteinase (TIMP)-1 messenger RNA (mRNA) to those expressing MMP-1 and MMP-3 mRNA in untreated celiac disease (41). Expression of MMP-1 and MMP-3 mRNA is mainly localized to subepithelial fibroblasts and macrophages. It is likely that the increased expression of metalloproteinases is related to activation of mucosal T cells.

PEPTIDE BINDING TO THE CELIAC DISEASE—ASSOCIATED DQ2 AND DQ8 MOLECULES

Both HLA class I and class II molecules bind peptides in a groove located in their membrane distal part (42). Stable binding is achieved by multiple hydrogen bonds between amino acids of HLA and peptide main chain atoms. There are many polymorphic variants of HLA molecules. Amino acid residues, which differ between the polymorphic variants, are clustered around the peptide binding site, where they contribute to the formation of specific binding pockets. Side chains of amino acids of the peptide (so-called anchor residues) fit into these pockets, and their interaction with HLA contributes to the binding of the peptide. The binding site of HLA class II molecules, in contrast to class I molecules, is open at both ends, allowing the bound peptides to protrude. The class II peptide ligands thus vary in length. The interactions with HLA mainly take place in a core region of nine residues. Within this region, side chains of amino acids in positions P1, P4, P6, P7, and P9 dock into pockets of the class II binding site. The chemistry and size of the various pockets vary between the different class II alleles, so that some amino acids are preferred and some are not.

DQ2 has a unique preference for binding peptides with negatively charged side chains at the three middle anchor positions (Fig. 2) (43–45). The binding motif of DQ8 is different from that of DQ2, but DQ8 also displays a preference for binding negatively charged residues at several positions (that is, P1, P4, and P9) (46–48). Hence, both the DQ2 and DQ8 molecules share a preference for negatively charged residues at some of their anchor positions.

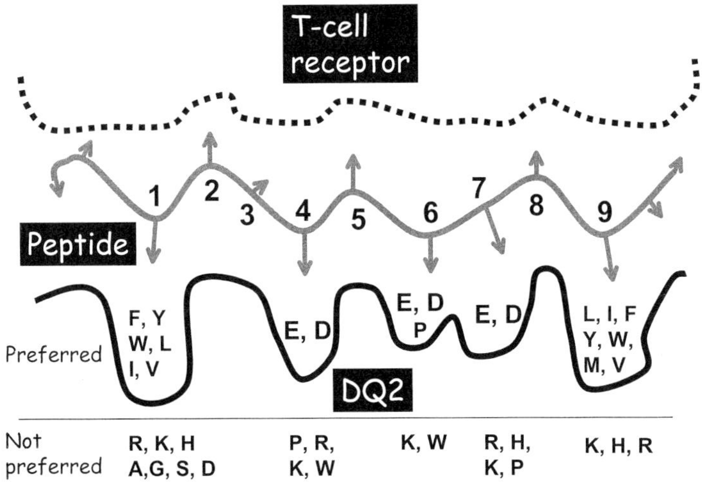

FIG. 2. Schematic depiction of the peptide binding groove of HLA-DQ2 (that is, HLA-DQ(α1*05, β1*02) with the peptide binding motif displayed with the one letter code for the amino acids. A bound peptide and a T cell receptor (TCR) recognizing the peptide/HLA complex are also indicated. Reprinted with permission from the *Annual Review of Immunology*, vol 18, 2000. (Available at www.AnnualReviews.org.).

PREFERENTIAL T-CELL RECOGNITION OF GLUTEN PEPTIDES PRESENTED BY DQ2 AND DQ8

The observation that gluten-reactive $CD4^+$ TCRαβ T cells can be isolated and propagated from intestinal biopsies of patients with celiac disease has been instrumental in recent achievements (49). Strikingly, such T cells in patients carrying the DR3-DQ2 haplotype were found to recognize gluten fragments presented by the DQ2 molecule rather than by other HLA molecules of the patients (49,50). Both DR3-DQ2–positive and DR5-DQ7/DR7-DQ2–positive antigen-presenting cells (that is, those carrying the DQA1*05 and DQB1*02 genes in *cis* or in *trans* configuration) are able to present the gluten antigen to these patient T cells. Likewise, T cells isolated from small-intestine biopsies of DQ2-negative, DR4-DQ8–positive patients predominantly recognize gluten-derived peptides when presented by the DQ8 molecule (49,50). Taken together, these results allude to presentation of gluten peptides in the small intestine as the mechanism whereby DQ2 and DQ8 confer susceptibility to celiac disease. HLA molecules are also important for determining the repertoire of peripheral T cells during maturation in the thymus. A thymic effect of the same DQ molecules on the TCR repertoire selection is, however, not excluded by these results.

IMPORTANCE OF GLUTEN DEAMIDATION FOR T-CELL RECOGNITION

Wheat gluten is a mixture of a large number of gliadin and glutenin polypeptides. Generally, gluten proteins are rich in proline and glutamine residues, whereas many other amino acids—including glutamic acid and aspartic acid—are unusually rare. Proteins of the gliadin fraction can be subdivided according to their sequence into the α, γ, and ω gliadins (51). Initially it was difficult to reconcile the DQ2 (and DQ8) binding motifs with presentation of gluten peptides, because gluten proteins have an unusual scarcity of negatively charged residues.

A clue that might help explain this paradox came from the observation that the stimulatory capacity of gliadin preparations for gliadin-specific intestinal T cells was significantly enhanced following treatment at high temperatures and low pH (52). These conditions are known to cause nonspecific deamidation of glutamines to glutamic acid and may thus convert gliadin from a protein with very few peptides with the potential to bind to DQ2/DQ8 into one with many. An important and general role for deamidation of gluten for T-cell recognition was sustained by analysis of the response pattern of a panel of polyclonal, gliadin-specific T-cell lines derived from biopsies (53). All the cell lines responded poorly to a gliadin antigen prepared under conditions of minimal deamidation (chymotrypsin digestion) when compared with the same antigen that had been further heat-treated in an acidic environment.

The characterization of gluten epitopes recognized by intestinal T cells has extended our knowledge of the importance of deamidation for T-cell recognition. Of the epitopes characterized up to now, most (DQ2-γ-gliadin-I, DQ2-α-gliadin-I, and DQ2-α-gliadin-II) fail to stimulate T cells in their native form, but are potent antigens when a single glutamine residue is exchanged with glutamic acid in certain positions. For one DQ8-restricted epitope (DQ8-α-gliadin-I), the T-cell recognition is augmented by introduction of negatively charged residues (54), whereas this is not seen for another DQ8-restricted epitope (DQ8-glutenin-I) (55). These data show that most, but not all, gluten-specific intestinal T cells from patients with celiac disease recognize gluten proteins only after they have undergone deamidation.

TISSUE TRANSGLUTAMINASE DEAMIDATES GLUTEN PEPTIDES *IN VIVO*

There is accumulating evidence that deamidation *in vivo* is mediated by the enzyme tTG (56,57). tTG is expressed in many different tissues and organs. In the small intestine, it is mainly expressed just beneath the epithelium in the gut wall (56). The activity of tTG in the small-intestine mucosa in patients with untreated celiac disease is increased compared with controls (58). The enzyme is present both intracellularly and extracellularly, and in the extracellular environment, tTG has been shown to play a role in extracellular matrix assembly, cell adhesion, and wound healing (59). The calcium-dependent transglutaminase activity of tTG catalyzes selective cross-linking or deamidation of protein-bound glutamine residues (60). In contrast to the nonenzymatically mediated deamidation that results in a near-random deamidation of the often numerous glutamine residues in gliadin peptides, tTG appears to carry out an ordered deamidation of some few specific glutamines.

In all the known major DQ2 and DQ8 restricted gluten epitopes recognized by gut T cells of adult patients, there are glutamic acid residues modified by tTG, which is important for T-cell recognition (53,54,61). Interestingly, the deamidation of glutamine residues that are not targeted by tTG (for example, by acid treatment) can be deleterious for T-cell recognition (57,62). This suggests that deamidation *in vivo* is mediated by tTG (Fig. 3). This idea is further supported by the results of experiments where T cell lines have been established from biopsies challenged with a minimally deamidated gliadin antigen (chymotrypsin-digested). In all but one of 18 adult patients, the established T-cell lines only barely responded to the chymotrypsin-digested gliadins, but efficiently recognized the *in vitro* tTG-treated variants of the same gliadins (63). Moreover, the addition of the tTG inhibitor cystamine during the gliadin challenge often resulted in T-cell lines with abolished or reduced responses to deamidated gliadin (63).

Normally, we do not mount immune responses to edible proteins. This phenomenon, which is termed *oral tolerance*, is believed to occur because of active tolerization toward edible proteins. In keeping with this thinking, oral tolerance to gluten in patients with celiac disease is either not established properly or disrupted. Given the preferential intestinal T-cell response to deamidated gluten fragments in patients with celiac disease, it may be that deamidation is central to the disturbance of oral gluten tolerance in such individuals. Deamidation increases the binding affinity of gliadin peptides for DQ2 from poor but significant binding to binding having reasonable but by no means exceptional affinity (53,61).

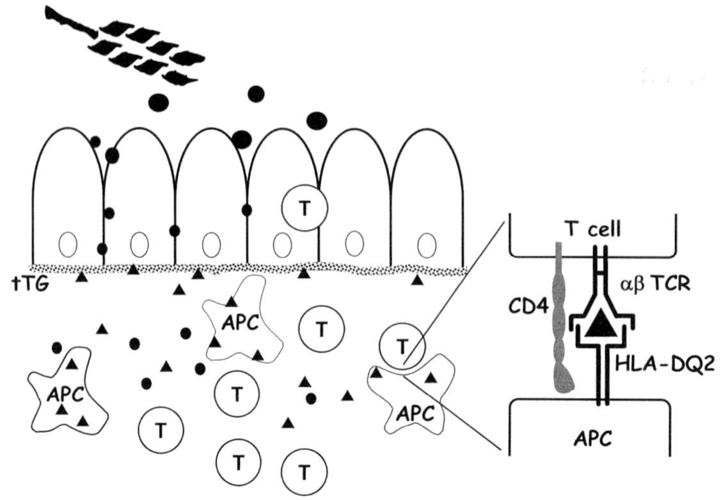

FIG. 3. CD4+ T lymphocytes in the lamina propria recognize deamidated gluten peptides presented by DQ2 or DQ8 molecules expressed by antigen-presenting cells (APC). The deamidation in vivo is likely to be mediated by the enzyme tissue transglutaminase, which in the gut mainly is expressed beneath the epithelium.

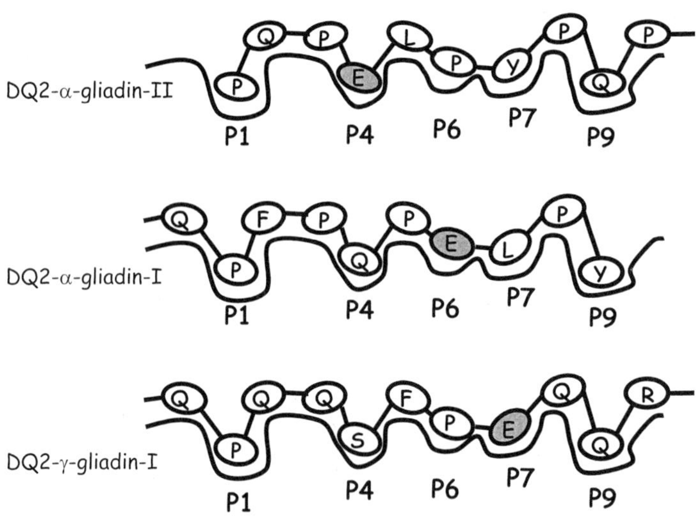

FIG. 4. A cartoon showing the amino acid sequence (one-letter code) and binding to DQ2 of the three known gluten epitopes recognized by intestinal T cells of patients with celiac disease. Two of the epitopes stem from α gliadins (DQ2-α-gliadin-II and DQ2-α-gliadin-I) and one from γ gliadins (DQ2-γ-gliadin-I). The T cell recognition of all three epitopes requires glutamine (Q) to glutamic acid (E) conversions in certain positions. This can be mediated by the enzyme tissue transglutaminase. (From the *Annual Review of Immunology*, Vol 18, 2000). (Available at www.AnnualReviews.org.).

The moderate binding affinity of these epitopes concurs with the finding that they do not carry optimal anchors in all the anchor positions. It is interesting that the modified glutamine residues for the three defined DQ2-restricted gliadin epitopes recognized by intestinal T cells occupy different pockets within DQ2 (Fig. 4). This suggests that the altered affinity of the gliadin peptides for DQ2 is a critical factor involved in loss of tolerance rather than recognition of a single "pathogenic" motif that binds to DQ2.

GLUTEN T-CELL EPITOPES IN CELIAC DISEASE: HOW MANY?

There are several epitopes in gluten that are recognized by small-intestine T cells of patients with celiac disease (52). Recent results from my laboratory (63a) and that of Koning in Leiden (63b) indicate that there may be more than ten distinct DQ2-restricted epitopes. The existence of multiple epitopes raises several interesting questions. Are only some of the epitopes pathogenic and thereby relevant to the HLA association? Are responses toward some of the epitopes generated during the early phases of disease development, whereas the responses to others are a result of epitope spreading? Are different epitopes recognized by distinct groups of patients (for example, children as opposed to adults)? Are some epitopes more relevant to disease because responses to them are found in most patients or because there is a higher precursor frequency of T cells in the lesion specific for these epitopes?

The answers to most of these questions must await further investigation. At present we know that for the DQ2-α-gliadin-I and DQ2-α-gliadin-II epitopes, intestinal T-cell reactivity is found in most, if not all adult DQ2$^+$ patients (61), whereas for the DQ2-γ-gliadin-I epitope, intestinal T-cell reactivity is found in fewer DQ2$^+$ patients (53). Less is known about the DQ8-restricted epitopes because few DQ8$^+$ patients have been tested so far. However, the DQ8-α-gliadin-I appears to be commonly recognized (54). What causes the variance in responsiveness to the different epitopes and whether this reflects qualitative or quantitative differences between the patients is presently unclear.

GLUTEN-SPECIFIC T CELLS MAY PROVIDE HELP FOR AUTOANTIBODY PRODUCTION

The serum IgG and IgA antibodies to tTG (also termed *antiendomysial antibodies*) are a hallmark of celiac disease, and detection of serum IgA tTG antibodies is used to predict the disease (64,65). Dieterich *et al.* (3) showed that tTG/gliadin complexes could be formed *in vitro* and argued that these complexes may contain neoepitopes that trigger mucosal T cells and break tolerance. The production of the anti-tTG IgA antibodies must be dependent on cognate T cell help to facilitate the isotype switching of the autoreactive B cells. Dieterich suggested that the necessary help for the B-cell production of anti-tTG IgA could be provided by autoreactive T cells specific for tTG. The existence of tTG-reactive T cells is doubtful. As tTG is expressed in the thymic epithelium, it is likely that tTG-reactive T cells are deleted during negative selection in the thymus (66). Moreover, the consequence of having autoreactive T cells specific for such a ubiquitously expressed enzyme as tTG would most likely be systemic autoimmunity.

As an alternative, tTG/gluten complexes may allow gluten-specific T cells to provide help to tTG-specific B cells by a mechanism analogous to the classical hapten carrier system (67). tTG-specific B cells will bind tTG/gluten complexes through their surface immunoglobulin. The complex will be processed, and gluten fragments may be produced that will bind to DQ2 or DQ8. Gluten-specific T cells may recognize this on the surface of the tTG-specific B cells and thereby provide cognate help for B cell maturation, isotype switching, and antibody secretion. This model can explain why tTG antibody levels in celiac disease are dependent on the presence of gliadin in the diet, as its removal will also abolish the T cell help needed for antibody production.

MECHANISMS INVOLVED IN FORMATION OF THE CELIAC LESION

The evidence discussed above provides strong evidence that $CD4^+$ $TCR\alpha\beta^+$ T cells in the lamina propria are central in controlling the immune response to gluten that produces the immunopathology of celiac disease (Fig. 3). The knowledge of the events downstream of T-cell activation is, however, still incomplete. Knowing that the immune system generally uses a multitude of effector mechanisms to fight its opponents, it is reasonable to assume that there may be multiple effector mechanisms involved in the creation of the celiac lesion. Adding to the complexity, recent *in vitro* organ culture studies have indicated that gluten exerts additional immune relevant effects that are independent of T cell activation (68,69). Some of these effects have rapid kinetics and conceivably the direct effects of gluten may facilitate subsequent T-cell responses.

Cytokines produced by lamina propria $CD4^+$T cells may be involved in the increased crypt cell proliferation and the increased loss of epithelial cells. IFN-γ induces macrophages to produce TNF-α. TNF-α activates stromal cells to produce keratinocyte growth factor (KGF), and KGF causes epithelial proliferation and crypt cell hyperplasia (70). IFN-γ and TNF-α can jointly have a direct cytotoxic effect on intestinal epithelial cells (71). It is also conceivable that IELs, and in particular $\gamma\delta$ T cells, play a role in the epithelial cell destruction by recognizing MIC molecules induced by stress (30).

Alterations to the extracellular matrix can also distort the epithelial arrangement, because the extracellular matrix provides the scaffold on which the epithelium lies. Enterocytes adhere to basement membrane through extracellular matrix receptors, so that modification or loss of the basement membrane can result in enterocyte shedding. There is evidence for increased extracellular matrix degeneration in celiac disease, and this may be an important mechanism for the mucosal transformation found in the disease (41). The increased production of metalloproteinases by subepithelial fibroblasts and macrophages is likely to be directly or indirectly induced by cytokines that are released from activated T cells.

We do not yet know whether the autoantibodies play a role in the pathogenesis of celiac disease. The tTG antibodies can inhibit the activity of tTG. This can cause villous atrophy by blocking interactions between mesenchymal cells and epithelial cells during the migration of epithelial cells and fibroblasts from the crypts to the tips of the villi (72). Moreover, the tTG antibodies may modulate the deamidating activity of tTG, in an inhibiting or a promoting fashion (73). Further research will tell us what role these antibodies play.

TRANSLATION OF THE NEW KNOWLEDGE INTO TREATMENT

The increasing insight into the molecular and cellular basis of celiac disease should yield benefit for patients. Knowledge about which gluten epitopes are recognized by gut T cells should allow improvements to the methods whereby gluten-free foods are assessed. Moreover, the new knowledge should uncover novel targets for treatment. There are already some attractive possibilities. Activation of $CD4^+$ gluten-specific T cells appears to be a critical checkpoint in the development of celiac disease, and interference with this step in the pathogenesis should be an effective way of controlling the disease. One possibility, which is basically an extension of today's treatment with a gluten-free diet, is to produce wheat seeds that are devoid of T-cell epitopes, either by breeding programs or transgenic technology.

tTG is a target for intervention because of the critical role it plays in generating gluten T-cell epitopes. Inhibitors of tTG activity exist and likely inhibitors suitable for use as drugs can be developed. The biggest problem with this approach is that tTG inhibitors may have unacceptable side effects. tTG is involved in many different physiologic processes including programmed cell death (59).

Another strategy would be to aim directly at the gluten-specific T cells. If patients with celiac disease have a normal oral tolerance to native gluten proteins but a disturbed tolerance to deamidated gluten peptides, exposing the gut immune system to deamidated peptides may establish oral tolerance to the deamidated gluten peptides as well. This approach would use the body's own mechanism to silence T cells. Alternatively, one could try to silence gluten-specific T cells directly by using soluble dimers of HLA/peptide complexes, which have been shown to induce antigen-specific apoptosis because of inappropriate T cell stimulation (74).

The central role of DQ2 and DQ8 in presenting gluten peptides offers yet another target for intervention. Blocking the binding sites of these HLA molecules would prevent presentation of disease-inducing gluten peptides. The challenge with this approach will be to find an efficient way of targeting and blocking the binding sites of DQ molecules, which are continuously synthesized by antigen-presenting cells. This approach—the blocking of peptide presentation—has also been suggested as treatment for other HLA-associated diseases. Celiac disease should be better suited to this approach because drug delivery in the gut is easy compared with, for example, the joints in rheumatoid arthritis or the pancreas in type 1 diabetes.

Whatever new therapeutic methods are introduced in celiac disease, they will have to be shown to be better than the current gluten-free diet regime. Long-term safety must also be taken into account when devising any new treatment. Although there are already interesting therapeutic principles with a good, rational basis that can be tested, it may for this reason take some years before any new treatments become a reality.

ACKNOWLEDGMENTS

Studies in the author's laboratory are funded by grants from the Research Council of Norway, the European Commission (BMH4-CT98-3087, QLRT-1999-00037, QLRT-2000-00657), Medinnova, the Jahre Foundation, and EXTRA funds from the Norwegian Foundation for Health and Rehabilitation.

REFERENCES

1. Fasano A, Catassi C. Current approaches to diagnosis and treatment of celiac disease: an evolving spectrum. *Gastroenterology* 2001; 120: 636–51.
2. Mäki M. The humoral immune system in coeliac disease. In: Howdle PD, ed. *Coeliac disease*. London: Bailliére Tindall, 1995: 231–49.
3. Dieterich W, Ehnis T, Bauer M, *et al.* Identification of tissue transglutaminase as the autoantigen of celiac disease. *Nat Med* 1997; 3: 797–801.
4. Ellis A. Coeliac disease: previous family studies. In: McConnell RB, ed. *The genetics of coeliac disease*. Lancaster: MTP Press, 1981: 197–9.
5. Risch N. Assessing the role of HLA-linked and unlinked determinants of disease. *Am J Hum Genet* 1987; 40: 1–14.
6. Petronzelli F, Bonamico M, Ferrante P, *et al.* Genetic contribution of the HLA region to the familial clustering of coeliac disease. *Ann Hum Genet* 1997; 61: 307–17.
7. Polanco I, Biemond I, van Leeuwen A, *et al.* Gluten sensitive enteropathy in Spain: genetic and environmental factors. In: McConnell RB, ed. *The genetics of coeliac disease*. Lancaster: MTP Press, 1981: 211–31.
8. Keuning JJ, Peña AS, van Leeuwen A, van Hooff JP, van Rood JJ. HLA-DW3 associated with coeliac disease. *Lancet* 1976; i: 506–8.
9. Ek J, Albrechtsen D, Solheim BG, Thorsby E. Strong association between the HLA-Dw3-related B cell alloantigen–DRw3 and coeliac disease. *Scand J Gastroenterol* 1978; 13: 229–33.
10. Mearin ML, Biemond I, Peña AS, *et al.* HLA-DR phenotypes in Spanish coeliac children: their contribution to the understanding of the genetics of the disease. *Gut* 1983; 24: 532–7.
11. Trabace S, Giunta A, Rosso M, *et al.* HLA-ABC and DR antigens in celiac disease. A study in a pediatric Italian population. *Vox Sang* 1984; 46: 102–6.
12. Tosi R, Vismara D, Tanigaki N, *et al.*Evidence that celiac disease is primarily associated with a DC locus allelic specificity. *Clin Immunol Immunopathol* 1983; 28: 395–404.
13. Sollid LM, Markussen G, Ek J, Gjerde H, Vartdal F, Thorsby E. Evidence for a primary association of celiac disease to a particular HLA-DQ α/β heterodimer. *J Exp Med* 1989; 169: 345–50.
14. Carrington M. Recombination within the human MHC. *Immunol Rev* 1999; 167: 245–56.
15. Lin L, Jin L, Kimura A, Carrington M, Mignot E. DQ microsatellite association studies in three ethnic groups. *Tissue Antigens* 1997; 50: 507–20.
16. Lin L, Jin L, Lin X, Voros A, Underhill P, Mignot E. Microsatellite single nucleotide polymorphisms in the HLA-DQ region. *Tissue Antigens* 1998; 52: 9–18.
17. Sollid LM, Thorsby E. HLA susceptibility genes in celiac disease: genetic mapping and role in pathogenesis. *Gastroenterology* 1993; 105: 910–22.
18. Greco L, Corazza G, Babron MC, *et al.* Genome search in celiac disease. *Am J Hum Genet* 1998; 62: 669–75.
19. Greco L, Babron MC, Corazza GR, *et al.* Existence of a genetic risk factor on chromosome 5q in Italian coeliac disease families. *Ann Hum Genet* 2001; 65: 35–41.
20. Djilali-Saiah I, Schmitz J, Harfouch-Hammoud E, Mougenot JF, Bach JF, Caillat-Zucman S. CTLA-4 gene polymorphism is associated with predisposition to coeliac disease. *Gut* 1998; 43: 187–9.

21. Naluai AT, Nilsson S, Samuelsson L, *et al.*The CTLA4/CD28 gene region on chromosome 2q33 confers susceptibility to celiac disease in a way possibly distinct from that of type 1 diabetes and other chronic inflammatory disorders. *Tissue Antigens* 2000; 56: 350–5.
22. Clot F, Fulchignoni-Lataud MC, Renoux C, *et al.* Linkage and association study of the CTLA-4 region in coeliac disease for Italian and Tunisian populations. *Tissue Antigens* 1999; 54: 527–30.
23. Kouki T, Sawai Y, Gardine CA, Fisfalen ME, Alegre ML, DeGroot LJ. CTLA-4 gene polymorphism at position 49 in exon 1 reduces the inhibitory function of CTLA-4 and contributes to the pathogenesis of Graves' disease. *J Immunol* 2000; 165: 6606–11.
24. Ligers A, Teleshova N, Masterman T, Huang WX, Hillert J. CTLA-4 gene expression is influenced by promoter and exon 1 polymorphisms. *Genes Immun* 2001; 2: 145–52.
25. Kagnoff MF, Austin RK, Hubert JJ, Bernardin JE, Kasarda DD. Possible role for a human adenovirus in the pathogenesis of celiac disease. *J Exp Med* 1984; 160: 1544–57.
26. Mahon J, Blair GE, Wood GM, Scott BB, Losowsky MS, Howdle PD. Is persistent adenovirus 12 infection involved in coeliac disease? A search for viral DNA using the polymerase chain reaction. *Gut* 1991; 32: 1114–16.
27. Kutlu T, Brousse N, Rambaud C, Le Deist F, Schmitz J, Cerf-Bensussan N. Numbers of T cell receptor (TCR) $\alpha\beta+$ but not of TcR $\gamma\delta+$ intraepithelial lymphocytes correlate with the grade of villous atrophy in coeliac patients on a long term normal diet. *Gut* 1993; 34: 208–14.
28. Spencer J, Isaacson PG, Diss TC, MacDonald TT. Expression of disulfide-linked and non-disulfide-linked forms of the T cell receptor γ/δ heterodimer in human intestinal intraepithelial lymphocytes. *Eur J Immunol* 1989; 19: 1335–8.
29. Halstensen TS, Scott H, Brandtzaeg P. Intraepithelial T cells of the TcR+ CD8− and Vδ1/Jδ1+ phenotypes are increased in coeliac disease. *Scand J Immunol* 1989; 30: 665–72.
30. Groh V, Steinle A, Bauer S, Spies T. Recognition of stress-induced MHC molecules by intestinal epithelial $\gamma\delta$ T cells. *Science* 1998; 279: 1737–40.
31. Groh V, Bahram S, Bauer S, Herman A, Beauchamp M, Spies T. Cell stress-regulated human major histocompatibility complex class I gene expressed in gastrointestinal epithelium. *Proc Natl Acad Sci USA* 1996; 93: 12445–50.
32. Lundqvist C, Melgar S, Yeung MM, Hammarström S, Hammarström ML. Intraepithelial lymphocytes in human gut have lytic potential and a cytokine profile that suggest T helper 1 and cytotoxic functions. *J Immunol* 1996; 157: 1926–34.
33. Halstensen TS, Farstad IN, Scott H, Fausa O, Brandtzaeg P. Intraepithelial TcR$\alpha/\beta+$ lymphocytes express CD45RO more often than the TcR$\gamma/\delta+$ counterparts in coeliac disease. *Immunology* 1990; 71: 460–6.
34. Kontakou M, Sturgess RP, Przemioslo RT, Limb GA, Nelufer JM, Ciclitira PJ. Detection of interferon gamma mRNA in the mucosa of patients with coeliac disease by in situ hybridisation. *Gut* 1994; 35: 1037–41.
35. Nilsen EM, Jahnsen FL, Lundin KEA, *et al.* Gluten induces an intestinal cytokine response strongly dominated by interferon gamma in patients with celiac disease. *Gastroenterology* 1998; 115: 551–63.
36. Nilsen EM, Lundin KEA, Krajci P, Scott H, Sollid LM, Brandtzaeg P. Gluten specific, HLA-DQ restricted T cells from coeliac mucosa produce cytokines with Th1 or Th0 profile dominated by interferon gamma. *Gut* 1995; 37: 766–76.
37. Nagashima R, Maeda K, Imai Y, Takahashi T. Lamina propria macrophages in the human gastrointestinal mucosa: their distribution, immunohistological phenotype, and function. *J Histochem Cytochem* 1996; 44: 721–31.
38. Scott H, Sollid LM, Fausa O, Brandtzaeg P, Thorsby E. Expression of major histocompatibility complex class II subregion products by jejunal epithelium in patients with coeliac disease. *Scand J Immunol* 1987; 26: 563–71.
39. Halstensen TS, Brandtzaeg P. Activated T lymphocytes in the celiac lesion: non–proliferative activation (CD25) of CD4+ α/β cells in the lamina propria but proliferation (Ki-67) of α/β and γ/δ cells in the epithelium. *Eur J Immunol* 1993; 23: 505–10.
40. Sturgess RP, Macartney JC, Makgoba MW, Hung CH, Haskard DO, Ciclitira PJ. Differential upregulation of intercellular adhesion molecule-1 in coeliac disease. *Clin Exp Immunol*1990; 82: 489–92.
41. Daum S, Bauer U, Foss HD, *et al.* Increased expression of mRNA for matrix metalloproteinases-1 and -3 and tissue inhibitor of metalloproteinases-1 in intestinal biopsy specimens from patients with coeliac disease. *Gut* 1999; 44: 17–25.
42. Madden DR. The three-dimensional structure of peptide-MHC complexes. *Annu Rev Immunol* 1995; 13: 587–622.

43. Johansen BH, Vartdal F, Eriksen JA, Thorsby E, Sollid LM. Identification of a putative motif for binding of peptides to HLA-DQ2. *Int Immunol* 1996; 8: 177–82.
44. van de Wal Y, Kooy YMC, Drijfhout JW, Amons R, Koning F. Peptide binding characteristics of the coeliac disease-associated DQ(α1*0501, β1*0201) molecule. *Immunogenetics* 1996; 44: 246–53.
45. Vartdal F, Johansen BH, Friede T, *et al.* The peptide binding motif of the disease associated HLA-DQ(α1*0501, β1*0201) molecule. *Eur J Immunol* 1996; 26: 2764–72.
46. Godkin A, Friede T, Davenport M, *et al.* Use of eluted peptide sequence data to identify the binding characteristics of peptides to the insulin-dependent diabetes susceptibility allele HLA-DQ8 (DQ 3.2). *Int Immunol* 1997; 9: 905–11.
47. Kwok WW, Domeier ML, Raymond FC, Byers P, Nepom GT. Allele-specific motifs characterize HLA-DQ interactions with a diabetes-associated peptide derived from glutamic acid decarboxylase. *J Immunol* 1996; 156: 2171–7.
48. Lee KH, Wucherpfennig KW, Wiley DC. Structure of a human insulin peptide-HLA-DQ8 complex and susceptibility to type 1 diabetes. *Nat Immunol* 2001; 2: 501–7.
49. Lundin KEA, Scott H, Hansen T, *et al.* Gliadin-specific, HLA-DQ(α1*0501,β1*0201) restricted T cells isolated from the small intestinal mucosa of patients with celiac disease. *J Exp Med* 1993; 178: 187–96.
50. Lundin KEA, Gjertsen HA, Scott H, Sollid LM, Thorsby E. Function of DQ2 and DQ8 as HLA susceptibility molecules in celiac disease. *Hum Immunol* 1994; 41: 24–7.
51. Shewry PR, Tatham AS, Kasarda DD. Cereal proteins and coeliac disease. In: Marsh M, ed. *Coeliac disease.* Oxford: Blackwell Science, 1992: 305–48.
52. Lundin KEA, Sollid LM, Norén O, *et al.* Heterogenous reactivity patterns of HLA-DQ-restricted small intestinal T-cell clones from patients with celiac disease. *Gastroenterology* 1997; 112: 752–9.
53. Sjöström H, Lundin KEA, Molberg Ø, *et al.* Identification of a gliadin T-cell epitope in coeliac disease: general importance of gliadin deamidation for intestinal T-cell recognition. *Scand J Immunol* 1998; 48: 111–15.
54. van de Wal Y, Kooy YM, van Veelen PA, *et al.* Small intestinal T cells of patients with celiac disease recognize a natural pepsin fragment of gliadin. *Proc Natl Acad Sci USA* 1998; 95: 10050–4.
55. van de Wal Y, Kooy YM, van Veelen P, *et al.* Glutenin is involved in the gluten-driven mucosal T cell response. *Eur J Immunol* 1999; 29: 3133–9.
56. Molberg Ø, McAdam SN, Körner R, *et al.*Tissue transglutaminase selectively modifies gliadin peptides that are recognized by gut-derived T cells. *Nat Med* 1998; 4: 713–17.
57. van de Wal Y, Kooy Y, van Veelen P, *et al.* Selective deamidation by tissue transglutaminase strongly enhances gliadin-specific T cell reactivity. *J Immunol* 1998; 161: 1585–8.
58. Bruce SE, Bjarnason I, Peters TJ. Human jejunal transglutaminase: demonstration of activity, enzyme kinetics and substrate specificity with special relation to gliadin and coeliac disease. *Clin Sci* 1985; 68: 573–9.
59. Aeschlimann D, Paulsson M. Transglutaminases: protein cross-linking enzymes in tissues and body fluids. *Thromb Haemost* 1994; 71: 402–15.
60. Folk JE. Mechanism and basis for specificity of transglutaminase-catalyzed ϵ-(γ-glutamyl) lysine bond formation. *Adv Enzymol Relat Areas Mol Biol* 1983; 54: 1–56.
61. Arentz-Hansen H, Körner R, Molberg Ø, *et al.* The intestinal T cell response to α-gliadin in adult celiac disease is focused on a single deamidated glutamine targeted by tissue transglutaminase. *J Exp Med* 2000; 191: 603–12.
62. Quarsten H, Molberg Ø, Fugger L, McAdam SN, Sollid LM. HLA binding and T cell recognition of a tissue transglutaminase-modified gliadin epitope. *Eur J Immunol* 1999; 29: 2506–14.
63. Molberg Ø, McAdam S, Lundin KEA, *et al.* T cells from celiac disease lesions recognize gliadin epitopes deamidated in situ by endogenous tissue transglutaminase. *Eur J Immunol* 2001; 31: 1317–23.
63a. Arentz-Hansen H, McAdam SN, Molberg Ø, *et al.* Celiac lesion T cells recognize epitopes that cluster in regions of gliadins rich in proline residues. *Gastroenterology* 2002; 123: 803–09.
63b. Vader W, Kooy Y, Van Veelen P, *et al.* The gluten response in children with celiac disease is directed toward multiple gliadin and glutenin peptides. *Gastroenterology* 2002; 122: 1729–37.
64. Dieterich W, Laag E, Schopper H, *et al.* Autoantibodies to tissue transglutaminase as predictors of celiac disease. *Gastroenterology* 1998; 115: 1317–21.
65. Sulkanen S, Halttunen T, Laurila K, *et al.* Tissue transglutaminase autoantibody enzyme-linked immunosorbent assay in detecting celiac disease. *Gastroenterology* 1998; 115: 1322–8.
66. Hällström O. Comparison of IgA-class reticulin and endomysium antibodies in coeliac disease and dermatitis herpetiformis. *Gut* 1989; 30: 1225–32.

67. Sollid LM, Molberg Ø, Mc Adam S, Lundin KEA. Autoantibodies in coeliac disease: tissue transglutaminase: guilt by association? *Gut* 1997; 41: 851–2.
68. Maiuri L, Picarelli A, Boirivant M, *et al.* Definition of the initial immunologic modifications upon in vitro gliadin challenge in the small intestine of celiac patients. *Gastroenterology* 1996; 110: 1368–78.
69. Maiuri L, Auricchio S, Coletta S, *et al.* Blockage of T-cell costimulation inhibits T-cell action in celiac disease. *Gastroenterology* 1998; 115: 564–72.
70. Bajaj-Elliott M, Poulsom R, Pender SL, Wathen NC, MacDonald TT. Interactions between stromal cell-derived keratinocyte growth factor and epithelial transforming growth factor in immune-mediated crypt cell hyperplasia. *J Clin Invest* 1998; 102: 1473–80.
71. Deem RL, Shanahan F, Targan SR. Triggered human mucosal T cells release tumour necrosis factor-alpha and interferon-gamma which kill human colonic epithelial cells. *Clin Exp Immunol* 1991; 83: 79–84.
72. Halttunen T, Mäki M. Serum immunoglobulin A from patients with celiac disease inhibits human T84 intestinal crypt epithelial cell differentiation. *Gastroenterology* 1999; 116: 566–72.
73. Sollid LM, Scott H. New tool to predict celiac disease on its way to the clinics. *Gastroenterology* 1998; 115: 1584–6.
74. Appel H, Seth NP, Gauthier L, Wucherpfennig KW. Anergy induction by dimeric TCR ligands. *J Immunol* 2001; 166: 5279–85.

DISCUSSION

Dr. Belli: What molecular evidence do we have for oat tolerance in celiac disease?

Dr. Sollid: For the time being the evidence is clinical. There are ongoing trials in Finland in which patients with celiac disease and dermatitis herpetiformis have been receiving oats for up to 5 years now, and there are no signs of immune activation or changes in mucosal architecture (1). What the long-term effects will be in terms of protection against osteoporosis and infertility and those more difficult endpoints is very hard to say, but I think the general understanding by scientists is that oats are tolerated. We now need to see whether we can find T-cell reactivity to any oats protein, as an independent line of evidence in support of oats being harmless.

Dr. B. Koletzko: You made a convincing case for HLA-DQ2 as providing the key to antigen presentation. My understanding was that the presence of the DR3 or DR5/7 genotype is really an indicator of increased risk but not an essential prerequisite for celiac disease. If that is so, are there other mechanisms involved in parallel to the antigen presentation by HLA-DQ2? Could there be a variety of pathophysiologies?

Dr. Sollid: I think this is the best argument for saying that there are non-HLA genes that contribute. Of the total genetic risk, HLA contributes perhaps as much as 30% to 50% and the rest of the effect is conferred by non-HLA genes, but the effect of each of these genes is modest—so small that they are hard to find by a genome-wide screen, where the power is relatively low. It could also be the case that different patients have different susceptibility genes, with disease heterogeneity. The chances of finding such genes then become much smaller. Despite the relatively gloomy data on the evidence of the non-HLA genes, I'm convinced that they exist, but we don't have the tools to locate them at this point.

Dr. B. Koletzko: My second question relates to the recommendations of Swedish pediatricians that gluten should be introduced earlier in the diet at the time of breastfeeding, based on a plausible argument that this would reduce the incidence of celiac disease (2). When one studies the data, however, there is an anomaly. It is apparent that while the recommendations of the Swedish pediatricians dated from 1996, the peak incidence of celiac disease was reported to be in 1994, and had declined by more than half by the time the recommendations were issued. Obviously one would expect that between the application of a new recommendation, if it were effective, and any effect on disease incidence, there would be a lag period, so it seems as if the disease prevalence anticipated the wisdom of the Swedish pediatricians. There does not seem to be a good case for a causal relation.

Dr. Sollid: I agree entirely. My Swedish colleagues say that there was a huge debate in Sweden on this topic, including several front-page articles in major newspapers on the subject before the national recommendations were changed. When the gluten consumption of these infants was measured, it was apparent that there was decline in the amount consumed that occured before the change in the national recommendations. This could be due to the public debate. But the data have also been used to argue for a role of breastfeeding in preventing celiac disease. The duration of the breastfeeding has been increasing over the years with the decline in celiac disease incidence, and from an immunologic point of view, there may be good reason to believe that breastfeeding is implicated in protection from development of celiac disease. Our model involves the creation of an inflammatory environment. We think the presentation of deamidated peptides will only occur if you already have inflammation, and such inflammation can be caused by an infection. Breastfeeding would protect you from that. So that could explain why you don't end up in a situation where you start to create dangerous peptides.

Dr. B. Koletzko: I certainly follow the argument that breastfeeding may be protective, but it doesn't really match the epidemiology. The reported change in the incidence rate does not parallel the change in breastfeeding habits.

Dr. Sollid: I agree. I think that the amount of gluten intake correlates much better than the change in the national recommendation, and this mostly reflects the fact that parents started to think for themselves. There was a major debate on this in Sweden.

Dr. Marquardt: I did understand the role of transglutaminase in the generation of peptides that then trigger the immune response, but why does one make antibodies against tissue transglutaminase?

Dr. Sollid: The basic answer is, we don't know. Tissue transglutaminase is expressed in almost all organs in the body, including the thymus, and I don't think there are any tissue-transglutaminase–specific T cells, so it's also very hard to see a role of gluten in this model. Tissue transglutaminase can cross-link proteins and a German group showed that it can cross-link glutamine residues of gluten proteins onto lysine, so you form a complex (3). Immunologists recognize this as a hapten–carrier complex. This complex is taken up by the B cell, which processes the antigens, and you end up with the gluten peptide bound to the class II molecule. We know that there are T cells specific for gluten peptide, and so this shortcut of the immune system can fool it into creating antibodies to self-protein. If you take away gluten from the diet, you basically take away this peptide. Although we don't have experimental data to prove this, I think it's the simplest model you can invoke to explain how antibodies to self antigens are driven by a T cell response to an exogenous antigen. I believe this is also relevant to other autoimmune diseases.

Dr. Fowler: I'm interested in the physiologic role of transglutaminase. Could you enlarge on that?

Dr. Sollid: Tissue transglutaminase is categorized as a wound-healing enzyme. If you induce a wound experimentally, you increase its expression at least ten-fold, and in the active celiac lesion, you also have increased expression. One problem we do not understand is what dictates cross-linking and what dictates deamidation. What our model requires is deamidation, and people who have worked with the biology of tissue transglutaminase say that the primary substrate for this enzyme is primary amines—lysine, spermidine, or spermine—and if those are present, you won't get deamidation. I would look at it from the other viewpoint. T cells basically record what is present, and what T cells recognize is the introduction of deamidated residues, which are very specific and which concur with the specificity of the enzyme. We do not really understand the microenvironment of the gut where these epitopes are formed. It could be that you deplete the storage of these primary amines when you have breakage of the epithelial barrier. You flood the lamina propria with gluten proteins and consume the primary amines, and you then start to create these epitopes. If you have a situation where you have more gluten, more substrate, and a higher enzyme level, you will reach that situation more rapidly. To try to mimic what is going on in this complex system in the gut is very difficult with *ex vivo* testing. We don't really know whether these peptides need to be taken up by the antigen-presenting cells and whether they can bind at the surface.

Dr. S. Koletzko: A couple of years ago a large epidemiologic study from Italy was published in *Gastroenterology* on more than 900 celiac patients diagnosed during childhood or adult life. The investigators showed that the earlier celiac disease was diagnosed and treated, the less problem there was with other autoimmune diseases later on. They showed a clear dose–response effect. Could you comment on that report? Do you believe you can prevent other autoimmune diseases by treating celiac disease early in life?

Dr. Sollid: What is clear is that there is an overrepresention of many autoimmune diseases in celiac disease—for example type 1 diabetes, thyroiditis, and some other diseases. The Italian investigators have not convinced me that celiac disease predisposes to the development of other autoimmune diseases. If you follow a cohort of patients with celiac disease, you will have an accumulation of other autoimmune diseases regardless of whether or not a gluten-free diet is in place. To prove that there is a difference, I think you would need larger numbers than were presented in that Italian study. However, this is not my field of expertise.

REFERENCES

1. Janatuinen EK, Kemppainen TA, Julkunen RJ, *et al.* No harm from five year ingestion of oats in coeliac disease. *Gut* 2002; 50: 332–5.
2. Ivarsson A, Persson LA, Hernell O, *et al.* Epidemic of coeliac disease in Swedish children. *Acta Paediatr* 2000; 89: 165–71.
3. Dietrich W, Ehnis T, Bauer M, *et al.* Identification of tissue transglutaminase as the autoantigen of celiac disease. *Nat Med* 1997; 3: 797–801.

Genetic Expression and Nutrition, edited by Claude Bachmann and Berthold Koletzko. Nestlé Nutrition Workshop Series, Pediatric Program, Vol. 50. Nestec Ltd., Vevey/Lippincott Williams & Wilkins, Philadelphia, © 2003.

Genetic Variations of Gliadins: New Approaches for the Treatment of Celiac Disease

Olin D. Anderson

Western Regional Research Center, Agricultural Research Service, United States Department of Agriculture, Albany, California, USA

The ingestion of a wide range of foods by humans is accompanied by an equally wide range of responses to those foods. While there are examples of raw foodstuffs containing major allergens or toxins requiring special processing before ingestion, such sources do not usually form part of the human diet. For the most common food sources, it is usual for only a small fraction of the populace to experience negative responses upon ingestion. In most cases, the contribution of a specific food item to the total diet is small enough for avoidance to be relatively straightforward, although diligence by the affected person is critical in the most severe cases. A significant exception is in the case of celiac disease, an enteropathy directly related to the ingestion of some cereal proteins, particularly those of wheat (1).

Celiac disease is a disorder in which ingestion of wheat gliadin, or related proteins, is followed by an immunologic-based injury to the lining and villi of the small intestine. Symptoms and associated disorders cover a wide range of demonstrated and suspected pathologies in children and adults, including villous atrophy, malabsorption with steatorrhea, weight loss, malnutrition coupled with multiple deficiency states, nonspecific abdominal pain, clotting disorders, juvenile diabetes, thyroid disease, dental anomalies, short stature, anemia, angina, osteopenic bone disease, lactose intolerance, infertility, hypoglycemia, osteoporosis, crypt hyperplasia, and dermatitis herpetiformis (1–4).

Currently, the only effective treatment of celiac disease is to refrain completely from eating any foods containing wheat, barley, or rye. This course of action almost always results in remission of symptoms. While such an effective remedy for a disease is seldom available and seems simple, its implementation can be extremely difficult, for two reasons. First, wheat is used in a wide array of foods that humans find highly appealing, so its avoidance needs lifelong self-control and diligence by the patient. Second, wheat products are so ubiquitous that wheat is present not only overtly, in products such as breads, pastas, cakes, and so on, but also as an ingredient in many processed foods. Thus, even if the patient carefully examines the contents of any such processed foods

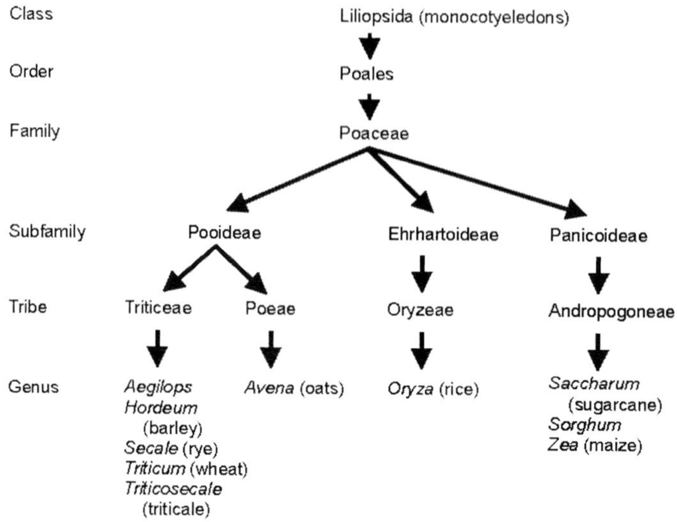

FIG. 1. Taxonomy of the major cereals and related grasses. Classifications are taken from the taxonomy database of the National Center for Biotechnology Information (NCBI) at http://www.ncbi.nlm.nih.gov.

and exercises firm diligence, the inclusion of wheat may not be adequately revealed, nor will the patient always be in a position to have access to all such information.

Although wheat is clearly identified as the most significant dietary factor in celiac disease, other cereals are also implicated, particularly other members of the *Triticeae* tribe of grasses (Fig. 1)—that is, barley, rye, and triticale (a hybrid of wheat and rye). The molecular agents eliciting the celiac response are certain classes of seed proteins—the prolamins—classically defined as seed proteins high in the amino acids prolamin and glutamine (hence the name) and separable from the other seed components by extraction in dilute alcohol mixtures (5). Prolamin-type proteins are found in all the major cereals shown in Fig. 1 (6), although the sizes, number of classes, and amino acid sequences vary greatly among the cereals. The potential for contributions to celiac disease of foods made from cereals outside the *Triticeae* tribe, such as rice, maize, and oats, is not completely defined. Rice and maize are considered safe (7), but the safety of oats is still not universally endorsed. For example, one study indicates that oat products are safe (8), whereas another (9) reports success using an oat peptide in a celiac detection kit. Clearly, more information is needed, including long-term effects and amounts of all suspect products that can safely be consumed.

The exact molecular elicitors of celiac disease are still not completely understood, although many studies have implicated the gliadin fraction of wheat seed prolamins—and more specifically some peptides of the α- and γ-gliadin protein classes described below (10–13), although there are some reports of reactions to the ω-gliadin and glutenin protein classes (14,15). There is also recent evidence that the active epitopes are not the original peptides within the gliadins, but specific deamidation products after tissue transglutaminase action on the ingested gliadins (16–18).

There are two basic approaches to helping the celiac patient, either direct, by improved medical treatment of the symptoms and the underlying immunologic disorder, or second, by making available a wider range of satisfying products devoid of the offending peptide motifs. In this chapter, I will address the second approach, and summarize the basic structure of the prolamins most often implicated in celiac disease. In addition, I will propose approaches to research and product development to eliminate the prolamin motifs that elicit celiac disease.

THE CEREAL PROLAMINS

Wheat white flour is made from the finely milled endosperm fraction of the wheat grain. The bulk of wheat flour is starch granules, comprising approximately 75% of the mass, and protein, which comprises another 10% to 15%. Most of the endosperm protein is composed of the prolamins, polypeptides high in proline and glutamine and divided into the glutenin and gliadin fractions. These fractions together comprise wheat gluten—an interacting network of proteins consisting of different types of disulfide, cross-linked polymers and gliadins (Fig. 2), which interact with the polymers through hydrogen bonds and other secondary forces (19,20). One of the original distinctions between the gliadin and glutenin fractions was physical—the glutenins form intermolecular cross-links to make enormous polymers that can reach tens of millions of Daltons (21), while the gliadins exist mainly as monomers. The glutenins are further divided into the high-molecular-weight glutenin subunits (HMW-GSs; MW, 60–100,000 kDa) and low-molecular-weight glutenin subunits (LMW-GSs; MW, 30,000–45,000 kDa). The gliadins are traditionally divided into α,

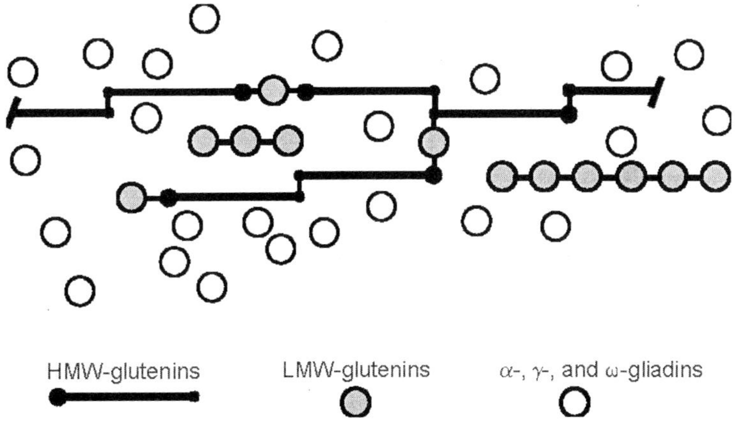

FIG. 2. Schematic view of wheat gluten structure. Polymers are formed through disulfide bonds, mainly by the HMW-glutenins plus LMW-glutenins, or by LMW-glutenins alone. Monomeric gliadins interact with each other and the polymers through hydrogen bonds. HMW-glutenins are indicated as rod-shaped molecules and the LMW-glutenins by filled circles, and α-, γ-, and ω-gliadins by empty circles. Intermolecular disulfide bonds are shown as links between glutenins in the polymers. HMW, high molecular weight; LMW, low molecular weight.

γ, and ω types by structure, with molecular weights of 30,000 to 35,000 kDa for the α- and γ-gliadins, and 30,000 to 45,000 kDa for the ω-gliadins. On an evolutionary basis, the LMW-GSs and various gliadin classes are members of the gliadin protein family, which itself is related to proteins found in all monocotyledonous cereals and in many dicotyledonous plants (6).

Wheat types are distinguished by the number and composition of their genomes: hexaploid bread and spelt wheats (AABBDD; the A, B, and D genomes), tetraploid wheats such as the durums (AABB), and diploid wheats such as einkorn (AA). Closely related are rye (RR) and barley (HH), along with triticale (AABBRR), a hexaploid derivative of tetraploid wheat and rye. The basic haploid genome complement of all these plants is seven chromosomes, the genes of which have maintained the same linear order among all of the genomes (22). These crops are all members of the *Triticeae* tribe (Fig. 1), and the similar genetics and genome structure have led to the group being considered a single genetic system for many research purposes. The gliadin genes are encoded at four complex loci on chromosomes 1 and 6: the γ- and ω-gliadins and the LMW-GS genes on chromosome 1, and the α-gliadins on chromosome 6 (23). This means that there are 8 and 12 gliadin loci in the tetraploid and hexaploid wheats, respectively.

The major cereals (barley, maize, oats, rice, rye, sorghum, triticale, wheat) are all members of the *Poaceae* family of monocotyledonous plants (Fig. 1), and all contain prolamin seed protein fractions. What is different among the cereals is the specific structure of the prolamins, the number of prolamin genes, and the proportion of the grain protein represented by the prolamins. In barley, maize, rye, triticale, and wheat, the prolamins are the major fractions of the grain proteins and typically account for over half the total protein. In contrast, the major grain storage proteins for oats and rice are the globulins, with prolamins comprising approximately 10% and 5%, respectively, of the total grain protein. Prolamin sizes vary from 10 to 100 kDa, and amino acid sequences can be almost completely dissimilar between some classes. The prolamins of wheat are from 40% to 75% proline+glutamine, and those amino acids are largely found in repetitive domains composed of a series of short repeats based on one or a few general motif patterns. Figure 3 shows the consensus repeat patterns for the major classes of *Triticeae* prolamins. From these basic motifs derive most of the repeat units found within the gliadin polypeptides and their related polypeptides in the other known *Triticeae* prolamins. The number of repeat units in a specific gliadin can range from eight to an estimated 60 or more for the largest ω-gliadins (none of the largest ω-gliadins is yet characterized).

The prolamins are related to various other seed protein families (6). The degree of similarity varies, and in some cases the only remnant of similarity is in the positions and spacing of conserved cysteine residues. However, in some cases, an evolutionary relation between two families can only be inferred by relations to third families—for example, the ω-gliadins are related to the γ-gliadins, and the γ-gliadins to some rice prolamins such as M23745 (GenBank accession no.)—but the similarity criteria find no direct relation between ω-gliadins and the same rice prolamin (not shown).

Prolamine Class	Repeat Motifs
LMW-glutenins	$P_{1-2}\,F\,\genfrac{}{}{0pt}{}{P}{S}\,Q_{2-6}$
B-hordeins	$P\ F\ P\ Q\ Q\ (PQ)_{0-1}$
α-gliadins	$P\,\genfrac{}{}{0pt}{}{F}{Y}\,P\ Q_{0-1}\ P\ Q_{1-2}$
γ-gliadins, γ-secalins, γ-hordeins	$P\ F\ P\ Q_{1-2}(PQQ)_{1-2}$
ω-gliadins, ω-secalins, C-hordeins	$P\ F\ P\ Q_{1-2}(PQQ)_{1-2}$
HMW-glutenins	PGQGQQGYYPTSPQQ $PGQGQQ(PQQ)_{0-1}$
LMW-gliadin 07h10	EPQHPSSPEHQPTPQPQEHPVPHQK
LMW-gliadin 09d3	QPQQSFPQQPQQKTVQSFFEQQ
LMW-gliadin 11dc7	SQGYGQCQQQPQQQ

FIG. 3. Prolamin classes and their repetitive domains. The major classes of *Triticeae* gliadin-related prolamins are listed along with the predominant repeat motif pattern. In the LMW-glutenin and α-gliadin classes, the second residues position is occupied approximately equally between phenylalanine and tyrosine. The first three amino acids are similar among the classes, but followed by variations of a glutamine (Q)-rich section. The short region of the LMW-gliadin types that is similar in structure and position to the major prolamin repetitive domain is given at the bottom. Examination of the DNA sequence of the LMW-gliadins reveals a closer relation than is evident in the amino acid sequences (24). LMW, low molecular weight.

This is because the ω-gliadins have lost the domains containing the conserved cysteine positions (see below) shared by the γ-gliadins and M23745, whereas the γ-gliadins and ω-gliadins have similar repeat domains and flanking, noncoding DNA sequences.

BASIC STRUCTURE OF THE GLIADINS

In this chapter, I will focus on the gliadins, and particularly the α-gliadins, because they have the clearest association with celiac disease (10–12). The basic structure of a gliadin polypeptide is shown in Fig. 4A and comprises a short N-terminal peptide followed by a repetitive domain composed of short proline+glutamine-rich motifs, and a C-terminal domain containing most or all of the cysteine residues. During the evolution of the various prolamin subfamilies, different domains have been either emphasized or reduced from this basic pattern. Panels B to D in Fig. 4 show the known gliadinlike polypeptides. The α-gliadins, γ-gliadins, and LMW-GSs share the generic gliadin structure, except that the LMW-GSs have one free cysteine in both N- and C- termini. These allow the formation of two intermolecular disulfide bonds and

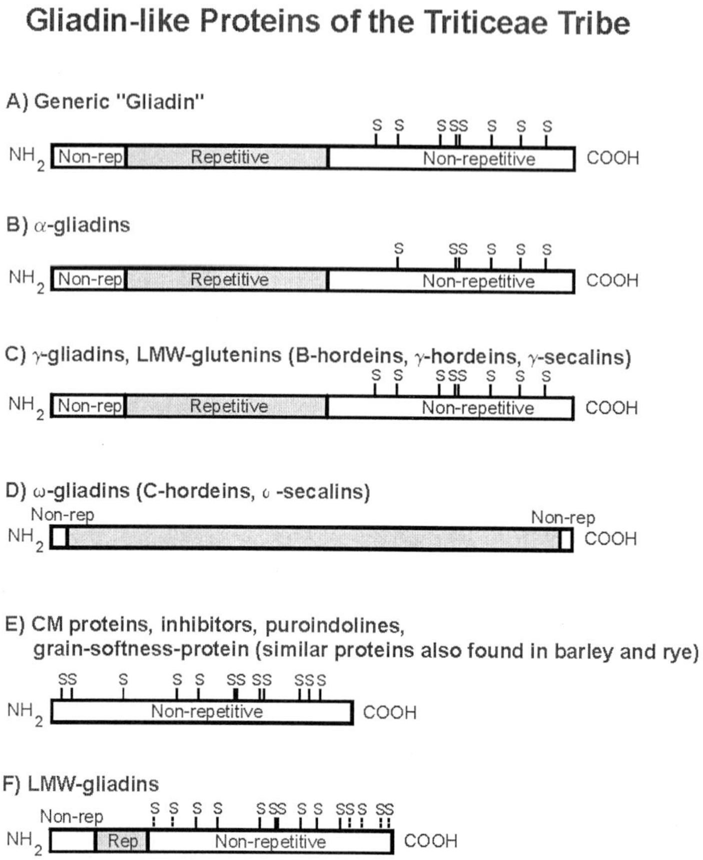

FIG. 4. Structure of the gliadin polypeptides of wheat and related cereals. The basic organization of the various gliadin and gliadin-related proteins of the wheat seed and flour is shown. Repetitive domains composed of repeats of short peptide motifs or related sequences are indicated by the gray domain. Positions of cysteine residues are indicated above the polypeptide by the S group available for intra- or intermolecular disulfide bond formation. Dotted lines between S groups and the polypeptide backbone indicate cysteine positions in only some LMW-gliadin subclasses. The N-terminal (NH_2) and C-terminal (COOH) ends of the polypeptides are indicated. Related prolamins from other cereals are in parentheses. **(A)** A "generic" wheat gliadin structure from which the specific groups can be conceptually derived. **(B)** α-Gliadins—unique to wheat. **(C)** γ-Gliadins of wheat and the related B- and γ-hordeins of barley, and the γ-secalins of rye. **(D)** ω-Gliadins of wheat and the related C-hordeins of barley and ω-secalins of rye. **(E)** A wide range of small seed proteins including classes identified as CM proteins, amylase/trypsin inhibitors, puroindolines, and grain-softness proteins. **(F)** A diverse group of LMW "gliadinlike" polypeptides found in wheat, barley, and rye. Characterization and distribution of the LMW-gliadin class is yet to be completed. LMW, low molecular weight.

explain the LMW-GSs participation in the glutenin polymer. The ω-gliadins differ in that almost all of the sequence is composed of the repetitive domain, and the polypeptides contain no cysteine residues. It is also known that other classes of endosperm proteins have a distant relation to the gliadins (for example, the CM/inhibitor proteins; Fig. 4E), but do not share the same physical/chemical properties or seem as relevant to dough properties. Recently it has been reported that there are several additional distinct classes of previously unrecognized gliadin-related endosperm proteins (24). These have been termed the *low-molecular-weight gliadins* and are composed of at least four classes, with molecular weights in the range of 17,000 to 20,000 dalton. These low-molecular-weight gliadins fit the basic gliadin model of Fig. 4A, but with a greatly reduced repetitive domain without a clearly discernable motif pattern. There is no evidence that these low-molecular-weight gliadins contribute to celiac disease, but their relation to the classical gliadin types and their sequence similarity mean that they must be considered as possible contributors. There is speculation that additional classes may well be found with more thorough analysis of the proteins in the wheat endosperm.

The high proline+glutamine amino acid residue composition of the prolamins reaches its maximum in the ω-gliadins, where 75% of the total amino acids are restricted to these two amino acids. The α-gliadins contain about 45% proline+glutamine, mainly in the repeat motifs of the repetitive domain. Glutamines are also found in the α-gliadins in two polyglutamine regions encoded by runs of CAG and CAA codons. These runs form microsatellites (25) that are particularly characteristic of the α-gliadin genes of wheat, and are the most variable region of any prolamin. These runs are also similar to the longer variable poly-CAG regions in genes associated with many human genetic disorders such as Huntington disease and spinocerebellar ataxia (26), but no connection between these runs and their products and celiac disease has been shown.

LEVEL OF KNOWLEDGE OF PROLAMIN STRUCTURE AND DISTRIBUTION AMONG THE CEREALS

In assessing potential contributions of various cereals to activate celiac disease, it would be useful to have as complete a picture as feasible of the protein composition and sequences of the seeds of those cereals. In this regard, the available evidence varies with the crop. The prolamins of wheat have been the most characterized, with as many entries in public databases for prolamin genes as all the other cereals combined (Table 1). Maize has a large number of entries, with fewer from barley, rice, and sorghum, but only four for oats and five for rye. Even some of these numbers are misleading—for example, in the case of the five rye sequences, three of the known sequences are from the same gene and one entire class of rye prolamins (the 40K secalins) has no representation in public DNA sequence databases. Clearly more information is needed on the prolamin fractions from all the major cereals.

In assessing the potential of a polypeptide for celiac toxicity, it is important to examine the amino acid sequence in detail. Some of the cereal prolamins have little

TABLE 1. *Number of cereal prolamine gene sequences in public databases*

Genus	Common name	Number of sequences
Avena	Oats	4
Hordeum	Barley	18
Oryza	Rice	24
Secale	Rye	5
Sorghum	Sorghum	11
Triticum	Wheat	149
Zea	Maize	49

convincing evolutionary relations, with most of the apparent similarity in sequence attributable simply to what would occur randomly for high glutamine+proline compositions. This is accentuated by the relative overutilization in cereal prolamin genes of the CCA codon for proline and CAA for glutamine. For example, the maize prolamin (zein) V01474 (GenBank accession no.) is 27% glutamine+proline, with 70% of the relevant codons being CCA and CAA. Similarly, γ-gliadin γ-13 is 46% glutamine+proline, and 61% of that is encoded by CCA plus CAA codons. A close examination of the two coding DNA sequences shows all significant apparent similarity to be within these glutamine/proline-rich regions and thus provides no evidence of evolutionary relatedness. In other cases, evolutionary relatedness is evident, but the sequence domains most probably related to celiac disease are quite different. For example, zein zX58197 (GenBank accession no.) would seem to have a similar structure to the γ-gliadin γ13 but the pattern of repeat motifs is completely different: VHLPPP and PHP(C/SPCQ) *versus* PFPQPQQ. Such different repeat structures may be useful in engineering less-toxic repeat domains (discussed below).

STRATEGIES TO ELIMINATE CELIAC-RELATED PEPTIDES

The approach to the treatment of celiac disease that I will now describe focuses on the elimination of the causative peptides from the food supply through combinations of breeding and genetic engineering. If the celiac causative motifs can be identified, can they be eliminated from wheat? The answer is yes, by at least two approaches. Before discussing those approaches, it must be emphasized that there is a critical problem with this strategy: its effect on the functionality of wheat doughs. The basis for the use of wheat in such a wide range of food products is the viscoelastic properties of doughs made from wheat flour. The wheat prolamins are the most important contributors to those properties. Thus a strategy that eliminates the α-gliadins, for example, may have a negative effect on dough properties to such a degree that the wheat cannot be used in traditional foods or traditional processing methods. This potential problem may not occur if the relative ratio of the gliadinlike proteins to glutenin polymer can be maintained. Thus, it will be important to test the effects of systematically removing specific subclasses of the prolamins from the doughs and to examine whether the resulting food products have beneficial effects for patients with celiac disease.

One strategy for eliminating proteins that cause celiac disease is the use of homology-dependent gene suppression. Both sense and antisense gene suppression are well-recognized phenomena in plants (27), although the basic molecular mechanisms are still not completely understood. It is currently believed that the RNA intermediates feedback on the homologous genes to shut off transcription, probably originating by rearrangement of the DNA sequence during the transformation and genome integration events. That wheat lines showing suppression can be generated has already been found in the reduction (28), or even elimination (Blechl AE, Anderson OD, June, 1998 unpublished data), of all HMW-glutenin subunits from lines of transgenic wheat. In these particular experiments, the objective was to generate a series of transgenic wheat lines expressing increased amounts of existing endogenous HMW-glutenin subunits by transforming wheat with intact HMW-glutenin genes. Although most lines did show increased levels of synthesis, a minority of lines had reductions in the expression of the five active endogenous HMW-glutenin genes, accompanied by lesser amounts of transgene expression (29). One line was found to express neither the transgene nor any of the endogenous HMW-glutenin genes (Fig. 5). The seeds were viable, and plants from these seeds developed apparently normally, showing that the presence of the HMW-glutenin subunits is not essential for seed viability and plant development.

In theory, the same method of suppression could be achieved with the gliadins. The difficulty is expected to come from the much larger number of gliadin genes and the closely related subfamilies of the gliadins. It is not known if the α-gliadin genes alone can be suppressed without also suppressing, either partially or completely, the γ- and ω-gliadins. Certainly, the complete lack of gliadins would be expected to have dramatically negative consequences on normal dough properties. It may still be possible to use such wheat to make some products, either from the gliadin-deficient dough itself, or from a mixture of such flours with flours from other cereals where the proteins could substitute to some degree for the gliadins.

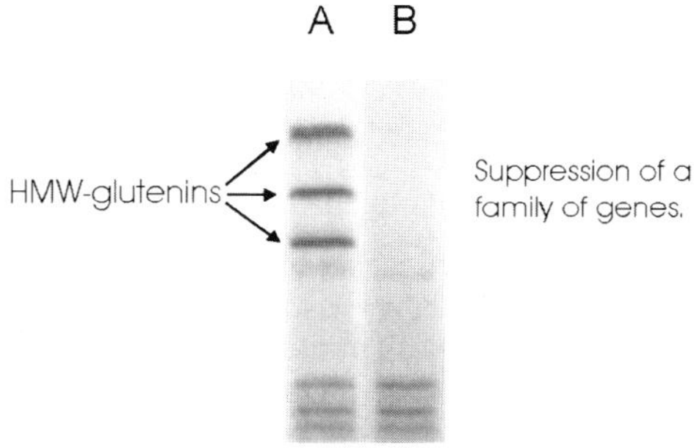

FIG. 5. Suppression of prolamin gene families. The complete suppression of a wheat prolamin gene family is exemplified by the HMW-glutenin genes. Shown are the portions of polyacrylamide gels displaying the gel region containing the wheat HMW-glutenin polypeptides (indicated by arrows). **(A)** Control wheat seed extract. **(B)** Extract from a transformed wheat line containing extra transgenic copies of one of the HMW-glutenin genes. HMW, high molecular weight. From Blechl AE, Anderson OD, unpublished data.

As previously mentioned, confounding factors in the strategies for eliminating α-gliadin gene expression by suppression are both the size of the gene family and the hexaploid nature of bread wheat. In a detailed examination of the wheat cultivar Cheyenne, it was estimated that there are 150 α-gliadin genes (30). Although the presence of a high percentage of pseudogenes seems to be a feature of the α-gliadin gene family, it was still estimated that there were 75 active α-gliadin genes. In addition, these genes are divided into the three homoeologous loci on the 6A, 6B, and 6D chromosomes. To effectively reduce α-gliadin content would require elimination of the activity of 75 genes (in this example) on three different chromosomes. Similar problems are presented by the other gliadin classes: The γ-gliadins and LMW-GS have approximately 20 to 40 genes each (31,32), and the ω-gliadins, 5 to 18 genes in durum and bread wheats (33). In addition to the large number of genes, the variable nature of the gene complements within different wheat lines makes it necessary to focus on the most conserved DNA sequences within a specific gliadin gene class. Wheat cultivars show a wide range of variability in gliadin composition, and the complex variation has been used to fingerprint individual cultivars (34).

Another strategy to eliminate the gliadins would be to use existing or newly developed wheat lines null for α-gliadin loci. There are several reports of wheat lines missing all or parts of the complex gliadin loci (35), including missing the entire locus on the 6A chromosome (36). Searches of existing wheat germplasm collections could uncover null lines for all three α-gliadin loci. These three null loci could then be combined in a single germplasm by a series of appropriate genetic crosses. If the α-gliadins were the main causative agent for celiac disease, then a wheat less toxic to patients with celiac disease might result. The advantage of this approach is that the resulting products would be produced by traditional breeding, thus avoiding controversy associated with the use of genetic engineering. The disadvantage is that each null locus would probably also be missing unknown neighboring portions of the genome surrounding the target prolamin loci. Such genetic deficiencies could have unpredictable negative effects on plant viability, crop yield, and flour quality. In addition, the transgenic approach has the significant advantage of being *trans*-dominant (affecting nonlinked genes), so that the single suppressing locus can be moved into a new genetic background and with the minimum number of back-crosses the desired parent variety could be reconstituted.

ENGINEERING NEW GLIADINS

A more ambitious approach to making wheat safe for patients with celiac disease is to construct entirely novel gliadin genes missing the offending peptide motifs. The assumption is that the core structural features that confer quality to doughs are not exactly the same as those causing celiac disease. The contributions of gliadins to dough functionality are as yet not well understood, but it is believed they are mediated largely through hydrogen bonding to the glutenin macropolymer. These interactions are thought to occur mainly among the repetitive domains, which also contain most, or perhaps all, of the toxicity for celiac disease. Because there is no reason to believe the celiac-causing peptides are important for dough quality, it should be possible to eliminate the suspect motifs while retaining dough functionality.

Figure 6 illustrates three approaches. For example, if it were determined that the PYP motif was part of the main causative peptides, then gliadin genes can be synthesized that maintain most of the original amino acid sequence but without the toxic motifs (Fig. 6A). A second possibility is to replace the entire repetitive domain with naturally occurring plant polypeptide domains having similar amino acid compositions to the gliadins but entirely different sequences (Fig. 6B). There would have to be some confidence that such replacement domains will not also trigger the celiac response. Finally, it is possible to synthesize completely novel genes instead

A) Direct a-gliadin modification

```
VRVPVPQLQPQNPSQQQPQEQVPLVQQQQFLGQQQPFPPQQPYPQPQPFPSQQPY
                                      Q   F       Q   F

LQLQPFPQPQLPYSQPQPFRPQQPYPQPQPQYSQPQQPISQQQQQQQQQQQQQ
        Q   FP          F

QQQILQQILQQQLIFCMDVVLQQHNIAHGRSQVLQQSTYQLLQELCCQHL
```

B) Substitution of non-gliadin high glutamine + proline domains

Oat domain :1 : QQQPILQQQQMLLQQQQQMLLQQQPLLQVLQQL

Oat domain :2: PFVQQQQPFVQQQQPFVQQQQMFLQPLLQQQL

C) Synthesis of novel repeat domains: HMW-glutenin example

```
PGQGQQ                  PGQGQQ GYYPTSPQQ
PGQGQQ GYYPTSPQQ        PGQGQQ GYYPTSPQQ
PGQWQQ                  PGQGQQ GYYPTSPQQ
PEQGQP RYYPTSPQQ        PGQGQQ GYYPTSPQQ
SGQLQQ                  PGQGQQ GYYPTSPQQ
PAQGQQ                  PGQGQQ GYYPTSPQQ
PGQGQQ GQQ              PGQGQQ GYYPTSPQQ
PGQGQP GYYPTSSQLQ       PGQGQQ GYYPTSPQQ
PGQLQQ                  PGQGQQ GYYPTSPQQ
PAQGQQ GQQ              PGQGQQ GYYPTSPQQ
PGQAQQ GQQ              PGQGQQ GYYPTSPQQ
PGQGQQ                  PGQGQQ GYYPTSPQQ
PGQGQQ GQQ              PGQGQQ GYYPTSPQQ
PGQGQQ                  PGQGQQ GYYPTSPQQ
PGQGQQ GQQ              PGQGQQ GYYPTSPQQ
LGQGQQ GYYPTSLQQ        PGQGQQ GYYPTSPQQ
   :
```

FIG. 6. Approaches to modifying α-gliadin repeat domains. Three examples are given for approaches to modifying the repeat domains of the α-gliadin genes and encoded proteins. **(A)** Substitution of individual amino acids to alter suspect polypeptide sequences. A section of an α-gliadin amino acid sequence is shown, with a region containing several celiac associated polypeptides underlined. Below that sequence are examples of amino acid changes not expected to alter this polypeptide's functionality. **(B)** Two oat avenin and one maize zein polypeptide domains are shown as examples of possible domains to substitute into the α-gliadins if such peptides can be shown to be innocuous for the celiac response. **(C)** Entirely synthetic repetitive domains are possible. Shown is a section of a completely synthetic repeat domain constructed for a synthetic HMW-glutenin gene (37). HMW, high molecular weight.

of replacing only portions of a prolamin gene. It has already been demonstrated that synthetic wheat prolamin genes, including long repetitive domains (Fig. 6C), can be constructed and will express new prolamin polypeptides (37). The identification of specific causative peptides would facilitate such engineering approaches, but it is likely that the complete replacement of the repetitive domains could accomplish the same purpose, even in the absence of detailed information on exactly which portion of the naturally occurring prolamin polypeptide sequences are most involved in the celiac response. It may ultimately be necessary to combine the above approaches by constructing new sets of gliadin genes and completely replacing the existing gliadin loci.

ENGINEERING OTHER CROPS TO REPLACE WHEAT IN FOOD PRODUCTS

Wheat is unique because of the properties of the viscoelastic dough resulting from wheat flour plus water mixtures (Fig. 1). Attempts to substitute other cereals—such as rice, maize, and oats—in similar products are as yet unsatisfactory. Can other plants (cereals and noncereals) be engineered to produce wheatlike products? There is no theoretical barrier to this, although the complexity of the engineering required is not clear. The wheat HMW-glutenin fraction is most critical for those viscoelastic properties unique to wheat doughs and responsible for the wide range and specific food products made from wheat. The HMW-glutenin genes are also attractive targets for genetic engineering, because the typical bread wheat cultivar only has five active HMW-glutenin genes. Engineering with such a small gene family is expected to be easier than with the much larger gliadin gene families. In the simplest experiment, HMW-glutenin genes can be transformed into nonwheat plants (cereals and noncereals), and effects on product flour processing and functionality tested. Wheat HMW-glutenin genes have been successfully transformed into barley and oats (Anderson OD, Blechl AE, and Somers DA, January, 1998, unpublished data) and the genes expressed normal levels of the wheat proteins in the new genetic background. Similar experiments are under way for maize and rice in several sites worldwide. Although functional testing of the resulting flours is yet to be reported, the expectation is that the properties of the resulting doughs will take on at least some of the characteristics of wheat doughs. Complete duplication of wheat dough functionality is not expected, as that functionality is the summation of interactions of HMW- and LMW-GS, gliadins, wheat starch granules, and other components. Nevertheless, even a modest improvement in functional properties provides opportunities for novel food products and some level of substitution for wheat. Longer term, a completely functional dough could be engineered with enough effort.

Assuming that such product development is feasible, it would also have to be determined if the HMW-glutenin enhanced non–wheat-based products are safe for patients with celiac disease. While the HMW-glutenins are also wheat prolamins, high in the amino acids proline and glutamine and possessing a long repetitive domain of short peptide motifs (Fig. 3), they are not considered a primary cause of the celiac condition, although there are reports of their involvement in mucosal T-cell responses (15). Thus even if the HMW-glutenins do not initiate the enteropathy, it must be confirmed that they do not make an ongoing contribution in sensitized patients through mechanisms such as epitope spreading (38). A combination of genetic engineering of novel HMW-glutenins in alternate cereals is a more complex option if necessary.

Reduction in celiac-related proteins in the diet may be helped by genetic engineering both to reduce the culprit peptides and to confer new functionality on cereals not currently suited to food products favored by humans. The specter of GMO (genetically manipulated organism) foods is currently a sensitive public issue, and it is expected that a combination of appropriate safeguards, better product labeling for both GMO and gluten composition, and greater public awareness of the details of the technology will lead to increasing acceptance of GMO-based foods in the human diet. While various societal elements are cautious about GMOs, many recognize the potential benefits, including such organizations as the Coeliac Society of Great Britain: "For coeliacs, genetic modification may not be all bad news—for it is possible that modification of wheat may, in time, result in a new product with all the excellent baking properties of wheat but without the harmful gluten content." (www.coeliac.co.uk)

RESEARCH NEEDS AND QUESTIONS

Accomplishing the objective of producing food products with the functionality, taste, and general consumer appeal, as found in traditional wheat-based products, while eliminating the celiac disease–eliciting peptides, would require or be facilitated by more detailed information in several areas.

(i) In-depth information is needed on seed proteins of all cereals; as the determination of DNA sequences is more efficient with today's technologies, information of protein sequences is more efficiently obtained through gene cloning and DNA sequencing. The most expeditious method is through EST (expressed sequence tag) sequencing. Such bulk DNA sequence data can be used as a basis for confirmatory protein chemistry analyses among all the cereals. EST programs are under way in all the major cereals, but no such programs are known for oats, rye, or the lesser cereals not considered here.

(ii) More information is needed on which peptides are related to celiac sensitivity. For example, what initiates the response? What peptides are or become reactive? How are levels of these peptides in foods related to the celiac response? What is the distribution of these peptides among proteins and cereals? Are there different reactive motifs or a single major motif?

(iii) Can genetic engineering replace or suppress the functioning of large multigene families such as the α-gliadins that are encoded by as many as 75 active genes in some cultivars (25)?

(iv) What is the effect on plant viability of eliminating the different prolamin groups from the seeds? Will seed development and germination proceed normally?

(v) What is the effect on product functionality of eliminating or replacing prolamin groups? Because the wheat prolamins are critical for dough function, can sufficient functionality be maintained after removing or modifying the prolamin content?

(vi) What are the effects of adding proteins now unique to wheat products to other crops, particularly other cereals not now associated with celiac disease? In other words, what is the functionality of the resulting flour and what are the effects on patients with celiac disease?

SUMMARY

There are probably no technical barriers to eliminating the offending peptide motifs eliciting the celiac response from the raw material used as the basis for cereal products. This elimination might be accomplished by searches for null alleles of the relevant prolamin genetic loci, or through genetic engineering. The latter approach also has the potential to confer functionality to flours of nonwheat cereals to allow production of food products not otherwise possible without the use of wheat flour. However, the lack of such technical barriers does not mean that such objectives will be attained in the foreseeable future. Significant information is still lacking in identification of the peptide triggers of the celiac response. This limits our ability to develop rational research and development directions. Second, the resources and time to accomplish such objectives would not be trivial. Thus, questions must be addressed as to the desirability of such approaches and the likelihood of sufficient commitment to their accomplishment.

REFERENCES

1. Trier J. Celiac sprue. *N Engl J Med* 1991; 325: 1709–19.
2. Bhattacharyya A, Patel MK, Tymms DJ. Coeliac disease in adults: variations on a theme. *J R Soc Med* 1999; 92: 286–9.
3. Maki MPC. Coeliac disease. *Lancet* 1997; 349: 1755–9.
4. Murray JA. The widening spectrum of celiac disease. *Am J Clin Nutr* 1999; 69: 354–65.
5. Osborne TB. *The vegetable proteins.* 2nd ed. London: Longmans Green, 1924.
6. Shewry PR. Plant storage proteins. *Biol Rev Camb Philos Soc* 1995; 70: 375–426
7. Kasarda DD. Grains in relation to celiac disease. *Cereal Foods World* 2001; 46: 209–10.
8. Hoffenberg EJ, Haas J, Drescher A, *et al.* A trial of oats in children with newly diagnosed celiac disease. *J Pediatr* 2000; 137: 361–6.
9. Ribes-Koninckx C, Alfonso P, Ortigosa L, *et al.* A beta-turn rich oats peptide as an antigen in an ELISA method for the screening of coeliac disease in a paediatric population. *Eur J Clin Invest* 2000; 30: 702–8.
10. Anderson RP, Degano P, Godkin AJ, Jewell DP, Hill AV. In vivo antigen challenge in celiac disease identifies a single transglutaminase-modified peptide as the dominant A-gliadin T-cell epitope. *Nat Med* 2000; 6: 337–42.
11. Arentz-Hansen EH, McAdam SN, Molberg O, Kristiansen C, Sollid LM. Production of a panel of recombinant gliadins for the characterisation of T cell reactivity in coeliac disease. *Gut* 2000; 46: 46–51.
12. Marsh MN, Morgan S, Ensari A, *et al.* In vivo activity of peptides 31–43, 44–56, 56–68 of α-gliadin in gluten sensitive enteropathy (GSE) [abstract]. *Gastroenterology* 1995; 108: A871.
13. Sjostrom H, Lundin KE, Molberg O, *et al.* Identification of a gliadin T-cell epitope in coeliac disease: general importance of gliadin deamidation for intestinal T-cell recognition. *Scand J Immunol* 1998; 48: 111–15.

14. Ensari A, Marsh MN, Moriarty KJ, Moore CM, Fido RJ, Tatham AS. Studies in vivo of omega-gliadins in gluten sensitivity (coeliac sprue disease). *Clin Sci (Colch)* 1998; 95: 419–24.
15. van de Wal Y, Kooy YM, van Veelen P, *et al.* Glutenin is involved in the gluten-driven mucosal T cell response. *Eur J Immunol* 1999; 29: 3133–9.
16. Arentz-Hansen H, Korner R, Molberg O, *et al.* The intestinal T cell response to alpha-gliadin in adult celiac disease is focused on a single deamidated glutamine targeted by tissue transglutaminase. *J Exp Med* 2000; 191: 603–12.
17. Molberg O, McAdams S N, Korner R, *et al.* Tissue transglutaminase selectively modifies gliadin peptides that are recognized by gut-derived T cells in celiac disease. *Nat Med* 1998; 4: 713–17.
18. van der Wal Y, Kooy Y, van Veelen P, *et al.* Selective deamidation by tissue transgluteminase strongly enhances gliadin-specific T cell reactivity. *J Immunol* 1998; 161: 1585–8.
19. MacRitchie F. Physiochemical properties of wheat proteins in relation to functionality. *Adv Food Nutr Res* 1992; 36: 1–87.
20. MacRitchie F. Wheat proteins: characterization and role in flour functionality. *Cereal Foods World* 1999; 44: 188–93.
21. Field JM, Shewry PR, Miflin BJ. Aggregation states of alcohol-soluble storage proteins of barley, rye, wheat and maize. *J Sci Food Agric* 1983; 34: 362–9.
22. Devos KM, Gale MD. Genome relationships: the grass model in current research. *Plant Cell* 2000; 12: 637–46.
23. Payne PI. Genetics of wheat storage proteins and the effect of allelic variation on breadmaking quality. *Annu Rev Plant Physiol* 1987; 38: 141–53.
24. Anderson OD, Hsia CC, Adalsteins AE, Lew EJ-L, Kasarda DD. Identification of several new classes of low-molecular-weight wheat gliadin-related proteins and genes. *Theor Appl Genet* 2001; 103: 307–15.
25. Anderson OD, Greene FC. The alpha-gliadin gene family. II. DNA and protein sequence variation, subfamily structure, and origins of pseudogenes. *Theor Appl Genet* 1997; 95: 59–65.
26. Warren ST. The expanding world of trinucleotide repeats. *Science* 1996; 271: 1374–5.
27. Matzke AJ, Matzke MA. Position effects and epigenetic silencing of plant transgenes. *Curr Opin Plant Biol* 1998; 1: 142–8.
28. Blechl AE, Anderson OD. Expression of a novel high-molecular-weight glutenin subunit gene in transgenic wheat. *Nat Biotechnol* 1996; 14: 875–9.
29. Blechl AE, Anderson OD, inventors. Glutenin genes and their uses. US patent 5,650,558, 1997.
30. Anderson OD, Litts JC, Greene FC. The alpha-gliadin gene family, I: characterization of ten new wheat alpha-gliadin genomic clones, evidence for limited sequence conservation of flanking DNA, and Southern analysis of the gene family. *Theor Appl Genet* 1997; 95: 50–8.
31. Cassidy BG, Dvorak J, Anderson OD. The wheat low-molecular-weight glutenin genes: characterization of six new genes and progress in understanding gene family structure. *Theor Appl Genet* 1998; 96: 743–50.
32. Sabelli PA, Shewry PR. Characterization and organization of gene families at the *Gli-1* loci of bread and durum wheats by restriction fragment analysis. *Theor Appl Genet* 1991; 83: 209–16.
33. Tatham AS, Shewry PR. The S-poor prolamins of wheat, barley and rye. *J Cereal Sci* 1995; 22: 1–16.
34. Metakovsky EV, Gomez M, Vazquez JF, Carrillo JM. High genetic diversity of Spanish common wheats as judged from gliadin alleles. *Plant Breeding* 2000; 119: 37–42.
35. D'Ovidio R, Lafiandra D, Tanzarella OA, Anderson OD, Greene FC. Molecular characterization of bread wheat mutants lacking the entire cluster of chromosome 6A controlled gliadin components. *J Cereal Sci* 1991; 14: 125–9.
36. Lafiandra D, Colaprico G, Kasarda DD, Porceddu E. Null alleles for gliadin blocks in bread and durum wheat cultivars. *Theor Appl Genet* 1987; 74: 610–16.
37. Anderson OD, Kuhl JC, Tam A. Construction and expression of a synthetic wheat storage protein gene. *Gene* 1996; 174: 51–8.
38. Lehmann PV, Sercarz FE, Forsthuber T, Dayan CM, Gammon G. Determinant spreading and the dynamics of the autoimmune T-cell repertoire. *Immunol Today* 1993; 14: 203–8.

Genetic Expression and Nutrition, edited by Claude Bachmann and Berthold Koletzko. Nestlé Nutrition Workshop Series, Pediatric Program, Vol. 50. Nestec Ltd., Vevey/Lippincott Williams & Wilkins, Philadelphia, © 2003.

Current Uses and Future Perspectives of Recombinant Modifications in Food Production

Dennis M. Bier

Children's Nutrition Research Center, and Department of Pediatrics, Baylor College of Medicine, Houston, Texas, USA

Agriculture is the world's oldest and largest industry. For more than 10,000 years, humans have been using biologic technologies to select and improve desired characteristics in plants and animals for food and other uses. In fact, the biotechnology of vine propagation by grafting, nearly three millennia old, is responsible for our word "cloning," which is derived from the Greek word for twig.

The first genetically modified foods were the results of successful natural wild species crosses or were produced by active selective breeding, the form of genetic engineering then available to farmers. These manipulations led not only to the introduction of different genes, but also to the assimilation of entire chromosomes; several crop species—including wheat, potatoes, strawberries, bananas, sugar cane, and coffee—are polyploids.

The successes led eventually to transformation of the wild precursors of maize, wheat, barley, legumes, and numerous other plants into today's staple crop foods, which bear little resemblance to their wild ancestors (1). Table 1 shows some of the ways that selective breeding has changed crop plants. The tomato is a good example of traditional genetic engineering. Today's varieties contain the genes of nine *Lycopersicon* species, the results of many genetic manipulations including distant crosses to introduce pest- and disease-resistance and the selection of beneficial irradiated mutants (2).

Perhaps the best recent example of the power of traditional manipulations of the plant genome is the complete removal of the detrimental fatty acid, erucic acid, from rapeseed to produce canola. This was recognized in 1985 as a new species, producing edible oil in which erucic acid has been replaced by oleic acid (Table 2). Similarly, biotechnologies using microorganisms have been used since at least 6000 BC for the fermentation of beer and for other fermentation products such as leavened bread, cheese, and yogurt, which date to at least 2000 BC, while the techniques of marination, pickling, and curing have been known for nearly three millennia.

The success of traditional agricultural practices has been extraordinary and last century's dramatic increase in the human population could not have been sustained

TABLE 1. *How plants change with domestication*

Characteristic	Example
Loss of dispersal mechanisms	Corn, wheat
Loss of dormancy	Wheat, oats, rice
Perennial conversion to annual	Rice, rye, cassava
Loss of fruit production	Yam, sweet potato
Loss of seed production	Banana
Increase in size of seed	Bean
Increase in size of fruit	Squash
Increase in storage organ	Carrot, cassava

without the increased crop yields resulting from highly successful selective breeding programs (3). These started with the commercialization of hybrid maize in the first half of the last century, when the world's population was less than 2 billion, and continued with the "Green Revolution" using semidwarf rice in the second half. In 1960, the United States' food, feed, and fiber crop production was 252 million tons, a value that increased to 700 million tons in 1999, even though it was produced on 10 million fewer acres of land (4). Similarly, the introduction of hybrid wheat into India and Pakistan increased each country's yields from 12.3 and 4.6 million tons, respectively, in 1965 to 73.5 and 21 million tons, respectively, in 2000 (4). Nonetheless, the world's population now stands at more than 6 billion people and is expected to continue to increase substantially, at least into the first half of the present century.

Although it is abundantly clear that no one can predict future population increases with any accuracy or precision (5), there is also little doubt that a net increase in food as the result of augmented crop production or decreased crop losses will be needed to feed the world adequately (3). Furthermore, it is not necessary to quibble about the effects of uncertainties in estimating future population growth on global food demands, when nutritional inadequacy is a current problem of immense magnitude. Various international agencies estimate that 150 to 180 million children younger than the age of 5 have protein-energy malnutrition, and 10 million of these die annually. Vitamin A deficiency affects 100 to 140 million children, 250,000 to 500,000 of whom go blind annually. Iron deficiency affects about 3.5 billion people, 1.5 billion of whom are frankly anemic. Iodine deficiency is present in about 50 million individuals and the number of persons who are deficient in other vitamins, minerals, and micronutrients is unknown.

TABLE 2. *Composition of rapeseed and canola oils*

Fatty acid	Rapeseed	Canola
16:0	1.7	4.0
18:0	0.9	1.4
18:1	12.3	56.9
18:2	12.7	23.8
18:3	7.6	10.1
22:1	59.4	0.0
24:0	1.6	0.3

Nevertheless, in attempting to increase crop yields to enhance existing food resources and to meet future demands, traditional agricultural systems are now under significant pressure. Crops grow on about 12% of the world's total land area, and the vast bulk of high-quality cropland is now under cultivation (6). In 1961, 0.44 hectares *per capita* of cultivated land supported food production, a figure that declined to 0.26 hectares *per capita* in 1997 and is projected to decline even further to 0.15 hectares *per capita* in 2050 (6). Population increases ensure continued losses of arable lands to other human uses such as housing, commerce, and transportation. In addition, topsoil losses are estimated to be about 25 billion tons annually, equivalent to the entire wheat-growing area of Australia. Salinization affects about 10% of the world's irrigated land and the potential detrimental effects of increased greenhouse gases are substantial (6).

Thus, to avoid disastrous consequences to the food supply, all traditional agricultural approaches to ensure sustainability must be enhanced. These include energy-efficient farming, combating soil erosion, expanding efficient irrigation while improving water conservation, maintaining biodiversity, improving pest control while simultaneously reducing dependency on pesticides and herbicides, and developing new, pest-resistant crop strains with enhanced yields, as well as salt, drought, and temperature tolerance (3,6). Nevertheless, newer technologic approaches to enhancing agricultural productivity are necessary as well. Recombinant molecular genetic methods for introducing agronomically and nutritionally desirable genes into crop plants is one of the available tools (7).

RECOMBINANT BIOTECHNOLOGY

Newer recombinant approaches differ from traditional biotechnologies, which operate relatively slowly across a narrow range of closely related species, because recombination methods allow a relatively fast exchange of genes from distant species. Nonetheless, both the Food and Agricultural Organization/World Health Organization (8) and the National Research Council of the US National Academy of Sciences (9) agree that recombinant approaches represent a "continuum" (8) in the long history of agricultural biotechnologies and reflect "no conceptual distinction. . .(as) the same physical and biological laws govern the response of organisms modified by modern molecular and cellular methods and those produced by classical methods." (9) While this chapter deals principally with the nutritionally related aspects of agricultural biotechnology, it is important to realize that the bounds of this tool's applications extend far beyond the sphere of crop foods, as shown in Table 3 (10).

In the two decades since recombinant plants were first produced experimentally, progress has been dramatic. More than 100 plant species have been transformed, and Table 4 shows the recombinant biotechnology crops available by the beginning of this century. In 2000, 44.2 million hectares (109.2 million acres) of land were planted with transgenic crops (11). This area is approximately equal to the land area of Iraq and represents an overall 25-fold increase since 1996 and an 11% increase within the last year (11).

TABLE 3. *Potential contributions of recombinant biotechnologies to agriculture*

Advance fundamental plant science	• Determine biochemical and genetic control mechanisms • Understand the molecular basis of plant growth and development • Enhance knowledge of plant interactions with microbes, animals, and environments • Delineate the structure of agriculturally relevant microbial systems
Spare forest, wilderness and marginal lands	• Produce more food from the same or less land area
Displace resource and energy intensive inputs	• Reduce the use of fuel, fertilizers, pesticides • Alternate alternative fuel sources
Reduce environmentally damaging practices	• Allow conservation tillage and consequent topsoil preservation • Biosensor for environmental monitoring
Advance restoration ecology	• Reestablish structure, function, and nutrient balance of altered ecosystems
Bioremediation removal of toxic materials and other unwanted products in soils	• Phytostabilization to eliminate bioavailability of toxic metals in soils • Phytoextraction of toxic metals from soils • Rhizofiltration of metals from polluted waters via plant root systems • Microbial degradation of pollutants • Biosensors for biomediation monitoring
Expand food safety assurance methods	• Testing for chemical and biologic contaminants
New product development	• Biopolymers, pharmaceuticals, biocatalysts, biomimetics • Tissue engineering
Advance aquaculture and marine biology	• Basic biology of marine plants and ecologic systems • Marine crop plants and novel bioactive compounds • Aquatic pollution remediation and waste disposal

Adapted from reference 10.

TABLE 4. *Biotech crops 2000*

Canola	Potato
Cantaloupe	Rice
Chicory	Soy
Maize	Squash
Cotton	Sugar beets
Flax	Tobacco
Papaya	Tomato

TABLE 5. *Land area planted with transgenic crops*

Year	Hectares (million)	Acres (million)	Equivalent area (country)
1996	2.8	7.0	Switzerland
1997	12.8	31.5	Greece
1998	27.8	69.5	United Kingdom
1999	39.9	98.6	Japan
2000	44.2	109.0	Iraq

From reference 11.

TABLE 6. *Potential risks of transgenic crops*

Invasiveness	• Crop to weed gene transfer through introgression
Horizontal gene transfer to viruses/bacteria	• Marker gene transfer to intestinal bacteria
	• Novel viral diseases as a consequence of new new recombination events
Emergence of resistant "super plants"	—
Destruction of ecological systems	• Nontarget effects on other species that depend directly or indirectly on the transgenic plant or the controlled pest • Loss of biodiversity
Toxic secondary metabolites and/or marker genes	—
New allergens in old foods	—
Old allergens in new foods	—
Variable and unexpected results	• Consequences of poorly understood exosystem complexity

Approximately 85% of transgenic crops are grown in eight industrialized countries, but the proportion grown in five developing countries has increased steadily from 14% in 1997 to 24% in 2000 (11). In fact, in 2000, the area planted with transgenic crops in the developing world increased by 51% from the previous year, while the corresponding increase in industrialized countries was only 2% (11). Sixty-eight percent of transgenic crops are grown in the United States, 23% in Argentina, 7% in Canada, 1% in China, and 1% in the remaining nine countries (11).

Four crops account for virtually all the transgenic crops cultivated. Transgenic soybean occupies 58%, maize 23%, cotton 12%, and canola 6% of the recombinant crops planted, with herbicide tolerance genes accounting for 74% of the introduced traits, insect resistance for 19%, and 7% of crops transformed with both traits (11). In 2000, 36% of all soybeans planted were transgenic, as was 16% of all cotton, 11% of all canola, and 7% of all maize (11). Globally, these four crops are planted in 271 million hectares of land, with 16% of the area now occupied by transgenic crops (Table 5). However, two thirds of the 271 million hectares are located in "developing countries where yields are lower, constraints are greater, and the need for improved production of food, feed, and fiber crops is the greatest" (11).

Nonetheless, underlying concerns have surfaced about the practical and theoretical risks of recombinant plants. In the first case, these center principally on potential threats to human health, agriculture, and the ecology (Table 6). In the second, they focus more on fundamental issues regarding the nature of genes, the nature of species, natural/unnatural boundaries, and the ethical, societal, political, and economic implications.

ISSUES OF HUMAN, AGRICULTURAL, AND ECOLOGIC SAFETY

In the context of human safety, it is important to realize that humans have not only been eating plant DNA throughout their evolutionary history, but also the DNA sequences of innumerable viruses and other pathogens contained in the plants—for example the cauliflower mosaic virus DNA that is ingested whenever broccoli is eaten. Crude estimates suggest that humans consume more than 100,000 genes daily. Most DNA is rapidly digested in the gastrointestinal tract, and there is no evidence that ingested DNA is transferred to enterocytes or gut bacteria. Furthermore, while there has been some concern about the consequences of ingesting the antibiotic marker genes used in early molecular recombination approaches, this health concern will ultimately become irrelevant as the use of these markers has been, or is being, eliminated.

The potential for new antigens (or for the appearance of known antigens in foods previously free of them) is certainly both a theoretical and an established risk. Nonetheless, although more than 160 foods are known to cause sporadic allergic reactions, eight foods (peanuts, soybeans, milk, eggs, fish, crustacea, wheat, and tree nuts) account for more than 90% of all food allergies (12). There are known chemical and biologic characteristics of food antigens, including stability to protease digestion (13–15), and algorithms exist for the approaches to identifying these before crop foods are marketed (12–14). A highly publicized instance of detection of a Brazil nut antigen in transgenic soybeans illustrated the successful application of decision tree algorithms for antigen detection and control. The testing resulted in discontinuation of the transgenic crop program before commercialization (16). Finally, what is often not also emphasized in discussions of the antigen risks associated with recombinant foods is that the available technologies have also both the potential for inserting genes for antigenically beneficial effects (as discussed in detail later) and for *removing* known antigens from traditional foods containing them. In this latter context, Tada *et al.* reduced the antigen content of rice using antisense methods (17), and others have developed approaches to eliminating antigens in peanuts (18) and wheat (19).

Ecologic risks (Table 6) remain a debatable concern, with knowledgeable experts taking both sides—which usually means that the outcome lies between the positions of the opposed parties. Wolfenberger and Phifer have recently published a succinct assessment of the issues. They concluded that "(n)either the risks nor the benefits of (genetically engineered organisms) are certain or universal," and "key experiments on both the environmental risks and benefits are lacking"(20). Similar assessments by others have generally supported their balanced assessment (6,21–26), including the report by the Nuffield Council on Bioethics (26), which also addressed in thoughtful detail the unresolved complexities of ethical issues surrounding experiments that may have environmental effects.

GENES, GENOMES, SPECIES, AND MOTHER NATURE

Other areas of concern about recombinant biotechnology center around more fundamental issues of genes, biology, nature, and society (Table 7). Although the religious, social, and economic issues are outside the scope of this review, the basic genetic issues are more appropriately addressed here.

Over the course of the last 50 years, the most widely held, fundamental genetic "truths" have been found to be less than axiomatic. These include former "facts" about genome size and species complexity, genome uniqueness and species speci-

TABLE 7. *Biotechnology assaults on fundamental beliefs*

Against nature
Transgresses species barriers
Violates genome uniqueness or "purity"
Contaminates the "natural foods" given to us by Mother Nature
Shortcircuits natural historical process of evolutionary adaptation
Breaks religious or moral taboos
Agribusiness as an evil force
Elimination of small farmer
Overturns traditional social and political systems
Opposition to labeling and the right to know

ficity, absence of horizontal gene transfers, lack of nonrandom mutations, equality of genes transmitted from either parent, and the absence of epigenetic effects on gene expression. What has emerged is an understanding that DNA approaches a single gene pool, with remarkable sequence conservation and gene similarity across phylogenetic kingdoms, including conserved sequences from plants to man (27). DNA sequences are now known to have been far more mobile than formerly appreciated. Microbial gene transfers through transformation, conjugation, and transfection are common (28). The ubiquitous soil bacterium, *Agrobacterium tumefaciens*, routinely transfers some of its own genes into host plant cells. In fact it is this natural property that forms the basis for the use of *Agrobacterium* to achieve recombinant transformation of plant cells, a method that is the foundation of modern biotechnology. "Similarities between the genomes of some plant-infecting, vertebrate-infecting, invertebrate-infecting, and microbe-infecting viruses indicate that they have common ancestors . . . " (29), and the "genomes of modern humans are riddled with thousands of endogenous retroviruses . . . , the proviral remnants of ancient viral infections of the primate lineage" (30). *Drosophila mariner*–transposable DNA elements are widely distributed among insect and other invertebrate genomes, but are also identified in vertebrate genomes, including those of prosimians, monkeys, great apes, and man, whose DNA contains 109 *mariner*-like elements (31–34). In fact, only about 1.5% of human genomic DNA codes for human proteins, provoking the disturbing realization that, at the DNA level, we are no longer the center of the universe. In fact, we are no longer even the "center" of our own complement of DNA.

Traditional species boundary issues are also more complex than common intuitive grouping tendencies and classical taxonomy approaches. For animals, reproductive isolation is a reasonable but not infallible criterion. In botany, however,

> The production of polyploids creates individuals that are usually unable to produce fertile offspring when crossed with the diploid plants that gave rise to them. . . Since the 'parents' and their offspring are reproductively isolated, they would be considered different species using the common zoological definition. . . . A second problem with the use of reproductive barriers as the criterion for defining a plant species is the ease with which plants hybridize, making the phrase 'reproductively isolated' difficult to apply (35).

Many scientists have concluded that there is no single definition of species that is not subject to refutation.

Addressing the "natural" and "unnatural" boundary is equally problematic. Agriculture is a form of accelerated evolution, and humans are the "worlds greatest evolutionary force" (36). But traditional agriculture, perhaps considered the most "natural" of human occupational activities, is, from two perspectives, "unnatural." First, it requires human intervention—that is, it is not what happens when Mother Nature is left to her own devices. Second, it runs counter to natural selection, as natural selection operates for the organism's own benefit but agriculture functions for man's benefit. From a different vantage point, however, by viewing man as the pinnacle of nature, both traditional agriculture and recombinant agricultural biotechnology are perfectly natural. In this context, there is a "duty laid on humanity to 'cultivate and reorder nature.' God's gift is a grant of sweeping authority to use the raw materials of nature wisely, i.e. the stewardship principle. Indeed it would represent ingratitude for God's bounty to neglect the materials placed before us" (26). Thus

> The 'natural/unnatural' distinction is one of which few practicing scientists can make much sense. Whatever occurs. . .can, in principle, be explained in terms of natural science. When human abilities to transform the world are limited, the distinction between nature and artifice seems clear: is a plant acceptably natural . . . if it has been successfully bred to have a particular gene complement, but unnatural . . . if precisely the same gene complement has been arrived at through laboratory processes? We can see no reason in ethics to draw a distinction" (26).

Enhanced concern about the dangers of recombinant plants, based on the safety of traditional agricultural products because of long human evolutionary adaptation to their consumption, also does not survive scrutiny. Humans in most regions of the world now consume crops that did not originate in those regions. Thus, for example, only about 37% of the crops now consumed in Japan, 44% of those in Latin America, 2% of those in the Mediterranean basin, and essentially none of those in Australia, the United States, and Canada had their origins in these respective regions (37). While Europeans worry about the current second American agricultural biotechnology invasion, they tend to forget the first. Thus, many of the foods now considered fundamental components of various European cuisines were not present in Europe until they were brought there following the discovery of the Americas. Before that time

> You must forget the tomato–difficult in the case of Italian cooking–and you must eliminate the potato, that welcome refuge of French housewives, because it was only lately that these products came to the European continent from the Americas. For the same reason, you must banish peppers . . . to which Europe has grown used to only through a relatively recent acquaintance. . . . Your polenta will not be made of corn, and you will have no coffee to sip after dessert (38).

In his book, *Columbus Menu* (39), Milioni reminds us that "anyone . . . who has come into contact with Italy's gastronomy is convinced that the tomato is a fundamental and inalienable element of it. In reality, the tomato's entry into the Italian pantry occurred relatively recently. . . ." In fact, the first mention of the tomato in an Italian cookbook was in 1692 in Latini's *The Modern Steward*, but the first tomato

TABLE 8. *More food is produced*

Disease and pest resistance	• Insects, viruses, fungi
Herbicide resistance	—
Insecticide resistance	—
Enhanced nitrogen fixation	—
Environmental stress tolerance	• Heat, cold, drought, salt

recipe in an Italian cookbook, *Il Cuoco Galante*, were instructions for preparing Pomodori Alla Certosina by Corrado in 1781 (39). It was not until 1839 that a recipe for pasta with tomato sauce first appeared in Covalcanti's cookbook, *La Cucina Teorico-Practica*, within an appendix entitled "Cucina Casareccia in Dialetto Napoletano" (39). On the evolutionary scale, several hundred years is woefully insufficient for adaptation to the sudden introduction of the entire tomato genome into European diets.

RECOMBINANT FOODS AND THE NUTRITION OF CHILDREN

Modern agricultural biotechnology can lead to improved childhood nutrition in at least four ways: agronomic properties that allow the production of more food; agronomic properties that permit more food to reach the child after harvest; "intrinsic fortification" of foods for nutrient enhancement—that is, "the food is healthier"; and enhancement of food properties that maintain the child's health and prevent disease—that is, "the child is healthier."

Table 8 shows the agronomic traits that have been introduced into foods to allow increased crop yields by resisting pathogens (7,40–46) and improving responses to adverse environmental conditions (47). Table 9 shows approaches to permit more crops to reach the child after harvest by controlling ripening (48), reducing senescence (49), improving taste (50), and improving functional characteristics of the food (51). Critics of recombinant biotechnology are quick to point out that these traits are, and will be, profitable to the companies that produce the recombinant plants. It is disingenuous, however, to ignore the fact that financial and crop losses that are caused by the crop diseases and pests themselves both reduce the available food and increase its cost. Thus, for example, the financial losses resulting from rice gall midge, corn rootworm, and the European corn borer together amount to far more than $1 billion annually in the United States alone, and marketable yield losses of casava, potato, and other roots and tubers from various plant viruses can be as great as 50% to 80% (52).

TABLE 9. *More food reaches the child*

- Control of ripening
- Delay senescence
- Improve flavor and sweetness
- Improve functional characteristics

TABLE 10. *The food is healthier*

Increased energy density	
Modification of plant storage lipids	• Increase essential fatty acids • Increased long-chain polyunsaturated fatty acids • Decreased saturated fatty acids • Decreased *trans* fatty acids
Modification of seed proteins and amino acids	• Lysine in grains; methionine in legumes
Manipulation of carbohydrate partitioning	—
Enhancement of vitamins and micronutrients	• Increased carotenoids, antioxidants • Introduce vitamin B_{12} into plant products • Increase essential mineral content
Enhancement of disease-limiting compounds	• Introduce or enhance biologically active phytochemicals
Removal of unwanted proteins/allergens	—
Elimination of bioavailability inhibitors	• Chelators, phytates, other sequestrants
Reduction of other antinutritive compounds	—

Commercial transgenic plants that specifically enhance the nutrient content of foods have not yet reached human testing or marketing, although it is precisely these improved quality traits that will help convince nutrition scientists, consumers in both developed and developing countries, and public policy makers that recombinant foods have direct nutritional benefit. Nonetheless, many transgenic plants with improved nutritional qualities are in various stages of development, as shown in Table 10. These include nutritionally beneficial modifications of vegetable oils (53–57), tocopherol enhancement of oil seeds (58,59), increased provitamin A content of tomatoes (60), protein enhancement of potatoes (61), lysine enhancement of rice (62), canola, and soybeans (63), methionine-rich protein enhancement of rice (64), augmentation of essential amino acids in sweet potatoes (65), and enhancement of iron content in rice by reduction of phytate (66). Perhaps the most widely publicized recombinant food nutrient enhancement was the creation of "golden" rice by Potrykus and his colleagues (67–69). In a remarkable feat, these workers inserted the genes for seven enzymes into rice: four enzymes were required to allow rice to convert geranylgeranylpyrophosphate into carotene and three were inserted to enhance iron absorption by introducing the ferritin gene, a gene that codes for a metallothionein protein high in cysteine, and the phytase gene, to diminish the inhibitory effects of phytate on iron absorption. Despite this Herculean effort and the accompanying enthusiasm, to my knowledge there

TABLE 11. *The child is healthier*

Active immunization	• Plant delivery of vaccine antigens
Passive immunization	• Plant antibody storage • Milk antibody delivery
Induction of immune tolerance	• Autoimmune diseases; allergy immunotherapy
Production/delivery of defensive proteins	• Milk transgenic lactoferrin, lysozyme, defensins

TABLE 12. *Vaccine antigens expressed in plants*

Antigen	Plant
Hepatitis B surface antigen	Tobacco
Escherichia coli heat labile enterotoxin B subunit	Potato
Vibrio cholerae toxin B subunit	Potato
Rabies virus glycoprotein	Tomato
Norwalk virus capsid protein	Potato
Malaria epitopes/TMV capsid protein	Tobacco
Human rhinovirus/cowpea mosaic virus	Cowpea
HIV-1/cowpea mosaic protein	Cowpea

has yet not been a single bioavailability study testing this transgenic product in animals or man. Dr. Marion Nestle has concluded that golden rice will fail to correct vitamin A deficiency (70), although an alternative analysis suggests that it will be beneficial to some degree in selected populations (71).

Table 11 lists the various ways in which transgenic plants can improve children's health and consequently their nutritional status. Particularly exciting is the use of plants as delivery systems for vaccine antigens (72–74), and various different antigens important for preventing significant human diseases have now been expressed in plants (75–80) (Table 12). Important progress has been made in this field since "proof of principle" experiments have been conducted successfully both in mice and in humans. In the mouse study, animals immunized with transgenic cholera toxin B subunit transgenic potatoes developed protective serum and intestinal mucosal antibody titers against the cholera enterotoxin, inhibited toxin ganglioside binding, and showed a reduction in toxin-induced diarrhea (77). Excitingly, human volunteers also developed appropriate antibody responses to transgenic potatoes containing heat labile *Escherichia coli* enterotoxin (78) and to transgenic potatoes containing Norwalk virus capsid protein assembled into viruslike particles (80). Finally, it is important to emphasize the role that transgenic plants might play in the delivery of dietary antigens to induce oral tolerance (81), as two dramatic "proof of principle" experiments have already shown that it is possible to prevent autoimmune diabetes in mice (82,83).

CONCLUSIONS

Recombinant agricultural biotechnology is a scientific advance of immense magnitude. Historically, revolutionary scientific advances of this nature have not been turned back. The current magnitude of, and continued rise in, land area planted with transgenic crops suggests that this technology is here to stay. To date, despite theories and contentions, there has been little evidence of actual harm, and genuinely debatable evidence of potential harm. Agronomically, there is early practical evidence of benefit. In nutrition there is considerable expectation of future benefit.

At this critical juncture, we must avoid the "Catch 22" of insisting that no agricultural biotechnology be used until it is safe, because verification of the risk/benefit ratio requires data. In other words, one cannot satisfy the safety questions without experiments to determine the level of risk. Thus, because most plant scientists estimate that the risks are readily manageable in the context of traditional agricultural experiments of the past, competent scientists should be allowed to test legitimate hypotheses—that is, to provide the data to answer hypotheses currently advocated by individuals on both sides of the recombinant biotechnology controversy.

REFERENCES

1. Lev-Yadun S, Gopher A, Abbo S. The cradle of agriculture. *Science* 2000; 288: 1602–3.
2. Institute of Food Technologists Expert Report on Biotechnology and Foods. Introduction. *Food Technology* 2000; 54: 124–36.
3. Conway G. *The doubly green revolution*. Ithaca: Cornell University Press, 1997.
4. Borlaug N. We need biotech to feed the world. *Wall Street Journal*. December 6, 2000: A.
5. Cohen JE. *How many people can the earth support?* New York: WW Norton, 1995.
6. Kendall HW, Beachy R, Eisner T, Gould F, Herdt R, Raven PH. *Bioengineering of crops*. Report of the World Bank panel on transgenic crops. Washington, DC: The World Bank, 1997: 4–8.
7. Walden R. Cell culture, transformation and gene technology. In: Lea PJ, Leegood RC, eds. *Plant biochemistry and molecular biology*. Chichester: John Wiley & Sons, 1999: 335–55.
8. FAO/WHO. *Strategies for assessing the safety of foods produced by biotechnology*. Report of a joint FAO/WHO consultation. Geneva: Food and Agricultural Organization/World Health Organization; 1991.
9. National Research Council. *Field testing genetically modified organisms: framework for decisions*. Washington, DC: National Academy Press, 1989.
10. Committee on Fundamental Science. *Biotechnology for the 21st century*. Washington, DC: National Science and Technology Council; 1995.
11. James C. *Global status of transgenic crops 2000*. Ithaca: International Service for the Acquisition of Agri-biotech Applications; 2001. Brief 21.
12. FAO/WHO. *Evaluation of allergenicity of genetically modified foods*. Report of a joint FAO/WHO expert consultation on allergenicity of foods derived from biotechnology. Rome: Food and Agriculture Organization of the United Nations; 2001.
13. Institute of Food Technologists Expert Report on Biotechnology and Foods. Human food safety evaluation of rDNA biotechnology-derived foods. *Food Technol* 2000; 54: 53–61.
14. Fuchs RL, Astwood JD. Allergenicity assessment of foods derived from genetically modified plants. *Food Technol* 1996; 50: 83–8.
15. Astwood JD, Leach JN, Fuchs RL. Stability of food antigens to digestion in vitro. *Nat Biotechnol* 1996; 14: 1269–73.
16. Nordlee JA, Taylor SL, Townsend JA, Thomas LA, Bush RK. Identification of a brazil-nut antigen in transgenic soybeans. *N Engl J Med* 1996; 334: 688–92.
17. Tada Y, Nakase M, Adachi T, *et al*. Reduction of 14–15 kDa allergenic proteins in transgenic rice plants by antisense gene. *FEBS Lett* 1996; 391: 341–5.
18. Bannon GA, Shin D, Maleki S, Kopper R, Burks AW. Tertiary structure and biophysical properties of a major peanut antigen, implications for production of hypoallergenic protein. *Int Arch Allergy Immunol* 1999; 118: 315–16.
19. Buchanan BB, Adamidi C, Lozano RM, *et al*. Thioredoxin-linked mitigation of allergic response to wheat. *Proc Natl Acad Sci USA* 1997; 94: 5372–7.
20. Wolfenberger LL, Phifer PR. The ecological risks and benefits of genetically engineered plants. *Science* 2000; 290: 2088–93.
21. Crawley MJ, Brown SL, Hails RS, Kohn DD, Rees M. Transgenic crops in natural habitats. *Nature* 2001; 409: 682–3.
22. Butler D, Reichardt T. Long-term effect of GM crops serves up food for thought. *Nature*1999; 398: 651–6.
23. Firbank LG, Forcella F. Genetically modified crops and farmland biodiversity. *Science* 2000; 289: 1481–2.
24. Institute of Food Technologists Expert Report on Biotechnology and Foods. Benefits and concerns associated with recombinant DNA biotechnology-derived foods. *Food Technol* 2000; 54: 61–80.
25. Brown K. Seeds of concern. *Sci Am* 2001; 284: 52–7.

26. Nuffield Council on Bioethics. *Genetically modified crops: the social and ethical issues*. London: Nuffield Council on Bioethics, 1999.
27. Potocki L, Chen K-S, Lupski J. Subunit 3 of the COP9 signal transduction complex is conserved from plants to humans and maps within the Smith-Magenis syndrome critical region in 17p11.2. *Genomics* 1999; 57: 180–2.
28. Ochman H, Lawrence JG, Groisman EA. Lateral gene transfer and the nature of bacterial innovation. *Nature* 2000; 405: 299–304.
29. Gibbs MJ, Weiller GF. Evidence that a plant virus switched hosts to infect a vertebrate then recombined with a vertebrate-infecting virus. *Proc Natl Acad Sci USA* 1999; 96: 8022–7.
30. Johnson WE, Coffin JM. Constructing primate phylogenies from ancient retroviral sequences. *Proc Natl Acad Sci USA* 1999; 96: 10254–60.
31. Lohe AR, Moriyama EN, Lidholm D-A, Hartl DL. Horizontal transmission, vertical inactivation, and stochastic loss of *mariner*-like transposable elements. *Mol Biol Evol* 1995; 12: 62–72.
32. Robertson HM, Zampano KL. Molecular evolution of an ancient mariner transposon, Hsmar1, in the human genome. *Gene* 1997; 205: 203–17.
33. Robertson HM, Martos R. Molecular evolution of the second ancient human mariner transposon, Hsmar2, illustrates patterns of neutral evolution in the human genome lineage. *Gene* 1997; 205: 219–28.
34. Reiter LT, Liehr T, Rautenstrauss B, Robertson HM, Lupski JR. Location of mariner DNA transposons in the human genome by PRINS. *Genome Res* 1999; 9: 839–43.
35. Simpson BB, Ogorzaly MC. *Economic botany: plants in our world*. New York: McGraw-Hill, 1995: 55.
36. Palumbi SR. Humans as the world's greatest evolutionary force. *Science* 2001; 293: 1786–90.
37. Crispeels MJ, Sadava DE. *Plants, genes, and agriculture*. Boston: Jones and Bartlett, 1994: 318.
38. Redon O, Sabban F, Serventi S. *The medieval kitchen*. Chicago: University of Chicago Press, 1998: xiii.
39. Milioni S. *Columbus menu*. Rome: Italian Trade Commission, 1992: 11–18.
40. Fitchen JH, Beachy RN. Genetically engineered protection against viruses in transgenic plants. *Annu Rev Microbiol* 1993; 47: 739–63.
41. Shah DM, Rommens CMT, Beachy RN. Resistance to diseases and insects in transgenic plants: progress and applications to agriculture. *Trends Biotechnol* 1995; 13: 362–8.
42. Rommens CM, Kishore GM. Exploiting the full potential of disease-resistance genes for agricultural use. *Curr Opin Biotechnol* 2000; 11: 120–5.
43. Kramer KJ, Morgan TD, Throne JE, Dowell FE, Bailey M, Howard JA. Transgenic avidin maize is resistant to storage insect pests. *Nat Biotechnol* 2000; 18: 670–4.
44. Krauter CM, Schachermayr G, Potrykus I, Sautter C. Antifungal activity of a virally encoded gene in transgenic wheat. *Nat Biotechnol* 2000; 18: 446–9.
45. Kloti WJ, Burkhardt PK, Biswas GC, Launis K, Iglesias VA, Potrykus I. Transgenic Indica rice breeding line IR58 expressing a synthetic cryIA(b) gene from Bacillus thuringiensis provides effective pest control. *Biotechnology* 1996; 14: 171–6.
46. Ai-Guo G, Hakimi SM, Mittanck CA, *et al*. Fungal pathogen protection in potato by expression of a plant defensin peptide. *Nat Biotechnol* 2000; 18: 1307–10.
47. Apse MP, Aharon GS, Snedden WA, Blumwald E. Salt tolerance conferred by overexpression of a vacuolar Na+/H+ antiport in arabidopsis. *Science* 1999; 285: 1256–8.
48. Theologis A. Control of ripening. *Curr Opin Biotechnol* 1994; 5: 152–7.
49. Gan S, Amasino RM. Inhibition of leaf senescence by autoregulated production of cytokinin. *Science* 1995; 270: 1986–8.
50. Penarrubia L. Production of the sweet protein monellin in transgenic plants. *Biotechnology* 1992; 10: 561–4.
51. Shewry PR, Tatham AS. Biotechnology of wheat quality. *J Sci Food Agric* 1997; 73: 397–406.
52. Carpenter JE, Gianessi LP. *Agricultural biotechnology: updated benefit estimates*. Washington DC: National Center for Food and Agricultural Policy; 2001.
53. Kinney AJ. Designer oils for better nutrition. *Nat Biotechnol* 1996; 14: 946.
54. Kinney AJ. Genetic modification of the storage lipids of plants. *Curr Opin Biotechnol* 1994; 5: 144–51.
55. Kinney AJ. Genetic engineering of oilseeds for desired traits. In: Setlow, JK, ed. *Genetic engineering*, vol 19. New York: Plenum Press, 1997: 149–66.

56. Ohlrogge JB. Design of new plant products: engineering of fatty acid metabolism. *Plant Physiol* 1994; 104: 821–6.
57. Volker T. Transgenic manipulation of edible oilseeds. In: Johns T, Romeo JT, eds. *Functionality of food phytochemicals*. New York: Plenum Press, 1997: 223–36.
58. Shintani D, DellaPenna D. Elevating the vitamin E content of plants through metabolic engineering. *Science* 1998; 282: 2098–100.
59. Grusak MA, DellaPenna D. Improving the nutrient composition of plants to enhance human nutrition. *Annu Rev Plant Physiol Mol Biol* 1999; 50: 133–61.
60. Rmer S, Fraser PD, Kiano JW, *et al.* Elevation of the provitamin A content of transgenic tomato plants. *Nat Biotechnol* 2000; 18: 666–9.
61. Chakraborty S, Chakraborty N, Datta A. Increased nutritive value of transgenic potato by expressing a nonallergic seed albumin gene from Amaranthus hypochondriacus. *Proc Natl Acad Sci USA* 2000; 97: 3724–9.
62. Zheng A. The bean seed storage protein -phaseolin is synthesized, processed, and accumulated in the vacuolar type-II protein bodies of transgenic rice endosperm. *Plant Physiol* 1995; 109: 777–86.
63. Falco SC, Guida R, Locke M, *et al.* Transgenic canola and soybean seeds with increased lysine. *Biotechnology* 1995; 13: 577–82.
64. Momma K. Quality and safety evaluation of genetically engineered rice with soybean glycinin: analyses of the grain composition and digestibility of glycinin in transgenic rice. *Biosci Biotechnol Biochem* 1999; 63: 314–18.
65. Prakash CS. Molecular insights into the biology of the sweet potato (Ipomoea batatas). In: Flores HE, ed. *Radical biology: advances and perspectives on the function of plant roots* (*Current topics in plant physiology*). Rockville: American Society of Plant Physiologists, 1997: 207–19.
66. Lucca P. Genetic engineering approaches to improve the bioavailability and the level of iron in rice grains. *Theor Appl Genet* 2001; 102: 392–7.
67. Ye X, Al-Babili S, Klti A, *et al.* Engineering the provitamin A (-carotene) biosynthetic pathway into (carotenoid-free) rice endosperm. *Science* 2000; 287: 303–5.
68. Potrykus I. Golden rice and beyond. *Plant Physiol* 2001; 125: 1157–61.
69. Al-Babili S, Ye X, Lucca P, Potrykus I, Beyer P. Biosynthesis of beta-carotene (provitamin A) in rice endosperm achieved by genetic engineering. *Novartis Found Symp* 2001; 236: 219–32.
70. Nestle M. Genetically engineered "golden" rice unlikely to overcome vitamin A deficiency. *J Am Diet Assoc* 2001; 101: 289–90.
71. Robetson R, Unnevehr L, Dawe D. Golden rice: what role could it play in alleviation of vitamin A deficiency? Available at: http://www.agbioworld.org/articles/goldenrice/html.
72. Mason HS, Arntzen CJ. Transgenic plants as vaccine production systems. *Trends Biotechnol* 1995; 13: 388–92.
73. Arntzen CJ. Pharmaceutical foodstuffs–oral immunization with transgenic plants. *Nat Med Vaccine Suppl* 1998; 4: 502–3.
74. Ma J K-C. Genes, greens, and vaccines. *Nat Biotechnol* 2000; 18: 1141–2.
75. Richter LJ, Thanavala Y, Arntzen CJ, Mason HS. Production of hepatitis B surface antigen in transgenic plants for oral immunization. *Nat Biotechnol* 2000; 18: 1167–71.
76. McGarvey PB, Hammond J, Dienelt MM, *et al.* Expression of the rabies virus glycoprotein in transgenic tomatoes. *Biotechnology* 1995; 13: 1484–7.
77. Arakawa T, Chong DKX, Langridge WHR. Efficacy of a food plant-based oral cholera toxin B subunit vaccine. *Nat Biotechnol* 1998; 16: 292–7.
78. Tacket CO, Mason HS, Losonsky G, Clements JD, Levine MM, Arntzen CJ. Immunogenicity in humans of a recombinant bacterial antigen delivered in a transgenic potato. *Nat Med* 1998; 4: 607–9.
79. Ball JM, Graham DY, Opekun AR, Gilger MA, Guerrero RA, Estes MK. Recombinant Norwalk virus-like particles given orally to volunteers: phase I study. *Gastroenterology* 1999; 117: 40–8.
80. Tacket CO, Mason HS, Losonsky G, Estes MK, Levine MM, Arntzen CJ. Human immune response to a novel Norwalk virus vaccine delivered in transgenic potatoes. *J Infect Dis* 2000; 182: 302–5.
81. Strobel S, Mowat AM. Immune responses to dietary antigens: oral tolerance. *Immunol Today* 1998; 19: 173–81.
82. Ma S-W, Zhao D-L, Yin Z-Q, *et al.* Transgenic plants expressing autoantigens fed to mice to induce oral immune tolerance. *Nature Med* 1997; 3: 793–6.
83. Arakawa T, Yu J, Chong DKX, Hough J, Engen PC, Langridge WHR. A plant-based cholera toxin B subunit-insulin fusion protein protects against the development of autoimmune diabetes. *Nat Biotechnol* 1998; 16: 934–8.

DISCUSSION

Dr. Böhles: A view has grown up that our bodily responses are determined by very remote times. In view of what you advocated at the end of your talk, doesn't this mean that we need a very individualized approach to those types of response? In other words, one person may be able to tolerate something perfectly while another cannot, just based on their genetic background.

Dr. Bier: Yes, I agree. I think we have learnt that from gluten enteropathy. One of the promises of the genome project is that in theory, if we know everyone's genes, we will have some idea of the basis of their reactions to the environment. However, one of the things we have learned in genetics in the last 50 years is the role of epigenetic modification of gene expression—that is, the interaction of your genetic make up with modifications from the environment. So, my answer is yes, but presumably, the number of individual responses is going to be almost infinite.

Dr. Belli: Do we have any evidence of modification of the composition of drinking water by the use of recombinant modifications in food production?

Dr. Bier: Do you mean in the sense of genes entering the drinking water?

Dr. Belli: Genes and other factors. We know that the use of pesticides and other agricultural products changes the composition of drinking water.

Dr. Bier: First of all, I think that the world's record of taking care of its drinking water by traditional technologies is so abysmal that I'm not sure that we can fault biotechnology for that. Such a large proportion of the world's population has no clean water even by traditional public health standards. The issue of whether the biotechnology products alter the composition of water depends on the introduction of pesticides and other things into the water. In my field, it's very hard to get complete agreement over what is happening with things like pesticide use, insecticide use, and so on. Several critics have said that in selected crop areas and crop studies, there has actually been no change or even an increase in pesticide use. In other published work of a wide variety, there are data to support that view that there has been less pesticide, insecticide, and herbicide use. This would presumably improve the drinking water.

Dr. Bachmann: Selection over years of famine and disease has led to selection of foodstuffs according to beliefs about edibility. We now also have several kinds of diet that are recommended for patients with various disorders. Do you foresee that directed manipulation of foodstuffs might create diets that are more palatable for such patients? In this way, we might be able to make life easier for such patients.

Dr. Bier: I think those things are certainly possible, but I doubt whether they are going to happen in the immediate future. I referred to several papers where individuals have introduced transgenic plants as natural sweeteners to improve the taste, and I think that moves to improve special diets in this way are probable in the future, though not immediately, as the applications are not wide enough to divert resources from what is necessary at this stage.

Dr. Rey: I have a question about "golden rice." You may know José Bové who produces Roquefort cheese in France. He is against mondialization because of punitive US taxes against Roquefort imports. He published an article in the newspaper *Le Monde* in which he wrote that golden rice would be unable to provide enough retinol and β-carotene for children in developing countries. My question is, how much β-carotene and retinol is there in golden rice?

Dr. Bier: I have referred to a publication and a website discussion of that particular issue in my paper. Nestle clearly laid out the reasons why golden rice is probably not going to alleviate worldwide vitamin A deficiency. A series of subsequent calculations was done based on population data from a community in the Philippines, and these confirmed Nestle's view. How much vitamin A golden rice provides depends first, obviously, on how much rice is consumed, and second, on the level of provision of β-carotene in the rice. In individuals who don't eat very much rice, and where the fortification level is at the lower end of the range, the actual increase in their vitamin A intake is going to be very small, only a few per cent. In such populations this is not going to be an answer; in fact, the situation is not much different from the fortification of grain with folate which we discussed earlier—the fact is that significant numbers of people don't eat grain products to the extent that fortifying them with folate is going to change their folate status. On the other hand, if one looks at high rice consumers and the higher end of β-carotene fortification then the result is of course better, but still not dramatically better—around 17% of the population might achieve an adequate β-carotene status.

I think it's wrong to look on biotechnology as the answer to the world's agriculture and pollution problems. It's one tool in the context of all the other tools that are available. No single food is going to solve the problems, just as no single food will solve any nutritional problem. We need to view biotechnology in the context of all the things that are available to us, and hopefully to introduce all of those things in a way that maximizes agricultural production and the nutrient content of food, minimizes crop losses, and so on.

Dr. Endres: My question is also related to quantitative aspects of the genetic engineering of crops. We are continually stimulation or modulating our immune systems by eating prebiotics or probiotics. What are the implications of this? This continuous boosting of our immune system is probably not necessary for the creation of antibodies against bacteria and viruses. What are the implications for children of the provision of this relatively uncontrolled source of vaccines implanted in plants?

Dr. Bier: Not being an immunologist, I don't know the immunologic answer to that question. I am sure there are some antigens where one exposure is sufficient for life, tetanus for example, and other antigens that are not going to be as powerful and may require repeated immunizations. How to use crops that may provide protection is something that immunologists will have to tell us.

Dr. Böhles: I read a book recently relating blood groups to diet (1). I laughed at the idea at first, but the more I thought about it, the more I saw something in it. We were after all hunters–gatherers, and we developed a Neolithic approach to farming. Is it not therefore relatively likely that people with certain blood group antigens are prone to particular problems but also resistant to particular dangers?

Dr. Bier: I think we have examples of this already from the genomic information that is available now. For example, a wide variety of nutrient–gene interactions is known about, like the apo-E effects on plasma cholesterol levels and potentially on the development of Alzheimer disease. Once we learn more about plant and human genomes and how they interact, and how the environment interacts, then yes, I agree with what you say. But, the permutations are so extensive that this is going to take a long time to work them all out.

Dr. Sollid: Just a short comment on vaccination by food consumption. The immune response of the gut can go into "protection mode" or into "tolerance mode," and we don't really understand all the rules. We have to be careful about this. There is no guarantee that if we introduce vaccines into plants we will get the desired effect. You may actually get the opposite. You may be aware of the clinical trials where they tried to tolerize people against self-antigens type 1 diabetes and multiple sclerosis (2). What they did was to give the self-antigens orally and hope that tolerization occurred, but I think there is a lot to learn about the principles involved, and we should be careful not to introduce this before we know what the outcome will be.

Dr. Bier: There are several published studies in mice prone to develop diabetes in which oral tolerance was induced and prevented the development of diabetes. There are related studies of other examples of oral tolerance, some of which have not specifically involved transgenics. There is at least one published example of what was called gene therapy, which was the oral introduction of an antigen that produced gut immunity without involving a plant. One certainly has to understand the science behind what one is trying to do, and the fact that it's in a plant doesn't make it safer or better or worse. To get the data we have to do the experiments.

Dr. Koletzko: What's your perspective on the use of food plants for the production of substances that have pharmacologic actions and that could be used as drugs? Some of those components have already been evaluated. And how would you respond to concerns that have been raised in the lay population that such modifications would allow pharmacologically active agents to enter the food chain?

Dr. Bier: I listed various way in which biotechnology in plants could be of benefit, and one of these was the production of products that might be helpful to society, such as drugs. Much research is being done in that area. What precisely will happen with horizontal gene transfer is an issue that has to be solved on the basis of experimentation. In the popular press, the risk/benefit ratio is generally calculated by adding the risks, and the benefit part of the equation is never put into the calculation. In practice, we can't derive a risk/benefit ratio without knowing the benefits. The idea of eating genes has been dramatized, but we've been eating genes since we've been eating food, and calculations have been made that we eat around 100,000 genes a day. We eat virus genes and viral DNA all the time. Plants are riddled with viruses, and we eat them. We get doses of viral and bacterial DNA every single day, and we have no evidence that this DNA has caused any harm by introducing genetic material into the intestine and so on. The modern tomato contains the genes of around nine tomato species. When lay people around the world were asked the question, "Do the tomatoes you eat contain genes?", 20% to 40% of them said no, and another 20% to 40% said they didn't know. So only about 20% to 30% of people know that tomatoes contain genes. We need to understand that there is an education process involved here.

Dr. El-Baroudy: I have read about the production of recombinant antithrombin-3 in goat's milk. The antigen is introduced into the goat and the goat then secretes milk containing antithrombin-3. I haven't read of anything comparable in the plant kingdom.

Dr. Bier: There are many animal studies being done in which animals are used as factories to produce various antibodies or other products. The same sort of work is being done in plants.

REFERENCES

1. D'Adamo PJ, Whitney C. 4 Blutgruppen. Vier Strategien für ein gesundes Leben. *Piper Taschenbücher* 2001; 2811.
2. Faria AM, Weiner HL. Oral tolerance: mechanisms and therapeutic applications. *Adv Immunol* 1999; 73: 153–264.

Genetic Expression and Nutrition, edited by Claude Bachmann and Berthold Koletzko. Nestlé Nutrition Workshop Series, Pediatric Program, Vol. 50. Nestec Ltd., Vevey/Lippincott Williams & Wilkins, Philadelphia, © 2003.

Diet as a Determinant of Drug Response

David G. Bailey

Lawson Health Research Institute, London Health Sciences Centre; Department of Medicine, Physiology, and Pharmacology, University of Western Ontario, London, Ontario, Canada

Therapeutic drugs and food are often taken together. Linking drug administration to a regular event like a meal can improve compliance with a treatment regimen, especially in elderly people (1). However, concomitant drug and food intake creates the opportunity for an interaction that may increase or decrease drug effectiveness or toxicity.

The response to a drug is largely dependent on its concentration at specific cells in the body. An increased drug concentration generally causes an enhanced magnitude and duration of effect, whereas a decreased drug concentration produces the opposite result. The concentration of an orally administered drug at the cellular site of action is determined by the net result of the rate and extent of absorption of the drug from the gastrointestinal tract into the systemic blood circulation (that is, its oral bioavailability), the distribution of the drug from the circulation to the target cells, and the removal of the drug from the site of action by its elimination from the body.

The oral bioavailability of a drug is dependent upon several factors. Drug dissolution is the initial important event following administration of a solid dose form. A standard tablet usually disintegrates and dissolves in the aqueous acidic environment of the stomach. Drugs with sufficient lipid solubility can then diffuse across the gastrointestinal tract into the circulation. However, this transfer usually does not take place in the stomach. For most drugs, absorption from the gastrointestinal tract occurs in the proximal portion of the small intestine (duodenum), which possesses a much greater surface area than the stomach. The duodenum also has a greater splanchnic blood flow, which creates a larger concentration gradient across the gut than the stomach for drug diffusion. Thus, the initiation and rate of movement of drug from the stomach to the duodenum—that is, gastric emptying—can be an important determinant of the onset and magnitude of the drug effect.

Before gaining access to the systemic circulation, drugs must also pass through the gut wall, enter the portal blood circulation, and then pass through the liver. Both the gut wall and the liver are capable of reducing drug bioavailability, either by enzymatically converting drugs to inactive derivatives (that is, presystemic or first-pass drug metabolism), and/or by transporting drugs back to the gastrointestinal tract where they may be reabsorbed (enterohepatic recirculation) or excreted in the feces.

Oral drug bioavailability is evaluated by measuring the drug concentration–time profile in blood or plasma. Changes in either the rate, as indicated by peak drug concentration (C_{max}) and time to C_{max} (t_{max}), or the extent, as determined by the area under the drug concentration–time curve (AUC), can have important implications for pharmacotherapy.

Food produces a wide range of physiologic changes to gastrointestinal function, some of which are relevant to drug bioavailability (2,3). In general, a meal increases the secretion of gastric fluid, which may facilitate drug dissolution. Importantly, food can also alter gastric emptying. For example, a hot or solid meal or a viscous solution delays and slows the passage of drug from stomach to duodenum. This can prolong the period of drug absorption and result in lower C_{max} and longer t_{max}, with no change in AUC of the drug. The effects are analogous to what occurs with a sustained-release drug formulation, although the mechanism is different (4). Thus, it is practical to take drugs with food to reduce the risk of adverse drug events from transiently high peak drug concentrations. In contrast, cool aqueous fluids accelerate gastric emptying. Thus drugs may be given on an empty stomach with a glass or two of water if a more rapid onset and greater drug effect is desired (4).

Food can also affect the extent of drug bioavailability. It is generally recognized that a clinically important reduction of absorption of tetracycline antibiotics from the gastrointestinal tract can occur following complexation with calcium in dairy products or with iron in vitamins. On the other hand, certain formulations of the antiinfective agents metronidazole and nitrofurantoin have increased bioavailability when taken with a meal, owing to better drug dissolution with delayed gastric emptying.

Ten years ago, a chance observation by our group led to the finding that grapefruit juice can increase the oral bioavailability and effects—both adverse and advantageous—of the dihydropyridine calcium channel antagonists. These are drugs that are often used in the management of cardiovascular disorders, particularly hypertension (5,6). Grapefruit juice acted by an original and specific mechanism. Subsequently, this initial observation has proved to be particularly important because of the magnitude of the interaction and the involvement of a wide range of drugs and some other foods. Grapefruit juice—drug interactions have been the subject of several reviews (7–10). In this chapter, I will focus on the interaction by retracing its discovery, reviewing its mechanism and the drugs involved, and discussing the clinical implications of the interaction in terms of its potential adverse and beneficial consequences.

DISCOVERY

Originally, a study was designed to test for an interaction between ethanol and the dihydropyridine calcium channel antagonist, felodipine, an analog of nifedipine (5). Grapefruit juice was chosen to mask the taste of the ethanol following an assessment of every juice in a home refrigerator one Saturday evening. White grapefruit juice was the most effective. Results of the study showed that plasma felodipine concentrations did not differ according to whether patients received the drug with ethanol (in grapefruit juice) or as a placebo (grapefruit juice alone). However, felodipine concentrations were several-fold higher than those observed in other pharmacokinetic investigations with the same dose of the drug. A systematic examination for obvious possible causes, such as incorrect dose or drug assay problems, did not resolve this discrepancy. This eventually resulted in a pilot project in a single volunteer to judge the role of the juice. Plasma felodipine concentrations were more than five-fold greater with grapefruit juice than with water (10). This result was confirmed in a subsequent formal clinical study (6). The discovery of this interaction illustrates the importance of unexpected observations in research.

MECHANISM AND RELEVANCE

Felodipine has been the most extensively studied probe for grapefruit juice–drug interactions. Normally, felodipine is completely absorbed from the gastrointestinal tract following oral administration (11). Thus, grapefruit juice did not act by improving drug dissolution or other pharmaceutic properties of the drug formulation. However, felodipine undergoes high presystemic (first pass) drug metabolism in the gut wall and then the liver, resulting in a low bioavailability, averaging 15% (Fig. 1)

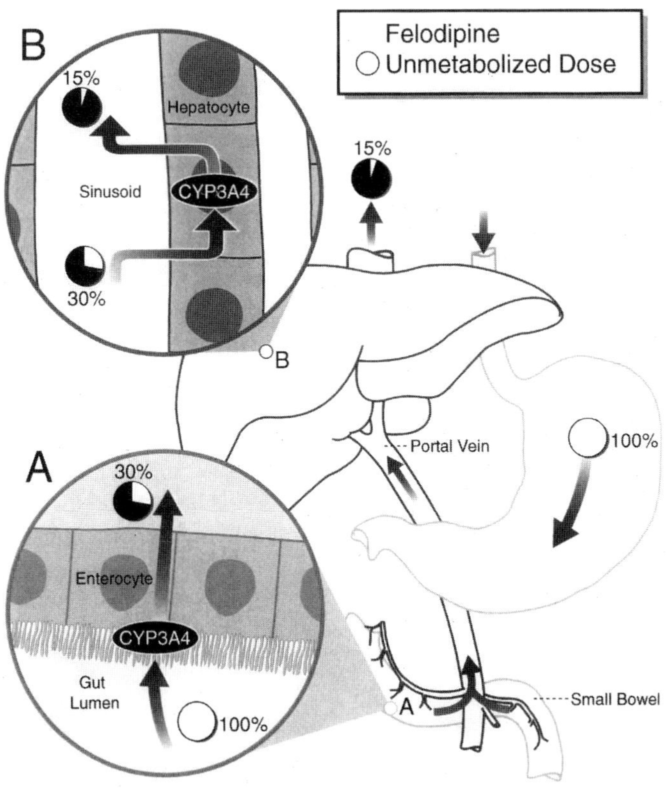

FIG. 1. Sequential presystemic felodipine metabolism by CYP3A4 in apical enterocytes of the small bowel **(A)** and then in the hepatocytes of the liver **(B)**. The percentage of unmetabolized felodipine is presented before and after passage through the gut wall and the liver.

but ranging from 4% to 36% among different individuals (11,12). It is metabolized by an enzyme belonging to a family known as cytochrome P450 (CYP) mixed function oxidases. CYPs account for most oxidative biotransformations of drugs and endogenous substances. These transformations usually produce substances (metabolites) that are more polar, less pharmacologically active, and more readily eliminated from the body in the urine. More than 40 different human CYPs have been identified. CYP3A4 metabolizes felodipine (13–15). CYP3A4 appears to be one of the most important human enzymes because approximately 60% of oxidized drugs are biotransformed, at least in part, by CYP3A4 (16). CYP3A4 drug metabolizing activity varies widely among individuals. Intestinal CYP3A4 is found in the apical enterocytes and its content varies 11-fold among individuals (17). However, CYP3A4 content is highest in the liver. Hepatic content has been shown to vary 20-fold and activity to range 10-fold among individuals. Intestinal and hepatic forms of CYP3A4 appear to be regulated independently of each other (18). The location of CYP3A4 in the small intestine and liver makes it well suited to playing a significant role in presystemic drug metabolism (10,19).

The first report of the interaction with felodipine revealed that grapefruit juice, but not orange juice, tripled the AUC or the extent of oral bioavailability of felodipine compared with water (6). Grapefruit juice also markedly increased the C_{max} or rate of absorption of felodipine but did not alter the plasma half-life (t 1/2) or the rate of elimination of felodipine from the systemic circulation. In addition, grapefruit juice did not change intravenous felodipine pharmacokinetics (20). Thus the interaction between grapefruit juice and felodipine appears to result from inhibition of presystemic drug metabolism mediated by CYP3A4.

The effect of grapefruit juice on drug-metabolizing enzymes of the small intestine and liver has been studied in humans. Grapefruit juice caused a 62% reduction in small intestinal enterocyte CYP3A4 protein content (21). This decrease was associated with a greater than three-fold increase in AUC and a greater than five-fold increase in C_{max} of felodipine. In addition, the magnitude of the interaction was highly variable among individuals, ranging from no change to six-fold greater plasma felodipine concentrations with grapefruit juice than with water. Subjects with the highest enterocyte content of CYP3A4 before grapefruit juice administration had the largest reduction of CYP3A4 and the greatest increase in felodipine C_{max} with the juice. In contrast, liver CYP3A4 activity, as measured by the erythromycin breath test, was not altered. Also, the content of other intestinal drug-metabolizing enzymes (CYP2D6, CYP1A1) was not affected. It was concluded that grapefruit juice acted by selective inhibition of intestinal CYP3A4 activity. The variability in the magnitude of the interaction with grapefruit juice among individuals was dependent upon inherent differences in baseline enterocyte CYP3A4 content.

These findings illustrate several important concepts. They show that the interaction with grapefruit juice is relevant only for orally administered drugs. They also suggest that drugs with high inherent presystemic metabolism are at risk of producing interactions of this type. Importantly, they demonstrate the significance of the gut as a site of drug metabolism.

Decreased expression of intestinal CYP3A4 by grapefruit juice implied that the interaction was not simple competition for substrate metabolism. As the content of enterocyte CYP3A4 messenger RNA (mRNA) was unchanged, it appears that grapefruit juice did not decrease CYP3A4 protein content by reducing its production (21). Rather, it indicates that the inhibitory effect results from enhanced degradation of the CYP3A4 protein. This effect could be caused by substances in grapefruit juice being metabolized by CYP3A4 to a reactive intermediate that covalently bonds to the enzyme to inactivate it, a process termed "suicide" or "mechanism-based" inhibition (10). The structurally modified CYP3A4 might then be expected to undergo rapid proteolysis within the cell. Consequently, the return of CYP3A4 activity would require *de novo* enzyme synthesis, suggesting prolonged effect of grapefruit juice.

The duration of activity of grapefruit juice has been evaluated. In one study, consumption of a single glass (200 ml) of juice at various time intervals before felodipine showed that the magnitude of the interaction was maximal between simultaneous administration and administration 4 hours previously (22). After that, the interaction declined slowly with increasing time interval. The half-life of disappearance of effect of grapefruit juice was estimated at 12 hours (10). Increased C_{max} of felodipine was still evident when grapefruit juice was consumed 24 hours beforehand. Thus a glass of grapefruit juice consumed yesterday has the potential to augment the bioavailability of drug administered today. Practically, this means that, when there is the potential for adverse effects from excessive drug concentrations, grapefruit juice consumption should best be avoided entirely during pharmacotherapy.

The amount of CYP3A4 present in the gut and liver can be increased (that is, induced) by certain drugs and environmental substances during repeated administration (17). As these substances can also initially inhibit this enzyme before induction, it is possible that the interaction between grapefruit juice and drugs could be attenuated under conditions of recurrent juice administration. The effect of routine grapefruit juice consumption has been evaluated. One glass (250 ml) of grapefruit juice augmented mean AUC and C_{max} of felodipine to 267% and 345%, respectively, of the values with water (21). Grapefruit juice three times daily with meals for 5 days further increased felodipine AUC and C_{max} to 345% and 538% of those with water. Thus there was no evidence of attenuation of the magnitude of the interaction or of possible induction of CYP3A4 activity. In contrast, it appears that there can be some cumulative increase in the effect of the juice under certain conditions of repeated administration.

INTERACTING DRUGS

Many drugs from a broad range of therapeutic categories are substrates for CYP3A4. Those that have been studied and found to have increased oral bioavailability with grapefruit juice are listed in Table 1 (5,15,20–77). These include several cholesterol-lowering drugs classified as hydroxymethylglutaryl-coenzyme A (HMG-CoA) reductase inhibitors (antilipemic agents) and the antihypertensive dihydropyridine calcium antagonist drugs (cardiovascular agents), both of which are widely used in

TABLE 1. *Drugs with increased oral bioavailability with grapefruit juice*

Antiinfective agents	*Central nervous system agents*
Artemether (23)	Buspirone (52)
Erythromycin (24)	Carbamazepine (53)
Saquinavir (25)	Diazepam (54)
Antiinflammatory agents	Midazolam (55)
Methylprednisolone (26)	Scopolamine (56)
Antilipemic agents	Sertraline (57)
Atorvastatin (27)	Triazolam (58,59)
Lovastatin (28)	*Estrogens*
Simvastatin (29,30)	Ethinylestradiol (60)
Cardiovascular agents	*Gastrointestinal agents*
Amiodarone (31)	Cisapride (61–63)
Carvedilol (32)	*Histamine H_1 antagonists*
Cilostazole (33)	Astemizole (64)
Felodipine (6,15,20–22,34–40)	Terfenadine (65–68)
Nifedipine (41–43)	*Immunosuppressive agents*
Nimodipine (44)	Cyclosporine (69–76)
NIcardipine (45)	Tacrolimus (77)
Nisoldipine (46,47)	
Nitrendipine (48,49)	
Verapamil (50,51)	

References in parentheses.

therapeutics. The gastrointestinal prokinetic drug cisapride and the antihistamines astemizole and terfenadine were also extensively used before their removal from the market because of serious adverse effects that could occur as a result of interactions with grapefruit juice.

Other drugs do not show increased oral bioavailability with grapefruit juice (78–96) (Table 2). Some of these belong to the same class as the interacting drugs and represent possible alternative agents when there is concern over a possible unwanted interaction. Such drugs include the HMG-CoA reductase inhibitor pravastatin and the dihydropyridine calcium antagonist amlodipine.

TABLE 2. *Drugs without increased oral bioavailability with grapefruit juice*

Antiasthmatic agents	*Antilipemic agents*
Theophylline (78)	Pravastatin (27)
Anticoagulants	*Cardiovascular agents*
Acenocoumarin (79)	Amlodipine (86,87)
Warfarin (80)	Diltiazem (88)
Antiinfective agents	Propafenone (89)
Clarithromycin (81)	Quinidine (90)
Indinivir (82)	*Central nervous system agents*
Itraconazole (83,84)	Alprazolam (91)
Quinine (85)	Clomipramine (92)
Antiinflammatory agents	Clozapine (93)
Prednisone (73)	Haloperidol (94)
	Phenytoin (95)
	Estrogens
	17β-estradiol (96)

References in parentheses.

We assessed interacting and noninteracting drugs for characteristics that might determine whether their pharmacokinetics were altered by grapefruit juice. As might be anticipated, drugs that interact with grapefruit juice have low to intermediate oral bioavailability (<5%–60%) because of presystemic metabolism mediated primarily by CYP3A4. Also, drugs with lower inherent bioavailability have an enhanced likelihood of greater interaction. Conversely, drugs that do not interact with grapefruit juice either have good innate oral bioavailability (amlodipine) or are not extensively metabolized by CYP3A4 (pravastatin).

ADVERSE CLINICAL CONSEQUENCES

When the clinical relevance of any interaction involving drugs is considered, the primary issue is safety. The following examples illustrate interactions that can result in important drug toxicities.

Torsades de Pointes

Torsades de pointes is a life-threatening cardiac ventricular arrhythmia that occurs in the setting of electrocardiographic QT interval prolongation. It has been shown to occur with several drugs that are CYP3A4 substrates. This adverse event was observed after these drugs were marketed. Drugs involved include the nonsedating antihistamines astemizole (97) and terfenadine (98–106), and the gastrointestinal prokinetic agent cisapride (107,108). These drugs appear to act in a concentration dependent manner to block the potassium rectifier current in cardiac conduction pathways (109–111). This delays repolarization of the ventricles, which is the basis for QT interval prolongation. Conditions in which plasma concentrations of these drugs are markedly increased are associated with a greater risk of developing this arrhythmia.

QT interval prolongation and torsades de pointes have been reported to occur during the concomitant administration of terfenadine, astemizole, or cisapride with grapefruit juice and other CYP3A4 inhibitors, including clarithromycin, erythromycin, itraconazole, and ketoconazole (97,101–103,106–108). In addition, formal pharmacokinetic/pharmacodynamic interaction studies have shown that these CYP3A4 inhibitors augment plasma substrate concentrations and prolong QT interval, supporting the predicted mechanism of action (61–68).

Rhabdomyolysis

HMG-CoA reductase inhibitors are an important class of cholesterol-lowering drugs. However, their use can be associated with diffuse myalgia and marked elevation of creatinine phosphokinase from skeletal muscle degeneration. There have been reports of severe rhabdomyolysis that precipitated acute renal failure (112–115). The mechanism by which these drugs produce this effect has not been clearly defined. However, the disorder appears to occur under conditions in which plasma concentrations of the parent drug and its metabolites are greatly increased, suggesting that it has a pharmacokinetic basis.

Lovastatin and simvastatin, which have very low oral bioavailability (<5%), both have profound (15-fold) increases in AUC when given with grapefruit juice (28–30). Atorvastatin, which has intermediate oral bioavailability, has a 2.5-fold higher AUC (27). Pravastatin has no interaction with grapefruit juice (27).

Symptomatic Hypotension

Hypotension is a dose-dependent side effect of many antihypertensive drugs. Several important antihypertensive agents, including most of the dihydropyridine calcium channel antagonists, are substrates of CYP3A4. Case reports of significant interactions resulting in dose-dependent adverse effects have been published for certain dihydropyridine calcium channel antagonists and CYP3A4 inhibitors (116–118). Grapefruit juice is the most extensively studied CYP3A4 inhibitor of this class of drug. It increases the AUC and C_{max} of the dihydropyridines between two- and threefold, except in the case of amlodipine, which normally undergoes low presystemic elimination and has a high oral bioavailability (19).

Sildenafil (Viagra) has recently become available for the treatment of erectile dysfunction. It is a relatively selective inhibitor of the cyclic guanosine monophosphate (cGMP)–specific phosphodiesterase type 5 (PDE5), thereby preventing degradation of cGMP in the corpus cavernosal smooth muscle (119). Nitric oxide (NO) is responsible for the production of cGMP in this tissue and in other systemic vascular beds. Under normal circumstances, no systemic vasodilatation is observed with sildenafil, because of its specificity for corpus cavernosal PDE5. However, under conditions of excessive sildenafil exposure, this selectivity can be lost. Sildenafil has 30% oral bioavailability from presystemic metabolism mediated by CYP3A4 (120). An initial report supports the possibility of enhanced oral bioavailability of sildenafil by grapefruit juice (121). This could be a particular problem with additional nitrate therapy because hypotension, myocardial infarction, or death may result (122–125).

BENEFICIAL CLINICAL CONSEQUENCES

While drug interactions should generally be avoided, inhibition of CYP3A4 may be used to achieve beneficial therapeutic goals. The following are creative examples of beneficial interactions.

Drug Cost Savings

Cyclosporine is a calcineurin inhibitor and immunosuppressive agent that is useful in preventing organ rejection following transplantation. It is metabolized by CYP3A4. It is also a substrate for P-glycoprotein (P-gp). P-gp is a drug efflux pump located on the luminal surface of epithelial cells of the small intestine and bile canalicular membrane of the liver (126). It affects the disposition of drugs by limiting their absorption from the gut and by facilitating their removal by secretion into the bile. CYP3A4 and P-gp in gut and liver appear to function cooperatively to facilitate drug elimination.

Cyclosporine is very expensive and must be taken on a daily basis for many years. It has only 30% to 40% oral bioavailability. The CYP3A4 and P-gp inhibitor ketoconazole more than doubles the oral bioavailability of cyclosporine (127). In view of the potential cost savings, two randomized controlled trials of cyclosporine with ketoconazole *versus* cyclosporine alone were undertaken (128,129). Plasma cyclosporine concentrations were routinely measured, and the dose of cyclosporine adjusted accordingly. This is particularly important in the case of cyclosporine because inadequate plasma drug concentrations can result in graft rejection, whereas excessive concentrations can cause nephrotoxicity. In both studies, combination therapy proved to be safe, effective, and considerably less costly than standard treatment over a 1- to 4-year period. This example shows that the concept of reduced drug dosage with equivalent drug efficacy is cost-effective.

Given the low cost and safety of grapefruit juice, it may be useful in this situation. Several investigations have shown that grapefruit juice can increase cyclosporine bioavailability (69–76). However, the effect was variable among studies and less than that observed with ketoconazole. There is another important difference between ketoconazole and grapefruit juice: In the case of grapefruit juice, there is additional concern over the consistency of the content of the active ingredient among batches and suppliers of the juice. This is likely to interfere with the uniformity of the effect on cyclosporine bioavailability, which is particularly crucial for this drug.

Maintenance of Drug Effectiveness

Artemether is a new and effective treatment for malaria because it has a fast onset of action, few side effects, and good activity against multidrug-resistant parasites. However, there is a high relapse rate during monotherapy. This may result from a change in the pharmacokinetics of artemether, because there is a marked reduction in plasma drug concentrations on repeated administration (130)—perhaps caused by induction of its own metabolism (autoinduction). Because artemether is a substrate for CYP3A4, this autoinduction may take place in the gut or liver, or in both. Previous work has shown that another inducer of CYP3A4, rifampin, has its primary effect in the gut (131–133). This is the site where grapefruit juice exerts its inhibitory effect on CYP3A4. Thus, grapefruit juice may be a useful therapeutic tool in this circumstance.

In a multidose study, grapefruit juice increased the oral bioavailability of artemether during single-dose administration compared with water (134). Grapefruit juice also produced higher plasma artemether concentrations after 5 days of concomitant administration compared with water. However, with both grapefruit juice and water, there was a time-dependent decrease in oral bioavailability of artemether. Thus, grapefruit juice consistently improved the oral bioavailability of artemether but did not totally prevent autoinduction. It was concluded that an oral treatment course with artemether is probably more effective when the drug is taken with grapefruit juice. Although grapefruit does not abolish the autoinduction of artemether, it may reduce the relapse rate by increasing and prolonging effective plasma drug concentrations. This example illustrates the concept that inhibition of CYP3A4 induction may be useful in maintaining drug efficacy.

Enhanced Drug Efficacy

The protease inhibitors are antiretroviral drugs used in the treatment of HIV-1 infection. Saquinavir, like cyclosporine, is a substrate for CYP3A4 and P-gp (135,136). It has very low bioavailability, in the region of 1% to 2% of the oral dose (137). Unlike cyclosporine, saquinavir does not produce important toxicity at high plasma drug concentrations. Thus any increase in bioavailability would produce only enhanced drug benefit.

Another protease inhibitor, ritonavir, is an inhibitor of CYP3A4 and P-gp. Its effect on saquinavir pharmacokinetics was studied in patients with advanced HIV disease. The addition of ritonavir resulted in a 58-fold greater AUC and 33-fold higher C_{max}, of saquinavir (138). In contrast, grapefruit juice doubled the oral bioavailability of saquinavir (25).

The findings from these pharmacokinetic studies have been put into clinical practice. Ritonavir is often coprescribed with saquinavir and other protease inhibitors. Grapefruit juice may be consumed with saquinavir to augment its bioavailability and therapeutic benefit. These examples illustrate the concept that enhanced bioavailability of drugs without dose-related toxicity can result only in greater drug efficacy, despite a large variability in the magnitude of the increase in plasma drug concentrations.

Oral Activity or More Consistent Dose Effect

Some drugs undergo complete first-pass metabolism and are active only when given by the intravenous route. As oral administration is much safer and more suitable for longer-term pharmacotherapy, it is preferable in most clinical situations. Thus, drugs that do not have oral activity are often discarded by pharmaceutical companies in the early phases of drug development.

Many marketed drugs have low as well as variable oral bioavailability. For example, felodipine has a mean 15% absolute oral bioavailability that ranges eight-fold (from 4% to 32%) among individuals (11,12). These pharmacokinetic shortcomings can have important clinical consequences. If a drug has inherently low oral bioavailability from high presystemic metabolism, concomitant administration of a food such as grapefruit juice, or of another drug that inhibits this metabolism, can be expected to produce a several-fold increase in drug bioavailability, even under single-dose conditions. The effect would be the same as if the individual took an accidental overdose of the drug. For example, administration of one tablet of lovastatin or simvastatin with a glass of grapefruit juice could have the same effect as giving 15 tablets of either of these drugs with a glass of water (28–30). Thus, high presystemic drug metabolism can be a risk factor for drug interactions.

If a drug has inherent low oral bioavailability, it generally means that it also has high variability in oral bioavailability among individuals. Clinically, this can complicate the determination of the correct dose for the clinical management of a patient's condition. Because it is currently not feasible to determine the extent of presystemic drug metabolism before treatment of the patient, the standard clinical approach is to start with the lowest dose of drug and to reassess its effect (adverse events, efficacy) after a certain period of treatment. At best, this only prolongs the time before the patient receives proper pharmacotherapy. On the other hand, a delay in receiving the correct dose of the drug could affect the outcome of more critical medical conditions. Also, the effect of a drug may not be readily measured. If a drug is used to lower blood pressure, its effect can be easily monitored; however, if the drug is used to prevent an epileptic seizure or an asthma attack, which can be irregular in occurrence, selection of the proper dose of drug can be much more difficult. In addition, drugs may also have nonlinear pharmacokinetics (that is, the bioavailability of the drug may increase disproportionately more than the increase in dose), or they may have a low therapeutic index (that is, the dose required to produce toxicity is not much greater than that required to produce efficacy). Thus, the selection of the correct dose of a drug at the onset of treatment can have a major impact on the management of the patient's medical condition. Consequently, if low and variable bioavailability of a drug could be transformed into complete and consistent bioavailability, this would result in improved treatment. This approach could also be a more viable therapeutic and financial option than starting at the beginning of the development process to find a drug with more desirable oral bioavailability characteristics.

One method that might be used to improve the oral bioavailability characteristics of certain drugs would be to incorporate into formulations a substance that inhibits presystemic metabolism—a process that appears to be mediated frequently by CYP3A4. The first criterion for an ideal bioavailability enhancer would be that it is a *nontoxic substance.* As grapefruit juice alone is generally recognized as a nutritional substance, it appears likely that any ingredient in the juice that is responsible for drug interactions is inherently safe. In practice this is important, as governmental agencies would not be likely to require extensive toxicologic testing before such an ingredient could be tested for activity as an enhancer of bioavailability in humans.

The second criterion is that the bioavailability enhancer can be *incorporated into the drug formulation as the pure substance*, to standardize and to have quality control over its effect. Although the active ingredient in grapefruit juice has not been definitely identified, several substances have been proposed on the basis of results from *in vitro* studies of the inhibition of CYP3A4 activity. These include the flavonoid naringin, which is probably metabolized to its more active aglycone, naringenin, after grapefruit juice consumption (5,139,140). However, commercial naringin in the same amounts as in grapefruit juice produced little or no interaction with felodipine or nisoldipine (36,46). Thus, naringin by itself does not appear to be capable of producing a clinical interaction like that observed with grapefruit juice.

Other proposed active ingredients include the furanocoumarins [bergamottin, 6′,7′-dihydroxybergamottin, and related dimers (141–145)]. For the assessment of their effects, it would be preferable to give them as pure compounds directly to humans. Unfortunately, these compounds are not yet approved for human intake. An alternative approach is to find another food that contains these substances in appropriate concentrations and then give this food with an established drug probe to determine whether it produces an interaction similar to that observed with grapefruit juice. It has been determined that Seville (sour) oranges contain bergamottin and 6′,7′-dihydroxybergamottin as well as naringin (40). Results showed that both Seville orange juice and grapefruit juice enhanced the oral bioavailability of the CYP3A4 substrate felodipine and altered the disposition of its metabolite dehydrofelodipine to a similar extent. Thus both bergamottin and 6′,7′-dihydroxybergamottin appear to be clinically important inhibitors of CYP3A4 in these juices, although other furanocoumarins and naringin may be also involved. This suggests that they may be "marker substances" to predict this type of interaction in other foods.

This study was important for an additional reason. A previous report had shown that, in comparison with grapefruit juice, Seville orange juice did not augment plasma concentrations of the CYP3A4 and P-gp substrate cyclosporine (146). This lack of effect of Seville orange juice occurred despite a marked decrease in enterocyte CYP3A4 content. Thus, inhibition of intestinal CYP3A4 cannot be the only mechanism whereby grapefruit juice enhances the oral bioavailability of cyclosporine. This supports the concept that grapefruit juice may inhibit both intestinal CYP3A4 and P-gp activities. Seville orange juice, on the other hand, may inhibit enteric CYP3A4 activity more selectively. Because bergamottin, 6′,7′-dihydroxybergamottin, and naringin are present in both Seville orange juice and grapefruit juice, these three substances are unlikely to be clinically important inhibitors of intestinal P-gp. Thus, there may be another substance in grapefruit juice that produces specific inhibition of P-gp.

The third criterion of an ideal bioavailability enhancer might be that it is a *selective inhibitor of the process determining oral drug bioavailability*. This could be important in preventing interactions with other drugs that are given concomitantly but which do not contain a bioavailability enhancer in their formulation. Because grapefruit juice has been shown not to interact with the CYP2D6 substrate propafenone (89), the CYP1A2 substrate theophylline (78), or the CYP2C9 substrate warfarin (80), this suggests that the inhibitory effect of the active ingredient in grapefruit juice may be limited to CYP3A4, at least within the cytochrome P450 family. However, many drugs are metabolized by CYP3A4 to some extent. Thus a drug interaction could still occur despite selective inhibition of CYP3A4.

The fourth criterion could be that the effect of the ideal bioavailability enhancer is *limited in duration of action to the period of drug absorption*. This property could further help to reduce the potential for an interaction with other drugs, perhaps by permitting their administration at another time in the day. As grapefruit juice has a long duration of action, the active ingredients in grapefruit juice may not meet this requirement (22,30,47).

The fifth and fundamentally important criterion for the enhancer is that it *produces complete or nearly complete oral drug bioavailability*. This effect would also satisfy the additional requirement for decreased variability of drug pharmacokinetics among individuals. The ability of grapefruit juice to meet this criterion was investigated in healthy elderly individuals who were at least 70 years old (38). This age group is particularly relevant as it comprises the individuals who are most often prescribed drugs.

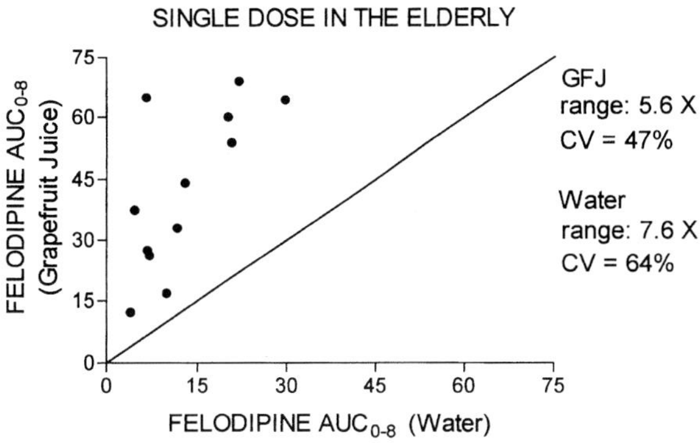

FIG. 2. Plasma felodipine area under the curve (AUC) for 12 healthy elderly subjects when a felodipine 5-mg extended release tablet was given with 250-ml water, plotted against when it was given with 250-ml grapefruit juice (GFJ). Diagonal represents the line of unity. CV, coefficient of variation.

Concomitant administration of a single glass of grapefruit juice (250 ml) produced a mean three-fold greater AUC of the pure CYP3A4 drug probe felodipine. Thus grapefruit juice caused a marked improvement in overall felodipine bioavailability.

Individually, the AUC of felodipine with water (the baseline condition) ranged 7.6-fold (coefficient of variation 64%) among individuals (Fig. 2). Grapefruit juice increased the AUC of felodipine in every individual. The magnitude of this increase was variable, ranging 1.2-fold to 8.2-fold. It might be anticipated that individuals with the lowest AUC of felodipine would show the greatest increase and *vice versa*, resulting in a more uniform AUC of felodipine with grapefruit juice among individuals. Results showed that the AUC of felodipine with grapefruit juice ranged 5.6-fold (coefficient of variation, 47%) among individuals. Thus grapefruit juice did make the oral bioavailability of felodipine more complete, but not sufficiently complete to produce a clinically relevant improvement in the consistency of oral bioavailability, at least under the conditions of the experiment. This may be because this amount of grapefruit juice contained only enough active ingredient to inhibit intestinal CYP3A4. Repeated administration and larger amounts of grapefruit juice appear capable of producing more complete drug bioavailability and inhibiting hepatic CYP3A4 (21,59). This approach may be necessary to improve the oral bioavailability enhancing characteristics of grapefruit juice and its active ingredients.

CONCLUSIONS

Foods can alter the response to drugs by affecting their oral bioavailability. This can occur because of non-specific changes in gastrointestinal function. A serendipitous observation led to the finding that grapefruit juice can produce inactivation of a drug-metabolizing enzyme, CYP3A4, in the gastrointestinal tract. This can lead to marked augmentation of the oral bioavailability of various drugs from a broad range of therapeutic categories. Clinically important adverse consequences of this interaction include torsades de pointes, rhabdomyolysis, and symptomatic hypotension with certain agents. Potentially beneficial clinical consequences are drug cost savings, maintenance of drug effectiveness, enhanced drug efficacy or oral activity, and a more consistent dose effect. The latter possibility requires the inclusion of a substance or substances into a drug formation to enhance oral drug bioavailability.

The ideal bioavailability enhancer should lack of toxicity, be available as a pure substance, be a selective inhibitor of the process determining oral drug bioavailability, have a duration of action limited to the period of drug absorption, and produce nearly complete oral drug bioavailability. The active ingredient or ingredients in grapefruit juice currently meet some of these requirements; however, further research is needed to determine the true practical value of this approach.

REFERENCES

1. Haynes R, Sackett D, Gibson E. Improvements of medication compliance in uncontrolled hypertension. *Lancet* 1976; i:1256–8.
2. Singh BN. Effects of food on clinical pharmacokinetics. *Clin Pharmacokinet* 1999;37: 213–55.
3. Winstanley PA, Orme ML. The effects of food on drug bioavailability. *Br J Clin Pharmacol* 1989; 28: 621–8.
4. Bailey DG. Food–drug interactions: emphasis on grapefruit juice effects. *Can J Clin Pharmacol* 1995; 2: 10–14.
5. Bailey DG, Spence JD, Edgar B, Bayliff CD, Arnold JMO. Ethanol enhances the hemodynamic effects of felodipine. *Clin Invest Med* 1989; 12: 357–62.
6. Bailey DG, Spence JD, Munoz C, Arnold JMO. Interaction of citrus juices with felodipine and nifedipine. *Lancet* 1991; 337: 268–9.
7. Bailey DG, Arnold JMO, Spence JD. Grapefruit juice and drugs: how significant is the interaction? *Clin Pharmacokinet* 1994; 26: 91–8.
8. Ameer B, Weintraub RA. Drug interactions with grapefruit juice. *Clin Pharmacokinet* 1997; 33: 103–21.
9. Fuhr U. Drug interactions with grapefruit juice: extent, probable mechanism of action and clinical relevance. *Drug Saf* 1998; 18: 251–72.
10. Bailey DG, Arnold JMO, Spence JD. Grapefruit juice–drug interactions. *Br J Clin Pharmacol* 1998; 46: 101–10.
11. Edgar B, Regardh CG, Johnsson G, *et al.* Felodipine kinetics in healthy man. *Clin Pharmacol Ther* 1985; 38: 205–11.
12. Blychert E, Edgar B, Elmfeldt D, Hedner T. A population study of the pharmacokinetics of felodipine. *Br J Clin Pharmacol* 1991; 31: 15–24.
13. Regardh CG, Baarnhielm C, Edgar B, Hoffman KJ. Pharmacokinetics and biotransformation of 1,4-dihydropyridine calcium antagonists. *Prog Drug Metab* 1990; 12: 41–86.
14. Guengerich FP, Brian WR, Iwasaki M, Sari MA, Baarnhielm C, Berntsson P. Oxidation of dihydropyridine calcium channel blockers and analogues by human liver cytochrome P450IIIA4. *J Med Chem* 1991; 34: 1834–44.
15. Bailey DG, Bend JR, Arnold JMO, Tran LT, Spence JD. Erythromycin–felodipine interaction: magnitude, mechanism, and comparison with grapefruit juice. *Clin Pharmacol Ther* 1996; 60: 25–33.
16. Guengerich FP. Cytochrome P-450 3A4; regulation and role in drug metabolism. *Annu Rev Pharmacol Toxicol* 1999; 39: 1–17.
17. Kolars JC, Schmiedlin-Rem P, Schuetz JD, Fang C, Watkins PB. Identification of rifampicin-inducible P450IIIA4 (CYP3A4) in human small bowel enterocytes. *J Clin Invest* 1992; 90: 1871–8.
18. Lown KS, Kolars JC, Thummel KE, *et al.* Interpatient heterogeneity in expression of CYP3A4 and CYP3A5 in small bowel–lack of prediction by the erythromycin breath test. *Drug Metab Dispos* 1994; 22: 947–55.
19. Dresser GK, Spence JD, Bailey DG. Pharmacokinetic-pharmacodynamic consequences and clinical relevance of cytochrome P450 3A4 inhibition. *Clin Pharmacokinet* 2000; 38: 41–57.
20. Lundahl J, Regardh CG, Edgar B, Johnsson G. Effects of grapefruit juice ingestion–pharmacokinetics and haemodynamics of intravenously and orally administered felodipine in healthy men. *Eur J Clin Pharmacol* 1997; 52: 139–45.

21. Lown KS, Bailey DG, Fontana RJ, *et al.* Grapefruit juice increases felodipine oral availability in humans by decreasing intestinal CYP3A protein expression. *J Clin Invest* 1997; 99: 2545–53.
22. Lundahl J, Regardh CG, Edgar B, Johnsson G. Relationship between time of intake of grapefruit juice and its effect on pharmacokinetics and pharmacodynamics of felodipine in healthy subjects. *Eur J Clin Pharmacol* 1995; 49: 61–7.
23. van Agtmael MA, Gupta V, van der Wosten TH, Rutten JP, van Boxtel CJ. Grapefruit juice increases the bioavailability of artemether. *Eur J Clin Pharmacol* 1999; 55: 405–10.
24. Kanazawa S, Ohkubo T, Sugawara K. The effects of grapefruit juice on the pharmacokinetics of erythromycin. *Eur J Clin Pharmacol* 2001; 56: 799–803.
25. Kupferschmidt HH, Fattinger KE, Ha HR, Follath F, Krahenbuhl S. Grapefruit juice enhances the bioavailability of the HIV protease inhibitor saquinavir in man. *Br J Clin Pharmacol* 1998; 45: 355–9.
26. Varis T, Kivisto KT, Neuvonen PJ. Grapefruit juice can increase the plasma concentrations of oral methylprednisolone. *Eur J Clin Pharmacol* 2000; 56: 489–93.
27. Lilja JJ, Kivisto KT, Neuvonen PJ. Grapefruit juice increases serum concentrations of atorvastatin and has no effect on pravastatin. *Clin Pharmacol Ther* 1999; 66: 118–27.
28. Kantola T, Kivisto KT, Neuvonen PJ. Grapefruit juice greatly increases serum concentrations of lovastatin and lovastatin acid. *Clin Pharmacol Ther* 1998; 63: 397–402.
29. Lilja JJ, Kivisto KT, Neuvonen PJ. Grapefruit juice–simvastatin interaction: effect on serum concentrations of simvastatin, simvastatin acid, and HMG-CoA reductase inhibitors. *Clin Pharmacol Ther* 1998; 64: 477–83.
30. Lilja JJ, Kivisto KT, Neuvonen PJ. Duration of effect of grapefruit juice on the pharmacokinetics of the CYP3A4 substrate simvastatin. *Clin Pharmacol Ther* 2000; 68: 384–90.
31. Libersa CC, Brique SA, Motte KB, *et al.* Dramatic inhibition of amiodarone metabolism induced by grapefruit juice. *Br J Clin Pharmacol* 2000; 49: 373–8.
32. SmithKline Beecham Pharma. Coreg Product Monograph. Oakville, Ontario; 1999.
33. Anonymous. Cilostazol for intermittent claudication. *Med Lett* 1999; 41: 44–6.
34. Bailey DG, Arnold JMO, Bend JR, Tran LT, Spence JD. Grapefruit juice–felodipine interaction: reproducibility and characterization with the extended release drug formulation. *Br J Clin Pharmacol* 1995; 40: 135–40.
35. Edgar B, Bailey DG, Bergstrand R, Johnsson G, Regardh CG. Acute effects of drinking grapefruit juice on the pharmacokinetics and pharmacodynamics of felodipine–and its potential clinical relevance. *Eur J Clin Pharmacol* 1992; 42: 313–7.
36. Bailey DG, Arnold JMO, Munoz C, Spence JD. Grapefruit juice–felodipine interaction: mechanism, predictability and effect of naringin. *Clin Pharmacol Ther* 1993; 53: 637–42.
37. Bailey DG, Kreeft JH, Munoz C, Freeman JD, Bend JR. Grapefruit juice–felodipine interaction: effect of naringin and 6′,7′-dihydroxybergamottin in humans. *Clin Pharmacol Ther* 1998; 64: 248–56.
38. Dresser GK, Bailey DG, Carruthers SG. Grapefruit juice–felodipine interaction in the elderly. *Clin Pharmacol Ther* 2000; 68: 28–34.
39. Bailey DG, Dresser GK, Kreeft JH, Munoz C, Freeman DJ, Bend JR. Grapefruit–felodipine interaction: effect of unprocessed fruit and probable active ingredients. *Clin Pharmacol Ther* 2000; 68: 468–77.
40. Malhotra S, Bailey DG, Paine MF, Watkins PB. Seville orange juice–felodipine interaction: comparison with dilute grapefruit juice and involvement of the furanocoumarins. *Clin Pharmacol Ther* 2001; 69: 14–23.
41. Rashid J, McKinstry C, Renwick AG, Dirnhuber M, Waller DG, George CF. Quercetin, an in vitro inhibitor of CYP3A, does not contribute to the interaction between nifedipine and grapefruit juice. *Br J Clin Pharmacol* 1993; 36: 460–3.
42. Rashid TJ, Martin U, Clarke H, Waller DG, Renwick AG, George CF. Factors affecting the absolute bioavailability of nifedipine. *Br J Clin Pharmacol* 1995; 40: 51–8.
43. Sigush H, Hippius M, Henschel L, Kaufmann K, Hoffmann A. Influence of grapefruit juice on the pharmacokinetics of a slow release nifedipine formulation. *Pharmazie* 1994; 49: 522–4.
44. Fuhr U, Maier-Bruggemann A, Blume H, *et al.* Grapefruit juice increases oral nimodipine bioavailability. *Int J Clin Pharmacol Ther* 1998; 36: 126–32.
45. Uno T, Ohkubo T, Sugawara K, Higashiyama A, Motomura S, Ishizaki T. Effect of grapefruit juice on the stereoselective disposition of nicardipine in humans: evidence for dominant presystemic elimination at the gut site. *Eur J Clin Pharmacol* 2000; 56: 643–9.
46. Bailey DG, Arnold JMO, Strong HA, Munoz C, Spence JD. Effect of grapefruit juice and naringin on nisoldipine pharmacokinetics. *Clin Pharmacol Ther* 1993; 54: 589–94.

47. Takanaga H, Ohnishi A, Murakami H, *et al.* Relationship between time after intake of grapefruit juice and the effect on the pharmacokinetics and pharmacodynamics of nisoldipine in healthy subjects. *Clin Pharmacol Ther* 2000; 67: 201–14.
48. Soons PA, Vogels BAPM, Roosemalen MCM, *et al.* Grapefruit juice and cimetidine inhibit stereoselective metabolism of nitrendipine in man. *Clin Pharmacol Ther* 1991; 50: 394–403.
49. Bailey DG, Munoz C, Arnold JMO, Strong HA, Spence JD. Grapefruit juice and naringin interaction with nitrendipine [abstract]. *Clin Pharmacol Ther* 1992; 51: 156.
50. Fuhr U, Harder S, Lopez-Rojas P, Muller-Peltzer H, Kern R, Staib AH. Increase of verapamil concentrations in steady state by coadministration of grapefruit juice [abstract]. *Naunyn Schmiedebergs Arch Pharmacol* 1994; 349[Suppl]: R134.
51. Ho PC, Ghose K, Saville D, Wanwimolruk S. Effect of grapefruit juice on pharmacokinetics and pharmacodynamics of verapamil enantiomers in healthy volunteers. *Eur J Clin Pharmacol* 2000; 56: 693–8.
52. Lilja JJ, Kivisto KT, Backman JT, Lamberg TS, Neuvonen PJ. Grapefruit juice substantially increases plasma concentrations of buspirone. *Clin Pharmacol Ther* 1998; 64: 655–60.
53. Garg SJ, Kumar N, Bhargava VK, Prabhakar SK. Effect of grapefruit juice on carbamazepine bioavailability in patients with epilepsy. *Clin Pharmacol Ther* 1998; 64: 286–8.
54. Ozedemir M, Aktan Y, Boydag BS, Cingi MI, Musmul A. Interaction between grapefruit juice and diazepam in humans. *Eur J Drug Metab Pharmacokinet* 1998; 23: 55–9
55. Kupferschmidt HHT, Ha HR, Ziegler WH, Meir PJ, Krahenbuhl S. Interaction between grapefruit juice and midazolam in humans. *Clin Pharmacol Ther* 1995; 58: 20–8.
56. Ebert U, Oertel R, Kirch W. Influence of grapefruit juice on scopolamine pharmacokinetics and pharmacodynamics in healthy male and female subjects. *Int J Clin Pharmacol Ther* 2000; 38: 523–31.
57. Lee AJ, Chan WK, Harralson AF, Buffum J, Bui BC. The effects of grapefruit juice on sertraline metabolism: an in vitro and in vivo study. *Clin Ther* 1999; 21: 1890–9.
58. Hukkinen SK, Varhe A, Olkkola KT, Neuvonen PJ. Plasma concentrations of triazolam are increased by concomitant ingestion of grapefruit juice. *Clin Pharmacol Ther* 1995; 58: 127–3.
59. Lilja JJ, Kivisto KT, Backman JT, Neuvonen PJ. Effect of grapefruit juice dose on grapefruit juice–triazolam interaction: repeated consumption prolongs triazolam half-life. *Eur J Clin Pharmacol* 2000; 56: 411–15.
60. Weber A, Jager R, Borner A, *et al.* Can grapefruit juice influence ethinylestradiol bioavailability? *Contraception* 1996; 53: 41–7.
61. Gross AS, Goh YD, Addison RS, Shenfield GM. Influence of grapefruit juice on cisapride pharmacokinetics. *Clin Pharmacol Ther* 1999; 65: 395–401.
62. Kivisto KT, Lilja JJ, Backman JT, Neuvonen PJ. Repeated consumption of grapefruit juice considerably increases plasma concentrations of cisapride. *Clin Pharmacol Ther* 1999; 66: 448–53.
63. Offman EM, Freeman DJ, Dresser GK, Munoz C, Bend JR, Bailey DG. Red wine–cisapride interaction: comparison with grapefruit juice. *Clin Pharmacol Ther* 2001; 70: 17–23.
64. Johnson & Johnson–Merck Consumer Pharmaceuticals of Canada. *Hismanal Product Monograph. Guelph, Ontario, 1999.*
65. Benton RE, Honig PK, Zamani K, Cantilena LR, Woosley RL. Grapefruit juice alters terfenadine pharmacokinetics, resulting in prolongation of repolarization on the electrocardiogram. *Clin Pharmacol Ther* 1996; 59: 383–8.
66. Honig PK, Wortham DC, Lazarev A, Cantilena LR. Grapefruit juice alters the systemic bioavailability and cardiac repolarization of terfenadine in poor metabolizers of terfenadine. *J Clin Pharmacol* 1996; 36: 345–51.
67. Rau SE, Bend JR, Arnold JMO, Tran LT, Spence JD, Bailey DG. Grapefruit juice–terfenadine single-dose interaction: magnitude, mechanism, and relevance. *Clin Pharmacol Ther* 1997; 61: 401–9.
68. Clifford CP, Adams DA, Murray S, *et al.* The cardiac effects of terfenadine after inhibition of its metabolism by grapefruit juice. *Eur J Clin Pharmacol* 1997; 52: 311–15.
69. Ducharme MP, Warbasse LH, Edwards DJ. Disposition of intravenous and oral cyclosporine after administration with grapefruit juice. *Clin Pharmacol Ther* 1995; 57: 485–91.
70. Yee GC, Stanley DL, Pessa JL, *et al.* Effect of grapefruit juice on blood cyclosporin concentration. *Lancet* 1995; 345: 955–6.
71. Ducharme MP, Provenzano R, Dehoorne-Smith M, Edwards DJ. Trough concentrations of cyclosporine following administration with grapefruit juice. *Br J Clin Pharmacol* 1993; 36: 457–9.
72. Herlitz H, Edgar B, Hedner T, Lidman K, Karlberg I. Grapefruit juice: a possible source of variability in blood concentration of cyclosporin A [letter]. *Nephrol Dial Transplant* 1993; 8: 375.

73. Hollander AAMJ, van Rooij J, Lentjes EGWM, *et al.* The effect of grapefruit juice on cyclosporin and prednisone metabolism in transplant patients. *Clin Pharmacol Ther* 1995; 57: 318–24.
74. Proppe DG, Hoch OD, McLean AJ, Visser KE. Influence of chronic ingestion of grapefruit juice on steady state blood concentrations of cyclosporine A in renal transplant patients with stable graft function. *Br J Clin Pharmacol* 1995; 39: 337.
75. Min DI, Ku Y-M, Perry PJ, *et al.* Effect of grapefruit juice on cyclosporine pharmacokinetics in renal transplant patients. *Transplantation* 1996; 62: 123–5.
76. Johnston A, Holt DW. Effect of grapefruit juice on blood cyclosporin concentration. *Lancet* 1995; 346: 122–3.
77. Westveer MK, Farquhar ML, George P, Mayers JT. Co-administration of grapefruit juice increases tacrolimus levels in liver transplant recipients. *Annu Meet Am Soc Transplant Physicians* 1996; 202: P115.
78. Fuhr U, Maier A, Keller A, Steinijans VW, Sauter R, Staib AH. Lacking effect of grapefruit juice on theophylline pharmacokinetics. *Int J Clin Pharmacol Ther* 1995; 33: 311–14.
79. van Rooij J, van der Meer FJM, Schoemaker HC, Cohen AF. Comparison of the effect of grapefruit juice and cimetidine on pharmacokinetics and anticoagulant effect of a single dose of acenocoumarol [abstract]. *Br J Clin Pharmacol* 1993; 35: 548P.
80. Sullivan DM, Ford MA, Boyden TW. Grapefruit juice and the response to warfarin. *Am J Health Syst Pharm* 1998; 55: 1581–3.
81. Cheng KL, Nafziger AN, Peloquin CA, Amsden GW. Effect of grapefruit juice on clarithromycin pharmacokinetics. *Antimicrob Agents Chemother* 1998; 42: 927–9.
82. Shelton MJ, Wynn HE, Hewitt RG, DiFrancesco R. Effects of grapefruit juice on pharmacokinetic exposure to indinavir in HIV-positive subjects. *J Clin Pharmacol* 2001; 41:435–42.
83. Kawakami M, Suzuki K, Ishizuka T, Hidaka T, Matsuki Y, Nakamura H. Effect of grapefruit juice on pharmacokinetics of itraconazole in healthy subjects. *Int J Clin Pharmacol Ther* 1998; 36: 306–8.
84. Penzak SR, Gubbins PO, Gurley BJ, Wang PL, Saccente M. Grapefruit juice decreases the systemic availability of itraconazole capsules in healthy volunteers. *Ther Drug Monit* 1999; 21: 304–9.
85. Ho PC, Chalcroft SC, Coville PF, Wanwimolruk S. Grapefruit juice has no effect on quinine pharmacokinetics. *Eur J Clin Pharmacol* 1999; 55: 393–8.
86. Josefsson M, Zackrisson AL, Ahlner J. Effect of grapefruit juice on the pharmacokinetics of amlodipine in healthy volunteers. *Eur J Clin Pharmacol* 1996; 51: 189–93.
87. Vincent J, Harris SI, Foulds G, Dogolo LC, Willavize S, Friedman HL. Lack of effect of grapefruit juice on the pharmacokinetics and pharmacodynamics of amlodipine. *Br J Clin Pharmacol* 2000; 50: 455–63.
88. Sigusch H, Henschel L, Kraul H, Merkel U, Hoffmann A. Lack of effect of grapefruit juice on diltiazem bioavailability in normal subjects. *Pharmazie* 1994; 49: 675–9.
89. Munoz CE, Ito S, Bend JR, *et al.* Propafenone interaction with CYP3A4 inhibitors in man [abstract]. *Clin Pharmacol Ther* 1997; 61: 154.
90. Min DI, Ku Y-M, Geraets DR, Lee H-C. Effect of grapefruit juice on the pharmacokinetics and pharmacodynamics of quinidine in healthy volunteers. *J Clin Pharmacol* 1996; 36: 469–76.
91. Yasui N, Kondo T, Furukori H, *et al.* Effects of repeated ingestion of grapefruit juice on the single and multiple oral-dose pharmacokinetics and pharmacodynamics of alprazolam. *Psychopharmacology* 2000; 150: 185–90.
92. Oesterheld J, Kallepalli BR. Grapefruit juice and clomipramine: shifting metabolic ratios [letter]. *J Clin Psychopharmacol* 1997; 17: 62–3.
93. Vandel S, Netillard C, Perault MC, Bel AM. Plasma levels of clozapine and desmethylclozapine are unaffected by concomitant ingestion of grapefruit juice. *Eur J Clin Pharmacol* 2000; 56: 347–8.
94. Yasui N, Kondo T, Suzuki A, *et al.* Lack of significant pharmacokinetic interaction between haloperidol and grapefruit juice. *Int Clin Psychopharmacol* 1999; 14: 113–18.
95. Kumar N, Garg SK, Prabhakar S. Lack of pharmacokinetic interaction between grapefruit juice and phenytoin in healthy male volunteers and epileptic patients. *Methods Find Exp Clin Pharmacol* 1999; 21: 629–32.
96. Schubert W, Cullberg G, Edgar B, Hedner T. Inhibition of 17β-estradiol metabolism by grapefruit juice in ovariectomized women. *Maturitas* 1994; 20: 155–63.
97. Tsai WC, Tsai LM, Chen JH. Combined use of astemizole and ketoconazole resulting in torsade de pointes. *J Formos Med Assoc* 1997; 96: 144–6.
98. Davies AJ, Harindra V, McEwen A, Ghose RR. Cardiotoxic effect with convulsions in terfenadine overdose. *BMJ* 1989; 289: 325.
99. Monahan BP, Ferguson CL, Killeavy ES, Lloyd BK, Troy J, Cantilena LR. Torsade de pointes occurring in association with terfenadine use. *JAMA* 1990; 264: 2788–90.

100. Matthews DR, McNutt B, Okerholm R, Flicker M, McBride G. Torsade de pointes occurring in association with terfenadine use. *JAMA* 1991; 266: 2375–6.
101. Zimmermann M, Duruz H, Guinand O, *et al.* Torsade de pointes after treatment with terfenadine and ketoconazole. *Eur Heart J* 1992; 13: 1002–3.
102. Pohjola-Sintonen S, Viitasalo L, Toivonen L, Neuvonen P. Torsade de pointes after terfenadine–itraconazole interaction. *BMJ* 1993; 306: 186.
103. Paris DG, Parente TF, Bruschetta HR, Guzman E, Niarchos AP. Torsade de pointes induced by erythromycin and terfenadine. *Am J Emerg Med* 1994; 12: 636–8.
104. Koh KK, Rim MS, Yoon J, Kim SS. Torsade de pointes induced by terfenadine in a patient with long QT syndrome. *J Electrocardiol* 1994; 27: 343–6.
105. Kamisako T, Adachi Y, Nakagawa H, Yamamoto T. Torsade de pointes associated with terfenadine in a case of liver cirrhosis and hepatocellular carcinoma. *Intern Med* 1995; 34: 92–5.
106. Spence JD. Drug interactions with grapefruit: whose responsibility is it to warn the public? *Clin Pharmacol Ther* 1997; 61: 395–400.
107. Wysowski DK, Bacsanyi J. Cisapride and fatal arrythmia. *N Engl J Med* 1996; 335: 290–1.
108. Sekkarie MA. Torsade de pointes in two chronic renal failure patients treated with cisapride and clarithromycin. *Am J Kidney Dis* 1997; 30: 437–9.
109. Woosley RL, Chen Y, Freiman JP, Gillis RA. Mechanism of the cardiotoxic actions of terfenadine. *JAMA* 1993; 269: 1532–6.
110. Yang T, Prakash C, Roden DM, Snyders DJ. Mechanism of block of a human cardiac potassium channel by terfenadine racemate and enantiomers. *Br J Pharmacol* 1995; 115: 267–74.
111. Rampe D, Wible B, Brown AM, Dage RC. Effects of terfenadine and its metabolites on a delayed rectifier K^+ channel cloned from human heart. *Mol Pharmacol* 1993; 44: 1240–5.
112. Ayanian JZ, Fuchs CS, Stone RM. Lovastatin and rhabdomyolysis [letter]. *Ann Intern Med* 1988; 109: 682–3.
113. Corpier CL, Jones PH, Suki WN, *et al.* Rhabdomyolysis and renal injury with lovastatin use. Report of two cases in cardiac transplant recipients. *JAMA* 1988; 260: 239–41.
114. Biesenbach G, Janko O, Stuby U, Zazgornik J. Myoglobinuric renal failure due to long-standing lovastatin therapy in a patient with pre-existing chronic renal insufficiency. *Nephrol Dialysis Transplant* 1996; 11: 2059–60.
115. Hino I, Akama H, Furuya T, *et al.* Pravastatin-induced rhabdomyolysis in a patient with mixed connective tissue disease. *Arthritis Rheum* 1996; 39: 1259–60.
116. Liedholm H, Nordin G. Erythromycin-felodipine interaction. *Drug Intell Clin Pharm* 1991; 25: 1007–8.
117. Neuvonen PJ, Suhonen R. Itraconazole interacts with felodipine. *J Am Acad Dermatol* 1995; 33: 134–5.
118. Tailor SA, Gupta AK, Walker SE, Shear NH. Peripheral edema due to nifedipine–itraconazole interaction: a case report. *Arch Dermatol* 1996; 132: 350–2.
119. Langtry HD, Markham A. Sildenafil: a review of its use in erectile dysfunction [review]. *Drugs* 1999; 57: 967–89.
120. Walker DK, Ackland MJ, James GC, *et al.* Pharmacokinetics and metabolism of sildenafil in mouse, rat, rabbit, dog and man. *Xenobiotica* 1999; 29: 297–310.
121. Lee M, Min DI. Determination of sildenafil citrate in plasma by high-performance liquid chromatography and a case for the potential interaction of grapefruit juice with sildenafil citrate. *Ther Drug Monit* 2001; 23: 21–6.
122. Webb DJ, Freestone S, Allen MJ, Muirhead GJ. Sildenafil citrate and blood-pressure-lowering drugs: results of drug interaction studies with an organic nitrate and a calcium antagonist. *Am J Cardiol* 1999; 83: 21–8C.
123. Cheitlin MD, Hutter AMJ, Brindis RG, *et al.* ACC/AHA expert consensus document. Use of sildenafil (Viagra) in patients with cardiovascular disease. American College of Cardiology/American Heart Association [review]. *J Am Coll Cardiol* 1999; 33: 273–82.
124. Nandwani R, Gourlay Y. Possible interaction between sildenafil and HIV combination therapy [letter]. *Lancet* 1999; 353: 840.
125. Hall MC, Ahmad S. Interaction between sildenafil and HIV-1 combination therapy [letter]. *Lancet* 1999; 353: 2071–2.
126. Cordon-Cardo C, O'Brien JP, Casals D, *et al.* Expression of the multidrug resistance gene product (p-glycoprotein) in human normal and tumor tissues. *J Histochem Cytochem* 1990; 38: 1277–87.

127. Gomez DY, Wacher VJ, Tomlanovich SJ, Hebert MF, Benet LZ. The effects of ketoconazole on the intestinal metabolism and bioavailability of cyclosporine. *Clin Pharmacol Ther*1995; 58: 15–9.
128. First MR, Schroeder TJ, Michael A, Hariharan S, Weiskittel P, Alexander JW. Cyclosporine–ketoconazole interaction. Long-term follow-up and preliminary results of a randomized trial. *Transplantation* 1993; 55: 1000–4.
129. Keogh A, Spratt P, McCosker C, Macdonald P, Mundy J, Kaan A. Ketoconazole to reduce the need for cyclosporine after cardiac transplantation. *N Engl J Med* 1995; 333: 628–33.
130. Ashton M, Hai TN, Sy ND, *et al.* Artemisinin pharmacokinetics is time-dependent during repeated oral administration in healthy male adults. *Drug Metab Dispos* 1998; 26: 25–7.
131. Fromm MF, Busse D, Kroemer HK, Eichelbaum M. Differential induction of prehepatic and hepatic metabolism of verapamil by rifampin. *Hepatology* 1996; 24: 796–801.
132. Holtbecker N, Fromm MF, Kroemer HK, Ohnhaus EE, Heidemann H. The nifedipine–rifampin interaction. Evidence for induction of gut wall metabolism. *Drug Metab Dispos* 1996; 24: 1121–3.
133. Fromm MF, Dilger K, Busse D, Kroemer HK, Eichelbaum M, Klotz U. Gut wall metabolism of verapamil in older people: effects of rifampicin-mediated enzyme induction. *Br J Clin Pharmacol* 1998; 45: 247–55.
134. van Agtmael MA, Gupta V, van der Graaf CAA, van Boxtel CJ. The effect of grapefruit juice on the time-dependent decline of artemether plasma levels in healthy subjects. *Clin Pharmacol Ther* 1999; 66: 408–14.
135. Eagling VA, Back DJ, Barry MG. Differential inhibition of cytochrome P450 isoforms by the protease inhibitors, ritonavir, saquinavir and indinavir. *Br J Clin Pharmacol* 1997; 44: 190–4.
136. Gutmann H, Fricker G, Drewe J, Toeroek M, Miller DS. Interactions of HIV protease inhibitors with ATP-dependent drug export proteins. *Mol Pharmacol* 1999; 56: 383–9.
137. Barry M, Gibbons S, Back D, Mulcahy F. Protease inhibitors in patients with HIV disease. Clinically important pharmacokinetic considerations. *Clin Pharmacokinet* 1997; 32: 194–209.
138. Merry C, Barry MG, Mulcahy F, *et al.* Saquinavir pharmacokinetics alone and in combination with ritonavir in HIV-infected patients. *AIDS* 1997; 11: F29–33.
139. Guengerich FP, Kim DH. In vitro inhibition of dihydropyridine oxidation and aflatoxin B_1 activation in human liver microsomes by naringenin and other flavonoids. *Carcinogenesis* 1990; 11: 2275–9.
140. Miniscalco A, Lundahl J, Regardh CG, Edgar B, Eriksson UG. Inhibition of dihydropyridine metabolism in rat and human liver microsomes by flavonoids found in grapefruit juice. *J Pharmacol Exp Ther* 1992; 261: 1196–9.
141. He K, Iyer KR, Hayes RN, Sinz MW, Woolf TF, Hollenberg PF. Inactivation of cytochrome P450 3A4 by bergamottin, a component in grapefruit juice. *Chem Res Toxicol* 1998; 11: 252–9.
142. Edwards DJ, Bellevue FH, Woster PM. Identification of 6′,7′-dihydroxybergamottin, a cytochrome P450 inhibitor, in grapefruit juice. *Drug Metab Dispos* 1996; 24: 1287–90.
143. Schmiedlin-Ren P, Edwards DJ, Fitzsimmons ME, *et al.* Mechanisms of enhanced oral availability of CYP3A4 substrates by grapefruit juice constituents: decreased enterocyte CYP3A4 concentration and mechanism-based inactivation by furanocourmarins. *Drug Metab Dispos* 1997; 25: 1228–33.
144. Kukuda K, Ohta T, Oshima Y, Ohashi N, Yoshikawa M, Yamazoe Y. CYP3A4 inhibitors in grapefruit juice: furocoumarin dimers as components of drug interaction. *Pharmacogenetics* 1997; 7: 391–6.
145. Guo LQ, Fukuda K, Ohta T, Yamazoe Y. Role of furanocoumarin derivatives on grapefruit juice-mediated inhibition of human CYP3A4 activity. *Drug Metab Dispos* 2000; 28: 766–71.
146. Edwards DJ, Fitzsimmons ME, Schuetz EG, *et al.* 6′,7′-Dihydroxybergamottin in grapefruit juice and Seville orange juice: effects on cyclosporine disposition, enterocyte CYP3A4, and P-glycoprotein. *Clin Pharmacol Ther* 1999; 65: 237–44.

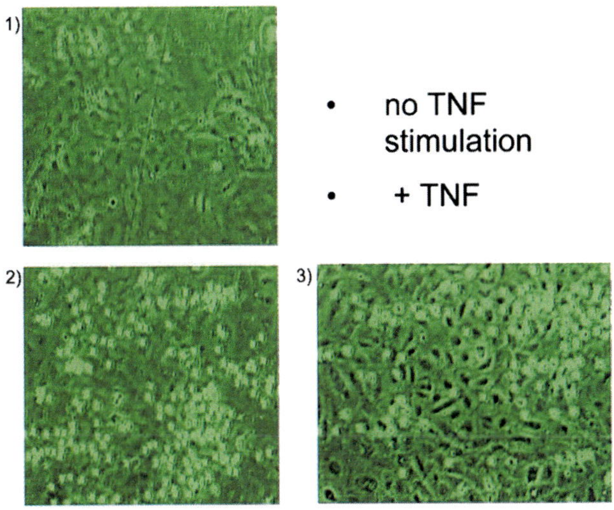

COLOR PLATE 1

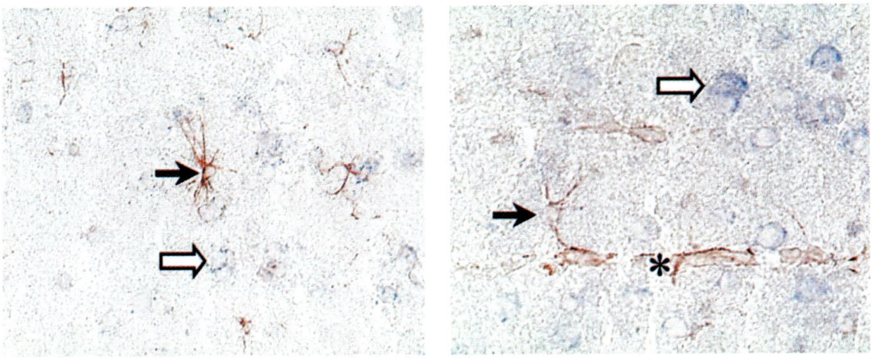

COLOR PLATE 2

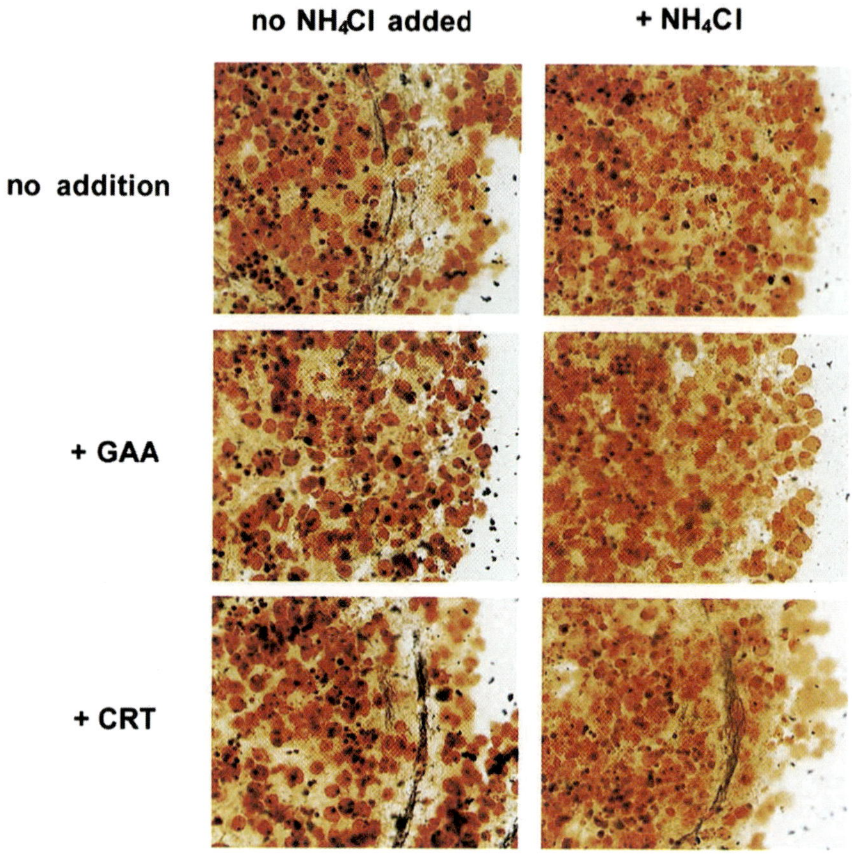

COLOR PLATE 3

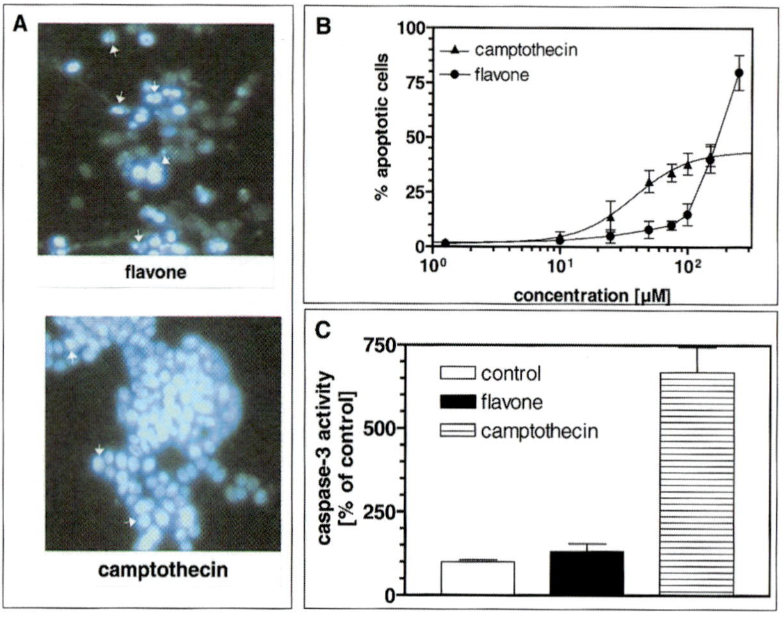

COLOR PLATE 4

Genetic Expression and Nutrition, edited by Claude Bachmann and Berthold Koletzko. Nestlé Nutrition Workshop Series, Pediatric Program, Vol. 50. Nestec Ltd., Vevey/Lippincott Williams & Wilkins, Philadelphia, © 2003.

Diet, Gene Expression, and Apoptosis: Clues to Cancer Prevention?

Hannelore Daniel and Uwe Wenzel

Department of Food and Nutrition Sciences, Molecular Nutrition Unit, Technical University of Munich, Freising-Weihenstephan, Germany

THE ROLE OF DIET IN THE INITIATION OF CANCER

The development of cancer results from the interplay between genetic factors and the environment, and dietary factors have been identified as modulators of almost every step in the process. The complexity of this interplay between genetic factors and the nutritional environment is summarized in Fig. 1. In its initial stage, the development of a tumor requires genetic damage, caused either by endogenous processes or by exogenous compounds that produce mutations or even gene deletions.

Free radicals are one of the endogenous mediators of oxidative DNA damage and mutations. They are produced by normal oxidative metabolism in cells, but their production is increased under pathologic conditions such as inflammation. The hydroxyl radical is considered to be the ultimate radical species responsible for DNA damage. There is evidence that the oxidative damage resulting, for example, in 8-oxoguanine and oxidized thymine bases, is involved in cancer induction and progression (1,2). Defense mechanisms protecting the DNA from the free radical attack include the antioxidant enzymes superoxide dismutase, catalase, and peroxidase, which require for their function trace elements such as selenium, zinc, and manganese. Other dietary constituents—for example, vitamin C and E, carotenoids, and flavonoids—also contribute to the antioxidative capacity of a cell or an organism.

In addition to those "classical" antioxidants, many other compounds, such as caffeine or diallyl polysulfides from aged garlic extracts (3,4), have been identified as antioxidants. However, under certain conditions and depending on a variety of factors, including the level of intake, antioxidants can also have pro-oxidant properties. In the presence of free transition metals such as iron or copper, oxidative DNA damage can be induced by, for example, ascorbic acid through acceleration of the Fenton reaction (5). The presence of metals has also been shown to accelerate DNA nicking by phenolic antioxidants (6).

Another important endogenous mechanism influencing the genetic events involved in tumor development is the biomethylation of DNA. About 3% of cytosine residues in mammalian DNA are methylated after replication, most probably to

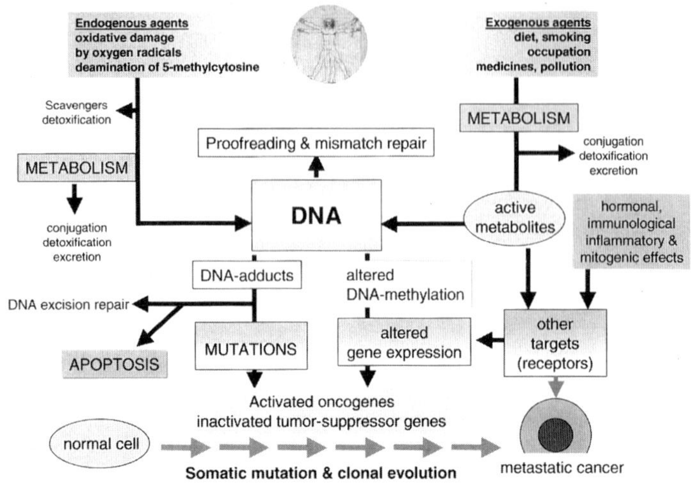

FIG. 1. Processes that affect the alteration of DNA and hence promote the transformation of a normal cell into a transformed cell and a tumor. A large variety of endogenous as well as exogenous factors can interact with the DNA and cause alterations such as mutations or DNA adduct formation. Protective nutritional factors that modulate those pathways are, for example, antioxidants scavenging reactive oxygen species or compounds that promote elimination of toxicants by the "excretory metabolism." DNA caretaking systems provide efficient repair mechanisms for maintaining genome stability. If oncogenes are activated and/or tumor suppressor genes are inactivated permanently, however, a normal cell can transform finally into a metastatic cancer. Essentially every step in these processes of tumor promotion and progression can be influenced by dietary factors (see text).

compartmentalize the genome into active and inactive regions. Changes in the pattern and intensity of methylation are found as the earliest and most consistent molecular abnormalities in human cancers (7,8). Methylation of "CpG islands"—that is, regions rich in cytosine–guanine doublets—in promoter regions of genes is a common mechanism of epigenetic silencing in somatic cells. If this silencing process takes place in tumor suppressor genes, cancer initiation and promotion are increased (9). It is not yet understood whether certain promotors are predisposed to gene silencing by this process or whether it occurs by chance (9). However, the inheritance of DNA methylation patterns in humans has been described (10), and the extent to which genes are silenced by this epigenetic mechanism appears to vary with age (11). It is possible, therefore, that inherited variation in the genes that regulate epigenetic silencing, as well as alterations in the activities of the corresponding gene products with age, contribute to an individual's susceptibility to cancer.

Dietary factors that might interfere with the processes of DNA methylation are the so-called lipotropes, which encompass the methyl group donors methionine and choline, and the methyl group transfer factors folic acid and vitamin B_{12}. Diets low in lipotropes are often associated with a higher risk for cancer development and for the altered reactions that are involved in DNA stability and carcinogenesis (12). Apart from hypomethylation of DNA, the disturbance of the nucleotide pools that is commonly associated with folate deficiency leads to uracil misincorporation into replicated DNA, and this might also contribute to mutagenesis (13,14). It must be stressed that antifolates used as chemotherapeutic agents act by generating a functional folate deficiency that is followed by enhanced mutagenesis, DNA damage, and apoptosis. An unfortunate consequence of this mode of action of antifolates and many other chemotherapeutic agents is that patients surviving this treatment are at increased risk of developing iatrogenic cancers because of enhanced rates of mutagenesis (15).

Exogenous genotoxic carcinogens encompass a wide range of chemically different compounds with the common property of forming chemical bonds with DNA, resulting in the generation of "DNA adducts" (16). Adduct formation can lead to base substitution, base deletions, or base additions and thus increased mutation frequencies. Mutations within a given gene often occur at "hot spots" determined by the sequence of DNA and by the chemical structure and reactivity of the adduct-forming agent (17). Although a single chemical will give rise to several different DNA adducts, it has to be assumed that a defined chemical has its individual "fingerprint" of DNA adducts generated. Molecular epidemiology tries to identify the mutational spectrum of certain cancers associated with an individual carcinogen to which subgroups of the population developing a particular kind of cancer have been exposed (18–20).

Most of the genotoxic carcinogens, such as polycyclic aromatic hydrocarbons, aromatic amines, and alkyl and arylnitrosamines, are activated by enzymes that are normally involved in the metabolism, detoxification, and elimination of potentially toxic compounds (21). The procarcinogens are activated by phase I enzymes, most of which belong to the superfamily of cytochrome P450–dependent monooxygenases. The normal function of phase I enzymes is to insert an oxygen molecule into relatively inert and usually nonpolar substrates. These intermediates formed by oxidative metabolism are then conjugated by phase II enzymes with endogenous polar substrates—such as glucuronides, glutathione or sulfate—to allow the excretion of the now more polar products (22). Phase II enzymes can therefore reduce the cellular exposure to carcinogens, whereas phase I enzymes can increase it. There is some evidence that the genetic susceptibility to cancer may in part be determined by polymorphisms and altered functions of these enzymes. For example, it has been postulated that polymorphisms in the genes that code for N-acetyltransferases (NATs) confer a differential susceptibility to the effect of red meat consumption on the development of colorectal cancer (23). The strongest association between red meat intake and cancer risk was found for individuals with the "rapid acetylator" genotype, in whom activation of the heterocyclic amines formed during the cooking of meats and hydroxylated by phase I metabolism is more rapid. Other examples include the association between cancer risks and certain polymorphisms in xenobiotic metabolizing enzymes such as glutathione-S-transferase (24) or cytochrome P450 E2 (25).

While polymorphisms may determine how efficiently procarcinogens are toxified or detoxified, dietary constituents also alter the phase I or phase II metabolism of potentially hazardous compounds, by an acute effect on the enzymes or by actions at the gene expression level. The isothiocyanate sulforaphane isolated from broccoli sprouts, for example, shows strong anticancer activity in a rat mammary tumor model by activating phase II enzymes (26), and isothiocyanates from cruciferous vegetables have been shown to increase the excretion of carcinogens in humans (27). Other isothiocyanates from *Brassica* vegetables or diallyl sulfone from garlic have been shown to inhibit phase I metabolism (28). Certain flavonoids identified in edible plants may serve as bifunctional inducers—that is, in addition to inducing phase II enzymes, they also can induce the procarcinogen activating phase I enzymes (29).

Based on their mechanism of action, compounds suppressing phase I enzymes or activating phase II enzymes can be regarded as "blocking agents"—that is, they are effective when given before or at the same time as the carcinogen. "Suppressing agents," in contrast, are also effective when given *after* the administration of a carcinogen, because they may induce apoptosis to eliminate transformed cells or cause growth arrest and promotion of differentiation in initiated cells.

Although these mechanisms that determine the occurrence of mutations in the genome are crucial, DNA that has undergone mutational alterations is normally subjected to repair processes. Here, the tumor suppressor p53 plays a very important role. The p53 protein is part of the gene damage survey system, which is also linked to the control of the cell cycle. Arrest of an initiated cell by this control system allows DNA repair mechanisms to proceed, and if DNA damage has exceeded the capacity for adequate repair, apoptosis will be induced to eliminate cells that have accumulated too many mutations. Various gene products are involved in the repair process for eliminating errors introduced during replication. Beyond the DNA care-taking systems, base excision repair (acting on DNA damage as induced by x-rays, oxygen radicals, alkylating agents, and spontaneous reactions), nucleotide excision repair (acting on DNA damage as induced by ultraviolet light or polycyclic aromatic hydrocarbons), recombinational repair (acting on DNA damage as induced by x-rays or antitumor agents), and mismatch repair (acting on replication errors) take place (30). Inherited genetic defects in any of these mechanisms generally appear to predispose to cancer development by allowing the accumulation of mutations within a cell.

Similarly, as already mentioned, mutations in tumor suppressor genes or protooncogenes also predispose to tumor development, because tumor suppressors as well as the protooncogenes are involved in the pathways regulating cell proliferation (Fig. 2). Tumor suppressors normally inhibit these pathways by inducing cell cycle arrest or apoptosis, whereas the protooncogenes—especially when mutated—promote cell growth. Accelerated cell division rates caused by loss-of-function mutations in tumor suppressors or gain-of-function mutations in protooncogenes (converting them to oncogenes) predispose the cell to accumulation of mutations because the time for DNA repair is shortened.

Paradigms for tumor suppressors and (proto-) oncogenes are the *p53* and *ras* genes, respectively. Both genes are found mutated in a large variety of human cancers with a high incidence. For example, mutations in the *ras* gene are found in 90% of pancreatic tumors and in 50% of colonic tumors. In lung tumors, as well as most other solid tumors, *ras* shows a mutation frequency of around 30% (31). The *p53*

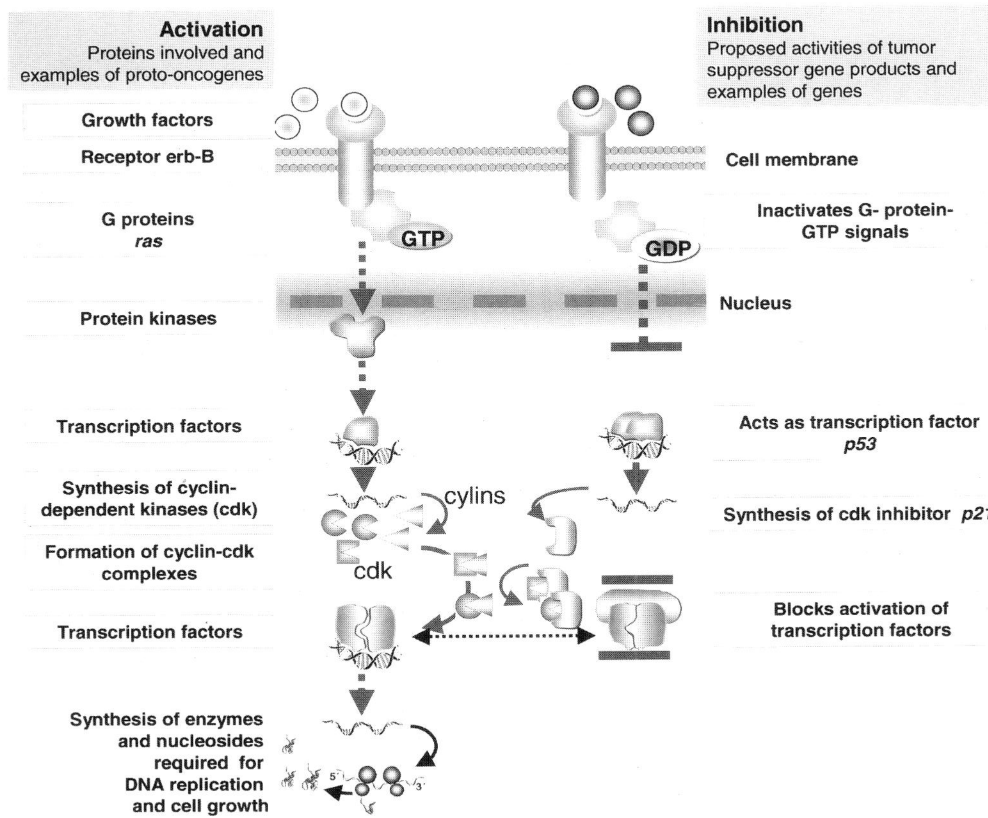

FIG. 2. Accelerated cell growth by loss-of-function mutations in tumor suppressors and/or gain-of-function mutations in protooncogenes. Both the ras oncogene and the tumor suppressor p53 show a high incidence of mutations in a large variety of human cancers. Ras carrying a gain-of-function mutation is permanently activated by losing its ability to hydrolyze guanosine triphosphate (GTP). The p53 with a loss-of-function mutation loses its ability to induce the expression of p21, which promotes cell cycle arrest, or of Gadd45, which enables DNA repair. Also, mutated p53 is no longer capable of inducing apoptosis in cells with excessive DNA damage.

gene bears mutations in 70% of all tumors (32). Ras belongs to the family of small cellular guanosine triphosphate (GTP) binding proteins (G proteins) and serves a signal transduction molecule with multiple functions including the stimulation of cell proliferation. The nonmutated ras hydrolyzes GTP after activating downstream signaling molecules, and becomes inactivated upon GTP hydrolysis. Gain-of-function mutations as found in *ras* prevent the hydrolysis of GTP and consequently keep ras in an activated state (33). The tumor suppressor p53 monitors genomic integrity and therefore has been called the "guardian of the genome"; p53 itself is involved in proofreading (34) and when DNA damage has occurred, it transcriptionally activates genes such as *p21*, coding for a cyclin-dependent kinase inhibitor that contributes to cell cycle arrest, or *Gadd45*, coding for a protein that increases the accessibility of damaged DNA to proteins involved in repair processes (35). The mechanisms whereby p53 induce apoptosis are poorly understood. However, it has been shown that p53 localizes to mitochondria and is involved in the onset of the apoptotic process by causing changes in mitochondrial membrane potential, leading to the release of cytochrome c, an important activator of the apoptotic cascade (36).

ROLE OF DIETARY COMPONENTS IN THE PROMOTION AND PROGRESSION OF CANCER

One of the mechanisms whereby proteins involved in controlling cell growth can acquire increased activity is by a sporadic gain in function mutations as described above. Increased activity, however, may also result from overexpression due to amplification or translocation of the corresponding genes. Gene amplification refers to increased copy numbers, usually as a result of end-to-end replication of a gene at the same chromosomal location. Such an amplification is found in almost 30% of breast carcinomas for the erb-B protein, a plasma membrane receptor for heregulin, which is a growth factor structurally related to epidermal growth factor (Fig. 2) (37). Overexpression of erb-B is associated with clinical outcome in cancer mainly because of its mitogenic effects, but also through an erb-B–induced increase in the production and secretion of vascular endothelial growth factor, which stimulates the angiogenesis necessary for progressive tumor growth (38). A paradigm for translocations associated with cancer development is Burkitt lymphoma. Here, chromosomal translocations fuse the *c-myc* gene with genes coding for immunoglobulins (39); accordingly, *c-myc* is placed under the control of promotors normally activated only during infections. Indeed, there is a strong association between the occurrence of infections with Epstein–Barr virus and Burkitt lymphoma (39). The normal cellular function of the c-myc protein is to act as a transcription factor participating in many cellular functions, such as replication, growth, metabolism, and differentiation. The constitutive expression of myc family members caused by such translocations seems to be a key event in the genesis of many cancers (40).

It was believed for a long time that genomic alterations—including mutations, amplifications, and translocations—must occur in selected genes in the initiation phase to provide a growth advantage for cells carrying these alterations in the promotion phase. In the worst case, this of course would allow the accumulation of further genomic alterations in the progression phase and hence promote cancer development. It becomes increasingly obvious, however, that many compounds are capable of inducing cancer—at least in animal models—without showing any genotoxicity. These compounds may act by stimulating proliferation, either by inducing regenerative proliferation because of their toxicity or by affecting proliferative pathways. These nongenotoxic substances, however, can enhance mutagenesis indirectly by reducing the time available for the cell to repair the replicated DNA. Consequently, cellular hyperproliferation is crucial in cancer development, especially in cells that already have suffered DNA damage (41,42).

A huge variety of food ingredients has been shown to affect cell proliferation rates in either direction. They therefore contribute in the promotion phase to cell transformation and may either reduce or increase the risk of cancer development and progression. Various studies—mainly in animal models—have shown that a high intake of dietary fat accelerates the development of breast cancer. In particular, the content of n-6 fatty acids such as linoleic acid is associated with increased rates of tumor formation, whereas the n-3 fatty acids such as eicosapentaenoic acid and docosahexaenoic acid do not affect tumor development but can inhibit the promoting effects of n-6 fatty acids (43). Another potent suppressor of mammary tumors in animal studies is conjugated linoleic acid, a good source of which is milk fat (44). In humans, however, there is evidence that increased body weight is a more critical variable for cancer development in the mammary gland than fat intake *per se* or the fatty acid pattern (45). Overnutrition in early life causes rapid growth, which results in early menarche, and consequently, the mammary tissue is confronted with high levels of growth-stimulating estrogens for a longer time. Overnutrition and a high fat consumption in later life results in a breast cancer promoting hormonal imbalance by increasing the levels of free steroid hormones associated with visceral obesity (46).

While certain dietary factors promote the proliferation of cells in mammary tissue by enhancing the circulating estrogen levels, there are also dietary ingredients that can suppress these effects. Phytoestrogens, found in vegetables such as soya, are weak estrogens but in competition with more potent endogenous estrogens show a net antiestrogenic activity. They also act as aromatase inhibitors, thus reducing the synthesis of estrogens (47). These actions of the phytoestrogens have been related to a reduced incidence of breast cancer associated with a diet rich in soya products (47).

It has been suggested that certain diets also act as tumor promoters in other types of cancer such as colon cancer, where high intakes of fat and phosphate have been linked to colonic hyperproliferation (48) and colon cancer development (49). Because there are dietary factors that increase the proliferation rate of colonic cells, there are also compounds that block uncontrolled tissue growth. Butyrate, produced by the gut microflora from fermentable dietary fiber, is one of the key factors shown to suppress growth of cancer cells by increasing the expression of proteins that permit cell cycle arrest. The induction of the cyclin-dependent kinase inhibitor p21, in particular, seems crucial for the growth-inhibiting effects of butyrate (50).

Selected flavonoids found in a large variety of vegetables and fruits can also cause cell cycle arrest. Their actions are linked to the increased expression of cell cycle regulators, such as p21, or the reduced expression of different cyclins or proteins affecting proliferation, such as cyclo-oxygenase-2 (COX-2) (51,52). The effects of these plant components on COX-2 expression could provide an important chemopreventive mechanism in colon cancer development because COX-2 is overexpressed in about 90% of all human colorectal cancers; moreover COX-2 inhibition by long-term treatment with non-steroidal anti-inflammatory drugs is considered to be a chemopreventive strategy (53,54). The mechanism whereby enhanced COX-2 expression promotes cancer development is probably through an increase in cell proliferation reflecting increased production of growth-promoting prostaglandins (55).

APOPTOSIS: THE CLUE TO CANCER PREVENTION?

Apoptosis is the process of programmed cell death that serves as a mechanism for regulating cell numbers during development, as well as in the adult organism, and for eradicating cells that have sustained damage to their DNA (56). Approximately 50 to 70 billion cells perish every day in an adult because of apoptosis. Cell death in self-renewing tissues, such as the skin, gut, and bone marrow, is necessary to maintain tissue and organ structure and functionality. Unlike accidental cell death caused by infarction or trauma, apoptosis represents a physiologic process culminating in the fragmentation of cells that are cleared by phagocytosis in neighboring cells without causing inflammatory reactions or tissue scarring.

Apoptosis of a cell is triggered by external plasma membrane–associated receptors as well as by internal sensors (Fig. 3). The ability of tumor cells to resist apoptosis in response to the death signals has become a central theme in the understanding of how tumors develop (57). For both chemoprevention and cancer treatment, the induction of apoptosis in precancerous or cancerous cells may be more important than inhibiting proliferation or inducing differentiation in such cells. This is because under certain unfavorable conditions, cell cycle arrest can provide an escape mechanism for apoptosis, as is often found in cancer treatments with cytostatic drugs (58,59).

On a morphologic basis, apoptosis is characterized by membrane blebbing, cell shrinkage, protein fragmentation, chromatin condensation, DNA degradation, and the engulfment of the generated corpses by neighboring cells (60). Most of the morphologic changes are caused by a set of proteases that are specifically activated in apoptotic cells (61). As these death-inducing proteases possess a cysteine in their active site and cleave their substrates C-terminally at an aspartate residue, they are called *caspases* (*c*ysteine proteases that cleave after *asp*artate). Caspases can autocatalytically cleave and activate themselves as well as other members of the caspase family. Once the first caspase in the apoptotic pathway has been activated, it processes downstream caspases and initiates a cascade of amplifying events that lead to the apoptotic death of a cell (62). The substrates of the downstream effector caspases are proteins that are either activated by the cleavage, such as the caspase-activated DNase, or inhibited, such as the inhibitor of caspase-activated DNase, the poly-ADP ribose polymerase or nuclear lamins.

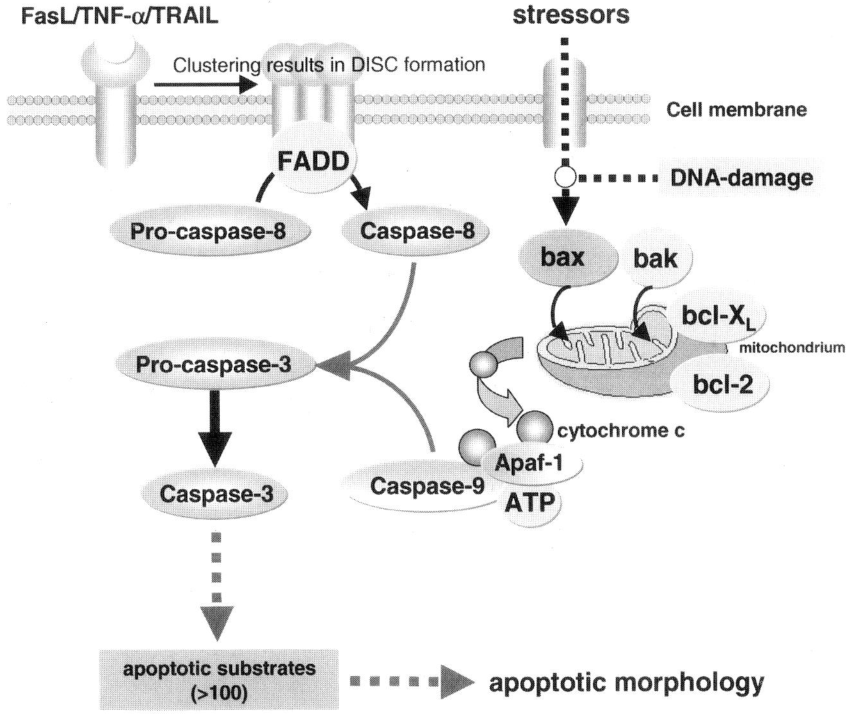

FIG. 3. Major pathways of apoptosis induction. Binding of Fas ligand (FasL), tumor necrosis factor α (TNF-α), or tumor necrosis factor related apoptosis-inducing ligand (TRAIL) leads to a clustering of the Fas/TNF-R1 death receptor in the cell membrane. As a result, the death inducing signaling complex (DISC) is formed that recruits, through the adaptor molecule FADD (Fas-associated death domain protein), multiple procaspase-8 molecules, resulting in the activation of this upstream caspase. Extracellular stressors or DNA damage activates the mitochondrial pathway by allowing proapoptotic proteins to migrate to the mitochondria. At the mitochondrial surface, the relation of pro- and antiapoptotic activities determines whether cytochrome c is released or not. Released cytochrome c contributes to the formation of the apoptosome, which, by activation of caspase-9, activates the effector caspase-3. At the level of caspase-3, both apoptosis pathways converge. Caspase-3 activation finally leads to the classical morphologic changes observed in cells undergoing apoptosis.

Various caspase substrates have been identified that induce each of the classic features of apoptosis (63). Two major apoptosis pathways have been defined in mammalian cells, the Fas/TNF-R1 death receptor pathway and the mitochondrial pathway (Fig. 3) (61). The first mechanism is triggered by the Fas ligand or tumor necrosis factor α (TNF-α) binding to the extracellular domain of the death receptors. Binding induces receptor clustering and the formation of a death inducing signaling complex (DISC). Through the adaptor molecule FADD (Fas-associated death domain protein), DISC recruits multiple procaspase-8 molecules, resulting in the activation of this upstream caspase. The mitochondrial pathway is triggered by intrinsic proapoptotic stimuli such as DNA damage and by extracellular or environmental factors such as cellular stress. The different triggers that converge in the mitochondrial pathway often act through the activation of proapoptotic members of the bcl-2 family, such as bax or bak, or through the inhibition of its antiapoptotic members, such as bcl-2 or bcl-X_L (64). Whereas bcl-2 and bcl-X_L remain constantly attached to the mitochondria, many proapoptotic members can shuttle between the cytosol and mitochondria (65). The cytosolic forms represent inactive proteins that are activated by proteolysis, dephosphorylation, and probably several other mechanisms, resulting in recruitment to the mitochondria in order to exert their proapoptotic effects.

The release of cytochrome c from the mitochondria is the critical step in the mitochondrial pathway, which is determined by the relation of the activities of pro- and antiapoptotic proteins at the mitochondrial surface. When released, cytochrome c associates with Apaf-1 (apoptotic protease activating factor-1) and then pro–caspase-9 is recruited and activated in the presence of adenosine triphosphate (ATP) or deoxyadenosine triphosphate (dATP). As a part of the so-called apoptosome, caspase-9 activates the effector–caspase-3, the point at which the mitochondrial pathway and the death receptor pathway converge (Fig. 3) (61). It has to be stressed that this model is an oversimplification of the apoptotic machinery—for example, caspase-3 activation can be antagonized by inhibitors of apoptosis (IAPs), which themselves are antagonized by Smac/DIABLO protein released from mitochondria. Moreover, downstream of caspase-3, the apoptotic program branches into a multitude of pathways. There is also a third apoptosis pathway that is initiated by stress in the endoplasmic reticulum, including disruption of calcium homeostasis. As a result caspase-12 is activated and most probably also converges on caspase-3 activation (66).

The complexity of apoptosis means that a multitude of proteins with impaired regulation during cancer development could serve as targets for treatment. Reduced sensitivity toward apoptotic stimuli, often observed in neoplasms, could result from loss-of-function mutations or reduced expression of proapoptotic proteins, or alternatively from gain-of-function mutations or enhanced expression rates of antiapoptotic proteins (67,68). Interestingly, tumor development is often associated with an overexpression of the Fas ligand in combination with a loss-of-function mutation in the Fas receptor of the transformed cells. In this way, the cancer cell acquires two advantages: resistance to apoptosis and, through Fas ligand overexpression, the killing of bystander cells such as immune cells providing an immune escape mechanism (69).

Other proteins involved in the apoptotic machinery and often found to be overexpressed in various tumor types are the antiapoptotic proteins bcl-2 and bcl-X_L, which exert their principal effects through stabilization of the mitochondrial membrane. Loss of Apaf-1 is also a relatively common event in malignant melanoma, and presumably also confers resistance to apoptosis (70). Alterations in the activities of proteins involved in the apoptotic machinery may not only enable precancerous cells to survive but may also play an important role in metastasis. Normally, detachment of cells from their neighbors or from the basal stroma is followed by spontaneous apoptotic suicide termed *anoikis.* The detachment deprives the cell of integrin- and cadherin-mediated survival signals, which, at least in part, account for the induction of apoptosis. Changes in integrin expression patterns and levels have been described during the progression of tumors from benign to malignant phenotypes and might contribute to the resistance of cells to anoikis (71).

MACRONUTRIENTS AND APOPTOSIS

It is becoming increasingly evident that diet has a major impact on cancer development, and that apoptosis—the dysregulation of which is crucial in allowing transformed cells to escape control—may directly or indirectly be affected by diet as well. One of the most striking findings is that apoptosis rates are increased by energy restriction (72). On a molecular basis, this increase in apoptosis is associated with an enhanced expression of p53 and a reduced expression of insulinlike growth factor 1 (IGF-1) (72,73). Besides the increased apoptosis, an increase in fidelity of DNA replication and enhanced DNA repair contribute to the maintenance of genomic integrity by low-energy diets. It is suggested therefore, that energy restriction retards tumor development as well as most physiologic indices of aging by maintaining the organism's ability to deal with stress (72).

Effects of diet on apoptosis, however, cannot be predicted *per se* and strongly depend on the protein pattern expressed in the cell. This is convincingly demonstrated by overexpression of the antiapoptotic proteins bcl-X_L and Akt (74). Cells overexpressing Akt show an increased glucose transporter expression level as well as enhanced glycolytic activity associated with the requirement of high levels of extracellular nutrients to support survival of the cells. Bcl-X_L–overexpressing cells, in contrast, survive in a more vegetative state, with reduced glycolytic activity, and are consequently less dependent on extracellular nutrients (74).

In addition to the effects on apoptosis of the macronutrients (carbohydrates, fat, and proteins) as sources of cellular energy or delivery systems for structural components, dietary constituents might also affect apoptosis by other mechanisms. Prebiotic fructans, for example, have been shown to increase apoptosis in the rat colon, and this contributes to the growing evidence that fructans may have cancer-preventing properties (75). In this case, the production of butyrate from the soluble fiber seems to be crucial for the inhibition of early and late events in colon tumorigenesis by controlling the transcription, expression, and activity of the key proteins involved in the apoptotic cascade (76).

Dietary fat affects apoptosis not only because it is a high-energy substrate but also because of its fatty acid constituents. It has recently been shown that a high-fat diet (rich in saturated fatty acids) increases eicosanoid production by increased COX-2 expression and causes a reduction in apoptosis. These effects are associated with increased colon cancer development and are not observed with a diet rich in n-3 fatty acids or a low-fat, corn oil–based diet (77). Further studies showed that consumption of a fish oil–containing diet is associated with increased apoptotic cell death, and it was suggested that n-3 polyunsaturated fatty acids can protect against carcinogenesis (78).

Among the amino acids, glutamine appears to play an important role in apoptosis. It serves as the preferred energy substrate in intestinal epithelial cells, and glutamine deprivation in the gut induces apoptosis, an effect not observed with cysteine or methionine deprivation (79). The effects of glutamine deprivation do not appear to be caused by lack of energy, because glutamine deprivation in leukemia and lymphoma cells is associated with apoptosis independently of the energy status, and glucose deprivation induces necrosis, not apoptosis (80).

The importance of nutritional factors in controlling cell proliferation and apoptosis rates has recently been shown in an exciting study in *Drosophila*. Both germline and somatic stem cells in the fruit fly adjust their proliferation rates to nutrient supply (81). Depending on the protein content of the diet, egg production rates varied almost 60-fold, reflecting a four-fold difference in proliferation rates and a 15-fold difference in apoptotic cell death (81). Moreover, it was found that an intact insulin pathway was necessary to fully up-regulate cell proliferation (81).

The finding that somatic stem cells can be drastically influenced by nutritional status may be of special importance in relation to cancer development, because somatic stem cells are probably the progenitors of transformed cells. In the case of colonic tissue, this was suggested on the basis of studies showing that the conversion of all cells in a human crypt to a radiation-induced mutant phenotype takes approximately 1 year (82). This is the time period necessary for a single stem cell to accumulate enough mutations to become transformed. On the other hand, this period is much longer than the lifespan of 2 to 3 days in a differentiated normal enterocyte. Intestinal stem cells are therefore very important, and it is suggested that the Wnt (Wingless-type) pathway may play a major role in the maintenance of the stem cell compartment in the colon (83). Wnt signaling stabilizes β-catenin, and alterations in this pathway are often found in adenomatous polyposis coli owing to the loss of heterozygosity in the adenomatous polyposis coli (APC) gene (84). Besides the loss-of-function mutation in APC, a stabilizing mutation in β-catenin can also activate the Wnt signaling cascade (84). B-Catenin transmits the signals of the Wnt pathway to the nucleus and amplifies the transcription of oncogenes (85). The importance of the Wnt pathway in the development of sporadic colorectal carcinomas implies that the maintenance of the stem cell compartment is crucial for the prevention of cancer.

An increase in stem cell numbers in response to nutritional factors may lead to the disruption of the spatial organization and control of cell proliferation, and consequently, proliferating cells are found in regions where no cell divisions normally occur. Subsequent mutations and clonal selection lead to a further increase in proliferation rates and reduced states of cell differentiation, allowing the accumulation of mutations and finally enabling cells to invade the mesenchyme and cause metastasis (83).

MICRONUTRIENTS AND APOPTOSIS

Vitamins and a limited number of minerals are essential for human life. They are necessary to maintain fundamental functions of the body such as growth, metabolism, and cellular integrity. Apoptosis is also dependent on the supply of micronutrients. It has been shown that the beneficial effects associated with decreases in energy intake described above only occur in the presence of sufficient micronutrient density (86). In the absence of proper nutrition, the sensitivity to carcinogens appears to be enhanced.

With regard to the effects on apoptosis, the most extensively studied micronutrients are vitamin A metabolites and selenium. Recently, retinoids selective for the retinoic acid receptor-α have been shown to induce expression of the membrane-bound tumor selective death ligand, TRAIL (tumor necrosis factor related apoptosis inducing ligand), in promyelocytic leukemia cells. This activates apoptosis by binding to the Fas/TNF-R1 death receptor (Fig. 3) (87). In other cancer cells, retinoic acid stimulates apoptosis by triggering the mitochondrial death pathway involving the activation of the proapoptotic bax and the release of cytochrome c, followed by caspase-9 activation (Fig. 3) (88).

Because selenium is the constituent metal in glutathione peroxidases, its availability contributes to the detoxification of reactive oxygen species. Thus selenium is a critical nutrient with regard to oxygen radical scavenging. However, it also plays an important role in the apoptotic process—its monomethylated forms in particular seem to promote its proapoptotic effects, and this can also be observed in cells with a p53-null phenotype (89). A novel chemopreventive mechanism is proposed involving the catalysis by selenium of the reversible cysteine/disulfide transformations that occur in a number of redox-regulated proteins, including transcription factors (89) that are also important in the apoptotic pathways.

Other micronutrients can affect apoptosis rates and could therefore interact in the process of carcinogenesis. For example, the calcium content of the diet has been shown to alter apoptosis rates, and a high calcium intake may reduce the formation of aberrant crypt foci in the colon (90). Antioxidant vitamins, such as ascorbic acid and tocopherol, have been shown to inhibit radiation-induced apoptosis (91), proving that apoptosis can be modulated by micronutrients in both directions. These antiapoptotic effects of antioxidants are important in anticancer treatment, where efficient apoptosis is crucial. Depletion of the nutritional antioxidants has been shown to be beneficial in a transgenic mouse brain tumor model by increasing the rate of apoptosis (92). However, not all antioxidants inhibit apoptosis *per se*, because vitamin E was shown to induce apoptosis in colon cancer cells by increasing the expression of p21 through a mechanism involving C/EBP-β (a member of the CCAAT/enhancer binding protein family of transcription factors), independent of p53 (93). It has to be assumed, therefore, that various different tissue-specific conditions determine the effect of antioxidant micronutrients on cell death and apoptosis.

NONNUTRITIVE FOOD INGREDIENTS AND APOPTOSIS

Epidemiologic studies consistently show that diets rich in vegetables—and to a lesser extent in fruits—are associated with a relative low cancer risk (94). However, trials to identify single compounds responsible for this effect have so far failed to reproduce the earlier findings (95). This is perhaps not surprising, given that vegetables and fruits contain a myriad of nonnutritive compounds—that is, substances without a direct nutritional value but with the ability to affect cancer development. One class of these dietary constituents is the flavonoids. These consist of an O-heterocyclic ring that is fused to an aromatic ring, and by a CMC bond to a second aromatic ring. The group of the flavonoids encompasses about 5,000 related chemical structures and very few of their biologic activities have been investigated (96).

Apart from the strong antioxidative properties of most but not all of the flavonoids investigated so far, they also possess a multitude of other biologic functions that could be relevant in cancer prevention. In particular, their inhibitory potency at various stages of tumor development in animal studies has attracted much attention (97). On a molecular basis, it has been found that selected flavonoids show pronounced effects on gene expression. For instance, individual compounds from the different flavonoid subgroups—such as the catechin epigallocatechin-3-gallate (EGCG) (98), the isoflavone genistein (99), the flavonol quercetin (100), and flavone (101)—have been shown to induce gene expression of the CDK inhibitor p21 (Fig. 4).

As pointed out above, expression of p21 is usually associated with cell cycle arrest, which allows control and repair of replicated DNA. We have shown recently that genes involved in the control of the apoptotic process are affected by the flavonoid flavone in colon cancer cells (51). The messenger RNA (mRNA) levels of the antiapoptotic factors bcl-X_L and NF-κB and of COX-2 were shown to be potently down-regulated by exposure of cells to flavone (Fig. 4). The expression of the antiapoptotic factors NF-κB and bcl-X_L is associated with colorectal tumorigenesis and disease progression (102,103). Thus, their down-regulation by flavone might prove to represent an effective form of chemoprevention by reducing the threshold for apoptosis. As already mentioned, overexpression of the COX-2 gene is consistently found during neoplastic development in a variety of tissues, and enhanced prostaglandin formation along the dysregulated COX pathways has been shown to mediate tumor promotion in animal models. Prostaglandins are considered to enhance tumor growth by promoting angiogenesis and metastasis and reducing immunosuppression (104). Consequently, COX-2 has become a key target of pharmacotherapy, especially in the prevention and treatment of colorectal cancers (105). Inhibition of COX-2 enzyme activity is achieved by nonsteroidal antiinflammatory drugs, including the newly developed COX-2 specific inhibitors, while the dietary flavonoid flavone prevents the gene expression of the enzyme.

The effects of flavone on gene expression in colonic tumor cells consistently lead to DNA fragmentation and chromatin condensation as important markers of the execution of the final stages of the apoptotic process (51). At flavone concentrations providing maximal effects, almost 80% of the colon cancer cells undergo apoptosis when exposed to flavone, whereas only 40% of cells are killed under the same conditions on exposure to camptothecin, a classical cytostatic drug (Figs. 5A and 5B; See Color Plate 4).

Flavone has been shown to induce apoptosis only in transformed colonocytes, and not in nontransformed primary murine colonocytes (Fig. 5C) (51). A similar selectivity with regard to apoptosis induction was also shown for the green tea polyphenol EGCG (106). The constitutive expression of NF-κB and the binding of NF-κB to DNA *cis*-regulatory elements was inhibited by EGCG at much lower concentrations in the cancer cells than in the non-transformed cells (107). Moreover, the activation of NF-κB by stress stimuli such as TNF-α or lipopolysaccharide was also inhibited by EGCG more prominently in cancer cells than in normal cells (107). The ability to trigger apoptosis in transformed cells but not in nontransformed cells is a prerequisite for effective chemoprevention, because enhanced apoptosis rates in normal cells also contribute to cancer development. This is shown by the increased cancer risks often found in patients undergoing chemotherapy. As induction of apoptosis by

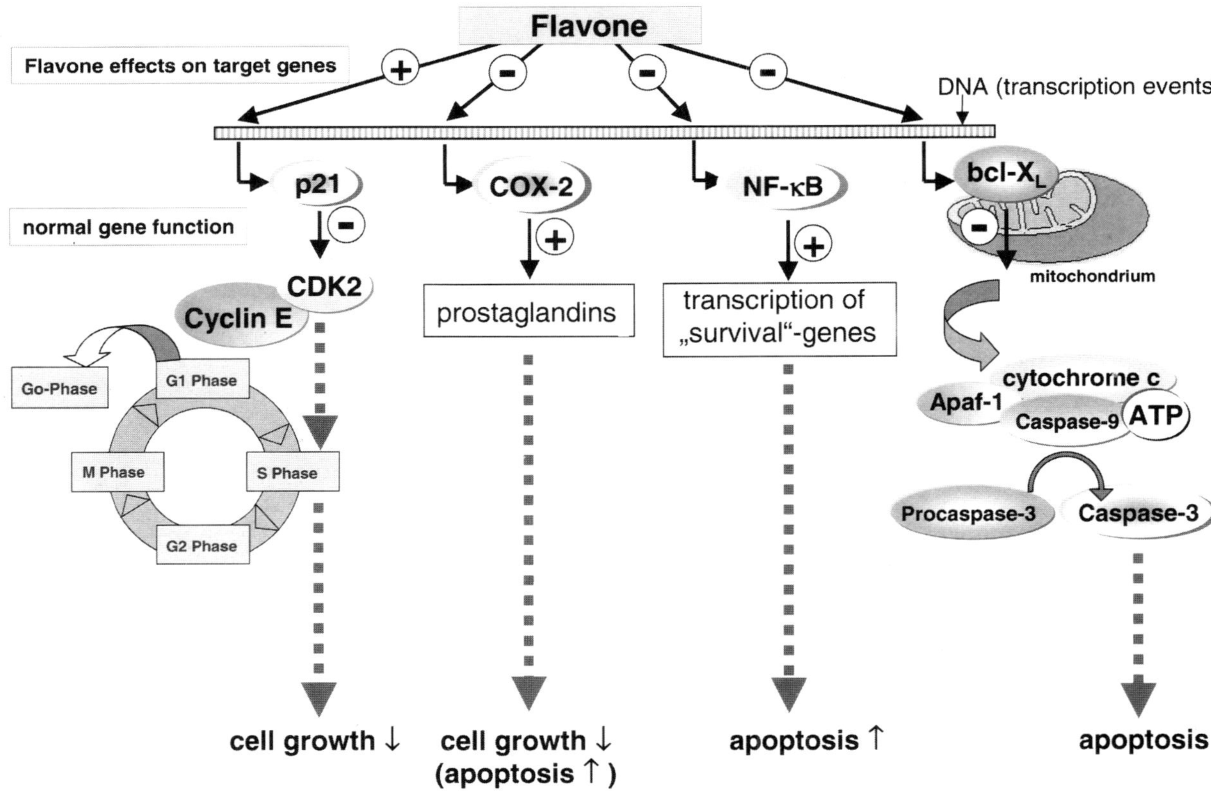

FIG. 4. Selected molecular effects of the flavonoid flavone in human cancer cells. Flavone increases the transcription of the cyclin-dependent kinase (CDK) inhibitor p21. Increased p21 levels cause inhibition of the CDK2–cyclin E complex that is crucial for S-phase entry in the cell cycle. As a result of increased p21 expression, flavone leads to cell cycle arrest. Growth inhibition may also be exerted by flavone through diminished cyclo-oxygenase-2 (COX-2) expression. COX-2 is responsible for the synthesis of prostaglandins. These have been shown to be involved in stimulation of cell growth and inhibition of apoptosis but may also play a role in angiogenesis and metastasis. The reduced mRNA levels of NF-κB observed by flavone exposure may inhibit the transcription of survival genes and thereby allow apoptotic signals to prevail. Moreover, flavone reduces the transcript levels of bcl-X_L, an effect that presumably also occurs in the apoptotic activities of flavone that are observed in cancer cells. +, stimulation; −, inhibition.

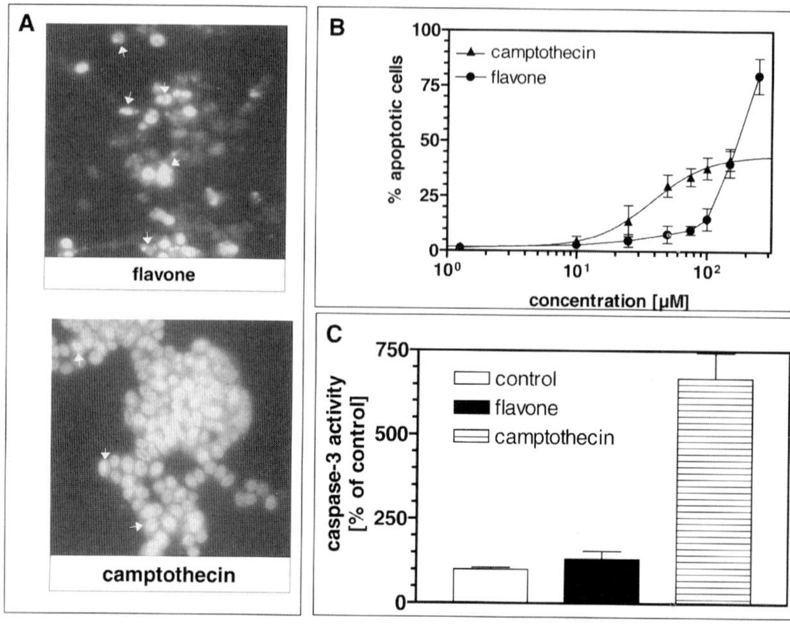

FIG. 5. (A,B) Apoptosis in human colon cancer cells as mediated by flavone or camptothecin. Cellular DNA was stained with Hoechst 33258 for the detection of apoptosis (*arrows*) after exposure of the cells to camptothecin or flavone. The fraction of apoptotic cells as a function of camptothecin or flavone concentrations, as measured after 24 hours of incubation, is given in **(C)**. **(D)** Effects of flavone and camptothecin on apoptosis in primary murine colonocytes. Caspase-3 activity, as a marker of apoptosis, was determined after 24 hours of incubation.

cytotoxic anticancer agents is not selective enough and thereby also affects nontransformed normal cells, the regenerative hyperproliferation may prime the normal cells to accumulate mutations that increase the risk of tumor recurrence (15).

FUTURE PERSPECTIVES: SINGLE NUCLEOTIDE POLYMORPHISMS AND COMPLEMENTARY DNA ARRAYS FOR BETTER CHEMOPREVENTION AND TREATMENT

Based on emerging new technologies, such as complementary DNA (cDNA) expression arrays or high-throughput analysis of single nucleotide polymorphisms (SNPs), new insights into the mechanisms of cancer initiation and progression are possible. Differential gene expression patterns between normal and transformed cells, and differences in amino acid sequences of proteins that are important for cell growth and apoptosis, are determined on a large scale (108–110). Simultaneous inspection of thousands of genes shows that cancers from different individuals differ extensively in their expression patterns. This result is, of course, not surprising because people differ in their genetic background and individual cancers in the same tissue differ in phenotype. This can only be explained by different activities of the proteins caused by their different expression levels or by mutations in the different genes. Nevertheless, gene expression patterns allow clustering to be detected, which indicates the existence of subgroups of genes that may be regulated simultaneously. By identification of these subgroup patterns it is possible, for example, to distinguish two forms of lymphoma that differ dramatically in their survival rates (109).

With chemoprevention as the ultimate goal, high-throughput screening arrays could also be helpful in providing clues to the mechanisms whereby dietary components alter gene expression and thereby affect cancer initiation and progression. In addition to investigating the gene expression patterns, it will be necessary to define important SNPs. As discussed above, SNPs occurring in the promotor can alter its expression level, and SNPs in the coding region of a gene affecting the catalytic or regulatory domains of the protein can critically alter its function within the orchestra of cellular proteins. This may be directly or indirectly associated with an individual's susceptibility to cancer. We are convinced that this will be one of the most important, most exciting, and most rewarding areas of postgenomic research.

REFERENCES

1. Reid TM, Loeb LA. Mutagenic specificity of oxygen radicals produced by human leukemia cells. *Cancer Res* 1992; 52: 1082–86.
2. Kasai H, Okada Y, Nishimura S, Rao MS, Reddy JK. Formation of 8-hydroxydeoxyguanosine in liver DNA of rats following long-term exposure to a peroxisome proliferator. *Cancer Res* 1989; 49: 2603–5.
3. Shi X, Dalal NS, Jain AC. Antioxidant behaviour of caffeine: efficient scavenging of hydroxyl radicals. *Food Chem Toxicol* 1991; 29: 1–6.
4. Horie T, Awazu S, Itakura Y, Fuwa T. Identified diallyl polysulfides from an aged garlic extract which protects the membranes from lipid peroxidation. *Planta Med* 1992; 58: 468–9.
5. Kang SA, Jang YJ, Park H. In vivo dual effects of vitamin C on paraquat-induced lung damage: dependence on released metals from the damaged tissue. *Free Radic Res* 1998; 28: 93–107.
6. Li AS, Bandy B, Tsang SS, Davison AJ. DNA-breaking versus DNA-protecting activity of four phenolic compounds in vitro. *Free Radic Res* 2000; 33: 551–66.
7. Bird A. The essentials of DNA methylation. *Cell* 1992; 70: 5–8.
8. Michalowsky LA, Jones PA. DNA methylation and differentiation. *Environ Health Perspect* 1989; 80: 189–97.
9. Baylin SB, Herman JG. DNA hypermethylation in tumorigenesis: epigenetics joins genetics. *Trends Genet* 2000; 16: 168–74.
10. Silva AJ, White R. Inheritance of allelic blueprints for methylation patterns. *Cell* 1988; 54: 145–52.
11. Issa JP, Ottaviano YL, Celano P, Hamilton SR, Davidson NE, Baylin SB. Methylation of the oestrogen receptor CpG island links ageing and neoplasia in human colon. *Nat Genet* 1994; 7: 536–40.
12. Christman JK, Sheikhnejad G, Dizik M, Abileah S, Wainfan E. Reversibility of changes in nucleic acid methylation and gene expression induced in rat liver by severe dietary methyl deficiency. *Carcinogenesis* 1993; 14: 551–7.
13. Branda RF, O'Neill JP, Brooks EM, Trombley LM, Nicklas JA. The effect of folate deficiency on the cytotoxic and mutagenic responses to ethylmethanesulfonate in human lymphoblastoid cell lines that differ in p53 status. *Mutat Res* 2001; 473: 51–71.
14. Ames BN, Gold LS. Endogenous mutagens and the causes of aging and cancer. *Mutat Res* 1991; 250: 3–16.
15. Fraser MC, Tucker MA. Second malignancies following cancer therapy. *Semin Oncol Nurs* 1989; 5: 43–55.
16. Poirier MC, Santella RM, Weston A. Carcinogen macromolecular adducts and their measurement. *Carcinogenesis* 2000; 21: 353–9.
17. Bigger CA, Flickinger DJ, St John J, Harvey RG, Dipple A. Preferential mutagenesis at G.C base pairs by the anti 3,4-dihydrodiol 1,2-epoxide of 7-methylbenz(a)anthracene. *Mol Carcinogen* 1991; 4: 176–9.

18. Aguilar F, Harris CC, Sun T, Hollstein M, Cerutti P. Geographic variation of p53 mutational profile in nonmalignant human liver. *Science* 1994; 264: 1317–19.
19. Rusin M, Butkiewicz D, Malusecka E, *et al.* Molecular epidemiological study of non-small-cell lung cancer from an environmentally polluted region of Poland. *Br J Cancer* 1999; 80: 1445–52.
20. Harty LC, Guinee DG, Travis WD, *et al.* p53 Mutations and occupational exposures in a surgical series of lung cancers. *Cancer Epidemiol Biomarkers Prev* 1996; 5: 997–1003.
21. Sheweita SA. Drug-metabolizing enzymes: mechanisms and functions. *Curr Drug Metab* 2000; 1: 107–32.
22. Coles B, Ketterer B. The role of glutathione and glutathione transferases in chemical carcinogenesis. *Crit Rev Biochem Mol Biol* 1990; 25: 47–70.
23. Chen J, Stampfer MJ, Hough HL, *et al.* A prospective study of N-acetyltransferase genotype, red meat intake, and risk of colorectal cancer. *Cancer Res* 1998; 58: 3307–11.
24. Sweeney C, McClure GY, Fares MY, *et al.* Association between survival after treatment for breast cancer and glutathione S-transferase P1 Ile105Val polymorphism. *Cancer Res* 2000; 60: 5621–4.
25. Lin DX, Tang YM, Peng Q, Lu SX, Ambrosone CB, Kadlubar FF. Susceptibility to esophageal cancer and genetic polymorphisms in glutathione S-transferases T1, P1, and M1 and cytochrome P450 2E1. *Cancer Epidemiol Biomarkers Prev* 1998; 7: 1013–18.
26. Fahey JW, Zhang Y, Talalay P. Broccoli sprouts: an exceptionally rich source of inducers of enzymes that protect against chemical carcinogens. *Proc Natl Acad Sci USA* 1997; 94: 10367–72.
27. Hecht SS, Carmella SG, Murphy SE. Effects of watercress consumption on urinary metabolites of nicotine in smokers. *Cancer Epidemiol Biomarkers Prev* 1999; 8: 907–13.
28. Yang CS, Brady JF, Hong JY. Dietary effects on cytochromes P450, xenobiotic metabolism, and toxicity. *FASEB J* 1992; 6: 737–44.
29. Yannai S, Day AJ, Williamson G, Rhodes MJ. Characterization of flavonoids as monofunctional or bifunctional inducers of quinone reductase in murine hepatoma cell lines. *Food Chem Toxicol* 1998; 36: 623–30.
30. Hoeijmakers JH. Genome maintenance mechanisms for preventing cancer. *Nature* 2001; 411: 366–74.
31. Bos JL. ras oncogenes in human cancer: a review. *Cancer Res* 1989; 49: 4682–9.
32. Levine AJ. p53, the cellular gatekeeper for growth and division. *Cell* 1997; 88: 323–31.
33. Quilliam LA, Zhong S, Rabun KM, *et al.* Biological and structural characterization of a Ras transforming mutation at the phenylalanine-156 residue, which is conserved in all members of the Ras superfamily. *Proc Natl Acad Sci USA* 1995; 92: 1272–6.
34. Huang P. Excision of mismatched nucleotides from DNA: a potential mechanism for enhancing DNA replication fidelity by the wild-type p53 protein. *Oncogene* 1998; 17: 261–70.
35. Ozbun MA, Butel JS. p53 tumor suppressor gene: structure and function. *Encyclopedia Cancer* 1997; 2: 1240–57.
36. Marchenko ND, Zaika A, Moll UM. Death signal-induced localization of p53 protein to mitochondria. A potential role in apoptotic signaling. *J Biol Chem* 2000; 275: 16202–12.
37. Ross JS, Fletcher JA. HER-2/neu (c-erb-B2) gene and protein in breast cancer. *Am J Clin Pathol* 1999; 112: S53–67.
38. Yen L, You XL, Al Moustafa AE, *et al.* Heregulin selectively upregulates vascular endothelial growth factor secretion in cancer cells and stimulates angiogenesis. *Oncogene* 2000; 19: 3460–9.
39. Hecht JL, Aster JC. Molecular biology of Burkitt's lymphoma. *J Clin Oncol* 2000; 18: 3707–21.
40. Dang CV, Resar LM, Emison E, *et al.* Function of the c-Myc oncogenic transcription factor. *Exp Cell Res* 1999; 253: 63–77.
41. Ames BN, Gold LS. Too many rodent carcinogens: mitogenesis increases mutagenesis. *Science* 1990; 249: 970–1.
42. Preston-Martin S, Pike MC, Ross RK, Jones PA, Henderson BE. Increased cell division as a cause of human cancer. *Cancer Res* 1990; 50: 7415–21.
43. Connolly JM, Gilhooly EM, Rose DP. Effects of reduced dietary linoleic acid intake, alone or combined with an algal source of docosahexaenoic acid, on MDA-MB-231 breast cancer cell growth and apoptosis in nude mice. *Nutr Cancer* 1999; 35: 44–9.
44. O'Shea M, Devery R, Lawless F, Murphy J, Stanton C. Milk fat conjugated linoleic acid (CLA) inhibits growth of human mammary MCF-7 cancer cells. *Anticancer Res* 2000; 20: 3591–601.
45. Rogers AE, Sullivan LM, Hafer LJ. Dietary fat, body weight, and cancer: contributions of studies in rodents to understanding these cancer risk factors in humans. *Toxicol Sci* 1999; 52: 66–71.
46. La Guardia M, Giammanco M. Breast cancer and obesity. *Panminerva Med* 2001; 43: 123–33.
47. Le Bail JC, Champavier Y, Chulia AJ, Habrioux G. Effects of phytoestrogens on aromatase, 3beta

and 17beta-hydroxysteroid dehydrogenase activities and human breast cancer cells. *Life Sci* 2000; 66: 1281–91.

48. Newmark HL, Lipkin M, Maheshwari N. Colonic hyperplasia and hyperproliferation induced by a nutritional stress diet with four components of Western-style diet. *J Natl Cancer Inst* 1990; 82: 491–6.
49. Lipkin M. Biomarkers of increased susceptibility to gastrointestinal cancer: new application to studies of cancer prevention in human subjects. *Cancer Res* 1988; 48: 235–45.
50. Archer SY, Meng S, Shei A, Hodin RA. p21(WAF1) is required for butyrate-mediated growth inhibition of human colon cancer cells. *Proc Natl Acad Sci USA* 1998; 95: 6791–6.
51. Wenzel U, Kuntz S, Brendel MD, Daniel H. Dietary flavone is a potent apoptosis inducer in human colon carcinoma cells. *Cancer Res* 2000; 60: 3823–31.
52. Casagrande F, Darbon JM. Effects of structurally related flavonoids on cell cycle progression of human melanoma cells: regulation of cyclin-dependent kinases CDK2 and CDK1. *Biochem Pharmacol* 2001; 61: 1205–15.
53. Hao X, Bishop AE, Wallace M, *et al.* Early expression of cyclooxygenase-2 during sporadic colorectal carcinogenesis. *J Pathol* 1999; 187: 295–301.
54. Fosslien E. Molecular pathology of cyclooxygenase-2 in neoplasia. *Ann Clin Lab Sci* 2000; 30: 3–21.
55. Sumitani K, Kamijo R, Toyoshima T, *et al.* Specific inhibition of cyclooxygenase-2 results in inhibition of proliferation of oral cancer cell lines via suppression of prostaglandin E2 production. *J Oral Pathol Med* 2001; 30: 41–7.
56. White MK, McCubrey JA. Suppression of apoptosis: role in cell growth and neoplasia. *Leukemia* 2001; 15: 1011–21.
57. Hanahan D, Weinberg RA. The hallmarks of cancer. *Cell* 2000; 100: 57–70.
58. St Croix B, Florenes VA, Rak JW, *et al.* Impact of the cyclin-dependent kinase inhibitor p27Kip1 on resistance of tumor cells to anticancer agents. *Nat Med* 1996; 2: 1204–10.
59. Waldman T, Zhang Y, Dillehay L, *et al.* Cell-cycle arrest versus cell death in cancer therapy. *Nat Med* 1997; 3: 1034–6.
60. Ferri KF, Kroemer G. Control of apoptotic DNA degradation. *Nat Cell Biol* 2000; 2: E63–4.
61. Hengartner MO. The biochemistry of apoptosis. *Nature* 2000; 407: 770–6.
62. Kumar S. Mechanisms mediating caspase activation in cell death. *Cell Death Differ* 1999; 6: 1060–6.
63. Chang HY, Yang X. Proteases for cell suicide: functions and regulation of caspases. *Microbiol Mol Biol Rev* 2000; 64: 821–46.
64. Kelekar A, Thompson CB. Bcl-2-family proteins: the role of the BH3 domain in apoptosis. *Trends Cell Biol* 1998; 8: 324–30.
65. Wolter KG, Hsu YT, Smith CL, Nechushtan A, Xi XG, Youle RJ. Movement of Bax from the cytosol to mitochondria during apoptosis. *J Cell Biol* 1997; 139: 1281–92.
66. Nakagawa T, Zhu H, Morishima N, *et al.* Caspase-12 mediates endoplasmic-reticulum-specific apoptosis and cytotoxicity by amyloid-beta. *Nature* 2000; 403: 98–103.
67. Todd R, Wong DT. Oncogenes. *Anticancer Res* 1999; 19: 4729–46.
68. Blandino G, Strano S. BCL-2: the pendulum of the cell fate. *J Exp Clin Cancer Res* 1997; 16: 3–10.
69. Muschen M, Warskulat U, Beckmann MW. Defining CD95 as a tumor suppressor gene. *J Mol Med* 2000; 78: 312–25.
70. Soengas MS, Capodieci P, Polsky D, *et al.* Inactivation of the apoptosis effector Apaf-1 in malignant melanoma. *Nature* 2001; 409: 207–11.
71. Frisch SM, Ruoslahti E. Integrins and anoikis. *Curr Opin Cell Biol* 1997; 9: 701–6.
72. James SJ, Muskhelishvili L, Gaylor DW, Turturro A, Hart R. Upregulation of apoptosis with dietary restriction: implications for carcinogenesis and aging. *Environ Health Perspect* 1998; 106: 307–12.
73. Dunn SE, Kari FW, French J, *et al.* Dietary restriction reduces insulin-like growth factor I levels, which modulates apoptosis, cell proliferation, and tumor progression in p53-deficient mice. *Cancer Res* 1997; 57: 4667–72.
74. Plas DR, Talapatra S, Edinger AL, Rathmell JC, Thompson CB. Akt and Bcl-xL promote growth factor-independent survival through distinct effects on mitochondrial physiology. *J Biol Chem* 2001; 276: 12041–8.
75. Hughes R, Rowland IR. Stimulation of apoptosis by two prebiotic chicory fructans in the rat colon. *Carcinogenesis* 2001; 22: 43–7.
76. Avivi-Green C, Polak-Charcon S, Madar Z, Schwartz B. Apoptosis cascade proteins are regulated in vivo by high intracolonic butyrate concentration: correlation with colon cancer inhibition. *Oncol Res* 2000; 12: 83–95.
77. Rao CV, Hirose Y, Indranie C, Reddy BS. Modulation of experimental colon tumorigenesis by types and amounts of dietary fatty acids. *Cancer Res* 2001; 61: 1927–33.

78. Latham P, Lund EK, Johnson IT. Dietary n-3 PUFA increases the apoptotic response to 1,2-dimethylhydrazine, reduces mitosis and suppresses the induction of carcinogenesis in the rat colon. *Carcinogenesis* 1999; 20: 645–50.
79. Papaconstantinou HT, Hwang KO, Rajaraman S, Hellmich MR, Townsend CM, Ko TC. Glutamine deprivation induces apoptosis in intestinal epithelial cells. *Surgery* 1998; 124: 152–9.
80. Petronini PG, Urbani S, Alfieri R, Borghetti AF, Guidotti GG. Cell susceptibility to apoptosis by glutamine deprivation and rescue: survival and apoptotic death in cultured lymphoma-leukemia cell lines. *J Cell Physiol* 1996; 169: 175–85.
81. Drummond-Barbosa D, Spradling AC. Stem cells and their progeny respond to nutritional changes during Drosophila oogenesis. *Dev Biol* 2001; 231: 265–78.
82. Campbell F, Williams GT, Appleton MA, Dixon MF, Harris M, Williams ED. Post-irradiation somatic mutation and clonal stabilisation time in the human colon. *Gut* 1996; 39: 569–73.
83. Taipale J, Beachy PA. The Hedgehog and Wnt signalling pathways in cancer. *Nature* 2001; 411: 349–54.
84. Laurent-Puig P, Blons H, Cugnenc PH. Sequence of molecular genetic events in colorectal tumorigenesis. *Eur J Cancer Prev* 1999; 8[Suppl 1]: S39–47.
85. Uthoff SM, Eichenberger MR, McAuliffe TL, Hamilton CJ, Galandiuk S. Wingless-type frizzled protein receptor signaling and its putative role in human colon cancer. *Mol Carcinogen* 2001; 31: 56–62.
86. Hart RW, Dixit R, Seng J, *et al*. Adaptive role of caloric intake on the degenerative disease processes. *Toxicol Sci* 1999; 52: 3–12.
87. Altucci L, Rossin A, Raffelsberger W, *et al*. Retinoic acid-induced apoptosis in leukemia cells is mediated by paracrine action of tumor-selective death ligand TRAIL. *Nat Med*2001; 7: 680–6.
88. Niu MY, Menard M, Reed JC, Krajewski S, Pratt MA. Ectopic expression of cyclin D1 amplifies a retinoic acid-induced mitochondrial death pathway in breast cancer cells. *Oncogene* 2001; 20: 3506–18.
89. Ganther HE. Selenium metabolism, selenoproteins and mechanisms of cancer prevention: complexities with thioredoxin reductase. *Carcinogenesis* 1999; 20: 1657–66.
90. Liu Z, Tomotake H, Wan G, Watanabe H, Kato N. Combined effect of dietary calcium and iron on colonic aberrant crypt foci, cell proliferation and apoptosis, and fecal bile acids in 1,2-dimethylhydrazine-treated rats. *Oncol Rep* 2001; 8: 893–7.
91. Ushakova T, Melkonyan H, Nikonova L, *et al*. Modification of gene expression by dietary antioxidants in radiation-induced apoptosis of mice splenocytes. *Free Radic Biol Med* 1999; 26: 887–91.
92. Salganik RI, Albright CD, Rodgers J, *et al*. Dietary antioxidant depletion: enhancement of tumor apoptosis and inhibition of brain tumor growth in transgenic mice. *Carcinogenesis* 2000; 21: 909–14.
93. Chinery R, Brockman JA, Peeler MO, Shyr Y, Beauchamp RD, Coffey RJ. Antioxidants enhance the cytotoxicity of chemotherapeutic agents in colorectal cancer: a p53-independent induction of p21WAF1/CIP1 via C/EBPbeta. *Nat Med* 1997; 3: 1233–41.
94. Department of Health. *Nutritional aspects of the development of cancer*. Report of the working group on diet and cancer of the Committee on Medical Aspects of Food and Nutrition Policy. London: HMSO, 1998.
95. Greenberg ER, Baron JA, Tosteson TD, *et al*. A clinical trial of antioxidant vitamins to prevent colorectal adenoma: Polyp Prevention Study Group. *N Engl J Med* 1994; 331: 141–7.
96. Middleton E, Kandaswamy C. The impact of plant flavonoids on mammalian biology: implications for immunity, inflammation and cancer. In: Harborne JB, ed. *The flavonoids: advances in research since 1986*. London: Chapman and Hall, 1994: 619–52.
97. Caltagirone S, Rossi C, Poggi A, *et al*. Flavonoids apigenin and quercetin inhibit melanoma growth and metastatic potential. *Int J Cancer* 2000; 87: 595–600.
98. Liang YC, Lin-Shiau SY, Chen CF, Lin JK. Inhibition of cyclin-dependent kinases 2 and 4 activities as well as induction of Cdk inhibitors p21 and p27 during growth arrest of human breast carcinoma cells by (−)-epigallocatechin-3-gallate. *J Cell Biochem* 1999; 75: 1–12.
99. Kuzumaki T, Kobayashi T, Ishikawa K. Genistein induces p21(Cip1/WAF1) expression and blocks the G1 to S phase transition in mouse fibroblast and melanoma cells. *Biochem Biophys Res Commun* 1998; 251: 291–5.
100. Iwao K, Tsukamoto I. Quercetin inhibited DNA synthesis and induced apoptosis associated with increase in c-fos mRNA level and the upregulation of p21WAF1CIP1 mRNA and protein expression during liver regeneration after partial hepatectomy. *Biochim Biophys Acta* 1999; 1427: 112–20.

101. Bai F, Matsui T, Ohtani-Fujita N, Matsukawa Y, Ding Y, Sakai T. Promoter activation and following induction of the p21/WAF1 gene by flavone is involved in G1 phase arrest in A549 lung adenocarcinoma cells. *FEBS Lett* 1998; 437: 61–4.
102. Wang CY, Cusack JC, Liu R, Baldwin AS. Control of inducible chemoresistance: enhanced anti-tumor therapy through increased apoptosis by inhibition of NF-kappaB. *Nat Med* 1999; 5: 412–17.
103. Maurer CA, Friess H, Buhler SS, *et al.* Apoptosis inhibiting factor Bcl-xL might be the crucial member of the Bcl-2 gene family in colorectal cancer. *Dig Dis Sci* 1998; 43: 2641–8.
104. Marks F, Furstenberger G. Cancer chemoprevention through interruption of multistage carcinogenesis. The lessons learnt by comparing mouse skin carcinogenesis and human large bowel cancer. *Eur J Cancer* 2000; 36: 314–29.
105. Elder DJ, Paraskeva C. COX-2 inhibitors for colorectal cancer. *Nat Med* 1998; 4: 392–3.
106. Ahmad N, Feyes DK, Nieminen AL, Agarwal R, Mukhtar H. Green tea constituent epigallocatechin-3-gallate and induction of apoptosis and cell cycle arrest in human carcinoma cells. *J Natl Cancer Inst* 1997; 89: 1881–6.
107. Ahmad N, Gupta S, Mukhtar H. Green tea polyphenol epigallocatechin-3-gallate differentially modulates nuclear factor kappaB in cancer cells versus normal cells. *Arch Biochem Biophys* 2000; 376: 338–46.
108. Alon U, Barkai N, Notterman DA, *et al.* Broad patterns of gene expression revealed by clustering analysis of tumor and normal colon tissues probed by oligonucleotide arrays. *Proc Natl Acad Sci USA* 1999; 96: 6745–50.
109. Alizadeh AA, Staudt LM. Genomic-scale gene expression profiling of normal and malignant immune cells. *Curr Opin Immunol* 2000; 12: 219–25.
110. Perou CM, Sorlie T, Eisen MB, *et al.* Molecular portraits of human breast tumours. *Nature* 2000; 406: 747–52.

DISCUSSION

Dr. Bachmann: You showed an increase in NF-κB activation. If you turn that around, statins are known to inhibit NF-κB activation. Do you think that an overlimitation of cholesterol synthesis or of the intermediates of the colon cholesterol synthesis pathway might explain claims that a limitation of cholesterol intake might lead to increased tumor formation?

Dr. Daniel: We have learned recently that the statins may be a two-edged sword, and it is too early to make any clear statement about their role. The idea that cholesterol biosynthesis, as well as elementary cholesterol, is involved in colon carcinogenesis was put forward a while ago, and there are good models to explain how that could happen, although again, we don't have any final proof. This raises an important question about the dietary factors that are now being introduced to reduce cholesterol absorption in the gut. These could result in a very high intestinal load of cholesterol, and even of some cholesterol metabolites that can be produced by bacteria in the colon—much higher than would normally be the case. We have been wondering whether this might have implications for colon carcinogenesis, but so far, this hypothesis has not been confirmed by animal data.

Dr. Micheli: You spoke about the effect of energy restriction on protein metabolism. I was puzzled that you said that energy restriction would increase protein turnover. Several of us here have measured the impact of energy intake on protein turnover in preterm babies and exactly the opposite occurs!

Dr. Daniel: I am aware of this experimental evidence. However, quite a number of the 6,500 genes in mouse muscle responded to energy restriction, and the mRNA levels and the enzymes involved in protein turnover were increased. There are other examples in which observed effects at the mRNA level do not match physiology and change opposite to the prediction. We can only say at present that the mRNA levels change but we cannot link this to the physiologic effects. We need to look additionally at the protein levels and, more important, at the metabolite levels. Consequently, the next step is metabolome analysis, and we now have the technology to undertake this on a high-throughput level. This will hopefully give us clearer answers. By this, we are generating enormous databases, and there is a need for people that can relate the data to the biology. These databases will grow even larger if we achieve high throughput metabolic profiling.

Dr. Pencharz: Glutamine seems to be a popular topic—we actually call it vitamin G! The trouble is that as a fuel for the enterocyte, it has to be converted to glutamate. I'm always concerned when people present data showing that *glutamine* does something, because if you're talking about it as a fuel for the enterocyte, it's actually glutamate. Reeds has shown that if you replace glutamine with glutamate, you see the same effects on growth in the small intestine. I'm not sure whether the caspase 2 and 3 data you showed refer to the glutamine molecule itself, or to the energetics, in which case it's glutamate. Also, did they do proper controls with glutamate *versus* glutamine?

Dr. Daniel: Clearly, the effect was much larger than could be accounted for just by glutamine, so glutamate was also involved. It all comes down to whether you believe the glutamine story or not. It is certainly an excellent energy fuel, and some of the findings with glutamine supplementation look convincing. However, I spent lots of money trying to find an approach for utilizing peptides rich in glutamine that could be provided from the luminal side in animal models of methotrexate treatment and so on, and it didn't work at all, so I'm a skeptic about all the glutamine effects. I agree that the effects shown in various studies could be mainly by glutamate.

Dr. Pencharz: You have to be careful about the orientation of cells in culture. Dr. Reeds has shown clearly that glutamine is taken up from the blood side and glutamate primarily from the luminal side, so unless you have the cells properly oriented, you cannot interpret the experiments.

Dr. Daniel: In the experiments I described, it was mainly the uptake from the basolateral side, as the cells do not differentiate so well. Clearly all the transporters for glutamine on the basolateral side were present, and the cells took up huge amounts of glutamine.

Dr. Batshaw: With the urea cycle disorders, even when affected children are stable, the glutamine levels are severely affected, and of course, there are also children with glutamate dehydrogenase deficiency who will have very low glutamine levels. There are also animal models with high glutamine levels. I wonder whether those various models could be helpful in answering some of the questions you posed.

Dr. Daniel: As I say, I'm a skeptic about the glutamine effects myself, but at least they look convincing at cell culture level, and I agree there are nice models around that could be explored.

Dr. Batshaw: I wonder if you or Dr. Pencharz would be able to predict the possible long-term effects of either very high or very low glutamine levels?

Dr. Pencharz: I cannot predict that at this moment. I think that the uptake of glutamine is primarily related to amino acid metabolism within the enterocyte, whereas glutamate is the fuel that is provided luminally. What happens physiologically, and how we interpret these *in vitro* studies with transformed cells, and how they relate to human and animal biologies is anyone's guess at the moment.

Dr. Bachmann: May I ask Dr. Pencharz, in addition, did you try your experiments with ornithine as a control, because when you start to look at glutamate you move into the proline and ornithine story as well? So is it glutamine, glutamate, ornithine, or proline? We have to look at these effects very critically.

Dr. Pencharz: You are absolutely right. We have not done that yet.

Dr. Daniel: Glutamine/glutamate deprivation has been looked at in comparison with other amino acids, and clearly the effect was on the glutamine-glutamate side, but they didn't look at ornithine in that study.

Dr. Clark: I think we have a responsibility to not lose sight of important biological questions. The calorie deprivation study (1) involved very severe energy deprivation, and the gene changes were relatively small. We must bear in mind that we have to ask good biologic questions and not simply jump on the bandwagon. The reported changes for mRNA levels were relatively small.

Dr. Daniel: You may have misunderstood me. I'm not saying that expression profiling necessarily provides the answer, but, as in other areas, the question is whether we want to explore it and whether we need to do it. You could ask the same questions about pharmacogenomics or toxicogenomics; so why not nutrigenomics? What I'm saying is that we should utilize these techniques to give us a better basis for understanding the effects of nutrition. I agree that a two-fold change in mRNA may not be considered significant, but there were ten-fold or 15-fold changes involved in the flavone effects I showed you.

Dr. Clark: I think you have to be very careful about -fold changes, because even with a many-fold change the effect may still be very small. We must also be wary of the severity of any restriction or intervention that is imposed experimentally. Can this really be achieved biologically? However, I agree with you that the techniques are exciting.

Dr. Bier: To follow up on your comments and those of Dr. Micheli, having lived through the era when we measured all sorts of enzymes in a pathway to try to determine the movement of material, we're now in the era of using changes in genes to estimate enzyme activity. We learned 30 years ago that just measuring the enzyme activity doesn't tell us the movement of material through a pathway unless we actually measure it in some kinetic way. We are a long way from doing that. My other comment is that when expression profiling is used in transgenic plants, the number of genes that respond is actually remarkably small considering the change this process induces.

Dr. Daniel: I agree, but here in plant biotechnology all those profiling techniques are used. As you know, we have a "novel food regulation" in the European Union (EU) where every transgenic material that goes into the food chain is evaluated for safety. One of the problems of this is the key statement that says "a novel food is a novel food if it is substantially different from a conventional one." When that was put in, I suppose nobody realized how difficult it would be to show "substantial equivalence." So, when is a transgenic tomato different from a normal tomato? There are hundreds of tomato varieties that change their content of nutrients, toxic alkaloids, and so on depending on where they grow, how much sunlight they get, and many other factors. As part of the safety assessment process, it is important to show what a *normal* tomato or potato is, although everybody thinks they know what a normal one is. Now that we have these new technologies, scientists are using them for safety assessments. We now verify "normality" in a potato, say, on the basis of its genome, of its proteome, and on the metabolome. I'm not saying that the answer should be based solely on this. It is just one of the approaches. If the transgenic procedure leads to expression of new proteins unexpectedly, you might need to worry about new allergens, for example, and then proteomics is a useful tool in safety assessment.

Dr. Bier: I'm afraid that these new methods will give us huge amounts of data but far less information, and certainly less precision.

Dr. Borum: You have nicely reviewed for us the correlations found between apoptosis and carbohydrate and amino acid metabolism. I would like to hear your comments about correlations of the ratios of other dietary components—not just energy—for example fatty acids.

Dr. Daniel: I cannot give you a scientific answer but I can talk about my beliefs. I stated initially that we still have the genome of a hunter and gatherer. The 8,000 years that have passed since the Neolithic revolution have not really made a huge impact on the genome, although we have some examples of recent genetic alteration, for example in salt tolerance acquired by mutations in the angiotensin-converting enzyme gene. However, if you analyze the diet of a hunter–gatherer you find a quite different ratio of n-6 and n-3 fatty acids compared to today's diet, and of course a much higher ratio of polyunsaturated to saturated fatty acids. The Neolithic revolution brought large quantities of saturated fat into our diet. Important fat components of the hunter's diet were found in wild game. If you analyze the meat of wild game, you find that the ratio of n-6 to n-3 almost matches present dietary recommendation. Thus, wild game has a quite different fatty acid pattern and ratio compared to domestic animals. There is also the matter of dietary fiber. In the United States and Europe we have been telling people they have to have 30 g of dietary fiber every day—it's good for your cholesterol and for protecting your colon from carcinogenesis. Then the large epidemiologic studies came out and told us that if there is an effect, it is only a moderate one. If you go back to the ancient diets of hunter–gatherers, or for example the diet of Aboriginal peoples of Australia—you find huge amounts of fiber, but not the fiber we are familiar with. The fiber in their diet is soluble fiber, composed of inulin, galacto- and fructo-oligosaccharides, and so on, which can very easily be hydrolyzed in the large intestine. Aborigines may eat up to 100 g a day of soluble fiber, whereas with the Neolithic revolution we replaced a large fraction of the soluble fiber in our diet by insoluble fiber. So my proposal is that we should look carefully into a moderate return to this earlier type of diet, which was responsible for molding our genome over thousands of years. Clearly, nutrition has been a very important genetic selector, so looking into the ancient diet could provide valuable information to guide us to a healthy diet for us now, even though the world has changed a lot.

Dr. Wanders: It has been shown that reactive oxygen species are involved in malignant cell transformation, and one hypothesis is that flavonoids are so potent because of their antioxidant properties. However, you mentioned that you tested various flavonoids and found no apparent relation between their antioxidant power and their effects in your tumor cell system. Could you explain that?

Dr. Daniel: What I can say is that the particular compound that had these very interesting effects on apoptosis is not an antioxidant. We also tested those with antioxidant activity, but those were much less effective. However, flavonoids have all kind of different cellular targets, which may have nothing to do with antioxidant activity. Moreover, there appears to be accumulating evidence that too many antioxidants, which are provided in all kinds of supplements, may not be desirable. We are talking about a balance here. For example, in cancer therapy, there is evidence that if the patient has a high intake of antioxidant compounds, the outcome of the treatment may not be as good as expected (2). This has its possible biochemical basis downstream in the cascade of apoptosis, where there is a step in which cells generate oxygen radicals within the apoptotic death pathway, and this might be important for the final execution of apoptosis. If you quench this process, you may have cells surviving that should not survive.

REFERENCES

1. Lee CK, Klopp RG, Weindruch R, Prolla TA. Gene expression profile of aging and its retardation by caloric restriction. *Science* 1999; 285: 1390–3.
2. Kong Q, Beel JA, Lillehei KO. A threshold concept for cancer therapy. *Med Hypotheses* 2000; 55: 29–35.

Genetic Expression and Nutrition, edited by Claude Bachmann and Berthold Koletzko. Nestlé Nutrition Workshop Series, Pediatric Program, Vol. 50. Nestec Ltd., Vevey/Lippincott Williams & Wilkins, Philadelphia, © 2003.

Gene Therapy: A Potential Solution for Nutrition-Dependent Disorders?

Mark L. Batshaw, *James M. Wilson, Xuehai Ye, Steven E. Raper

*Children's Research Institute, Children's National Medical Center, Department of Pediatrics, George Washington University School of Medicine and Health Sciences, Washington DC, USA; Departments of Surgery and *Molecular and Cellular Engineering, University of Pennsylvania School of Medicine, Philadelphia, Pennsylvania, USA; Institute for Human Gene Therapy, University of Pennsylvania School of Medicine, Philadelphia, Pennsylvania, USA*

Over the past few decades there has been significant progress in the treatment of inherited nutrition-dependent disorders, especially the inborn errors of amino acid and organic acid metabolism (1). In some instances, long-term nutritional modification—for example dietary restriction of phenylalanine in phenylketonuria—has led to normal development of affected children (2). In other cases, such as the urea cycle disorders, there has been some improvement in survival, but morbidity remains high (3); in these conditions, despite life-long treatment with both medicines and nutritional restriction, affected children remain at risk of hyperammonemic crises precipitated by intercurrent infections and other unanticipated events.

An alternative to medical/nutritional management in some of these disorders is enzyme replacement by organ transplantation. The effect of liver transplantation in urea cycle disorders is to replace a severe disorder by a less severe and more manageable one (that is, organ rejection and immunosuppression) (4). Gene transfer offers the potential for yet another approach to long-term correction, especially as these disorders may only require this to be done peripherally.

Although there has been great enthusiasm for the potential of gene transfer, the results to date have been for the most part discouraging. Ironically, two of the most well-publicized failures of gene transfer have involved inborn errors of urea synthesis. The first reported trial of gene transfer involved the use of the Shope papilloma virus to treat two subjects with arginase deficiency (5). This trial was found to have had no benefit and to have been ethically suspect, and an inactive papilloma virus may have been used (6–8). The second instance was a pilot study of adenovirus-mediated gene transfer in ornithine transcarbamylase (OTC) deficiency that resulted in the well-publicized death of a patient 4 days after he received an intrahepatic gene transfer (9). This trial has come under investigation for protocol lapses (see *www.fda.gov*). In this chapter, we will use this OTC deficiency study to discuss the risks and potential benefits of gene transfer for the treatment of nutrition-dependent inborn errors of metabolism. We also summarize two more-recent pilot gene transfer trials in hemophilia and severe combined immunodeficiency that have shown promise (10,11).

ORNITHINE TRANSCARBAMYLASE DEFICIENCY: CURRENT TREATMENT AND OUTCOME

Inborn errors of each of the enzymes in the urea cycle have been described, OTC deficiency being the most common, with a prevalence of 1:40,000 to 1:80,000 births. This X-linked disorder classically presents in males as hyperammonemic coma during the first days of life. Long-term survival of affected neonates has been distressingly poor despite acute treatment with hemodialysis and chronic treatment with alternative-pathway therapy (3,12–14). Mortality is lower in late-onset (partial) OTC deficiency in both male patients and symptomatic, heterozygous female patients, although both groups remain at risk of life-threatening hyperammonemic episodes (15). Several patients with OTC deficiency have received orthotopic liver transplants to provide enzyme replacement therapy, with a survival rate of 80% (4). Most surviving males with neonatal-onset OTC deficiency (and other urea cycle disorders) have mental retardation and other developmental disabilities (14,16).

CONSIDERATIONS IN DEVELOPING AN OTC GENE TRANSFER PILOT STUDY

Because of the high morbidity and mortality associated with this disorder, we became interested in developing a gene transfer approach to OTC deficiency. First, we needed to develop a vector that would permit gene transfer to the desired organ, the liver. We chose an adenovirus vector because of its rapid expression in the liver following intravascular infusion. Although its effects are transient, for individuals with OTC deficiency who are in hyperammonemic coma, the acute effects of adenovirus could be life saving. Next, an animal model was needed to test the vector to determine transduction and expression of the transgene. In the case of OTC deficiency, there are two rodent models available, the spf mouse and the spf^{ash} mouse (17,18). Once shown to be effective, the vector was then tested for toxicity in both rodents and nonhuman primates. This informed our decision about the dose to be used in a pilot safety study in human subjects.

Choice of Vector for Gene Transfer

Intravascular infusion of an adenovirus vector leads to specific and very efficient gene transfer into virtually 100% of hepatocytes (19–21). While this is more efficient than any other form of gene transfer, its effects are transient and subsequent readministration is ineffective. Before contemplating the application of adenovirus vectors for liver-directed gene transfer in humans, we studied the basic mechanisms responsible for the problems of its transient expression and induced inflammation.

Extensive studies in a variety of animals indicated that a major problem was the immune response to the virus and to virus-infected cells. Antigen-specific cellular responses were found to destroy or destabilize the genetically corrected cells, effectively eliminating the transgene and causing inflammation (22–24). Simply deleting the E1 (pathogenic) adenovirus gene does not ablate all viral protein expression. As a result, newly expressed viral proteins or the transgene product activate $CD8^+$ T cells to form cytotoxic T lymphocytes that target and destroy or destabilize the genetically corrected hepatocytes. Input viral proteins activate B cells as well as $CD4^+$ T cells (that is, T helper cells) that produce neutralizing antibodies to the virus (24). This has no effect on a primary challenge but can diminish the efficiency of gene transfer in the context of a second administration of virus.

The model described above suggests that further inactivating the adenovirus vector by diminishing its propensity to express viral proteins would improve its performance by diminishing cytotoxic T lymphocyte responses, leading to less inflammation and more extended gene expression. These findings resulted in our use of an E1,E4-deleted adenovirus for a clinical trial.

Genetic Correction of OTC Deficiency in spf and spf^{ash} Mice

In 1990, Stratford-Perricaudet *et al.* (25) explored the potential of an E1-deleted recombinant adenovirus for introducing and expressing the normal OTC gene in newborn spf^{ash} mice. In this study, liver OTC activity rose to normal levels in four of 15 mice, and persisted for 1 to 2 months. These mice also showed a decrease in urinary orotic acid excretion, suggesting metabolic correction.

In the following years, Grompe *et al.* (26) used a defective recombinant retrovirus to transduce primary hepatocytes from either spf or spf^{ash} mice. Transduction of primary cultures was highly efficient, with an average proviral copy number of 0.5 to 2 per cell in the population of transduced hepatocytes. Morsy *et al.* (27) subsequently used an adenoviral vector carrying human OTC complementary DNA (cDNA) under the control of a chimeric adenovirus SRα promoter (Ad.SRαhOTC) to infect Spf and OTC-deficient human hepatocytes. Normal levels of OTC activity were achieved in spf hepatocytes. Virus-mediated OTC expression was also detectable in OTC-deficient human hepatocytes.

In 1996, Morsy *et al.* (28) reported significant improvement of OTC deficiency in spf^{ash} mice using the same Ad.SRαhOTC vector. A 70% reduction in orotic aciduria was observed within 24 hours of treatment. Metabolic correction was later associated with phenotypic correction and a moderate increase in enzyme activity. In an effort to identify the level of gene expression required to achieve wild-type levels of enzyme activity, these investigators uncovered a dominant negative effect of the endogenous mutant protein on the activity of the delivered recombinant wild-type protein. We, however, could not replicate this finding (29). Kiwaki *et al.* (30), using Ad.SRαhOTC and another virus containing a CAG promoter (Ad.CAGhOTC), also showed hepatic expression of OTC by Western blot and the normalization of orotic aciduria for up to 60 days in spf^{ash} mice.

In adult animals, the efficacy of the first-generation, E1-deleted viruses in correcting the biochemical abnormalities such as orotic aciduria and glutaminemia appeared to be limited by a host-mediated immune response (31). Microscopic analysis of liver tissue from recipient spf mice revealed substantial inflammation that was characterized by degeneration of hepatocytes and mononuclear infiltrates (31).

Using the spf and spf^{ash} models, we studied the following: (i) whether *in vivo* gene transfer with an adenovirus vector would be successful in correcting the metabolic derangements of OTC deficiency; (ii) if so, how long the metabolic correction would persist; (iii) whether repeated injections could be given to provide long-term treatment; and (iv) whether gene transfer could be used to treat or prevent acute hyperammonemia.

By creating a second-generation virus with a temperature-sensitive mutation at E2a, we sought to limit the cytotoxic T lymphocyte response by decreasing viral protein expression. The level of late gene expression was significantly reduced at both RNA and protein levels in mice receiving this virus, compared with livers infected with identical doses of the first-generation OTC virus. Microscopic inspection of liver tissues infected with the second-generation OTC virus also revealed reduced histopathology. Expression of the late viral genes (for example hexon), as measured by RNA blot analysis and immunocytochemistry, was substantially reduced compared with the results with the first-generation virus. From the perspective of efficacy, histochemical staining for OTC activity in recipient livers showed significantly raised activity in most hepatocytes at day 7 after infusion, and persistence for at least 1 month—far longer than for the first-generation virus. We also found that there was prolonged metabolic correction of plasma glutamine and urinary orotic acid excretion, lasting 2 to 3 months (31).

A second concern was that repeated injections to achieve long-term therapy might be prevented by an anamnestic immune response to the virus. Indeed, we did find that repeated injections were not successful in replicating the correction we had observed after the initial infusion of recombinant adenovirus. To address this problem we injected an anti-CD4 monoclonal antibody (GK1.5) on the days immediately preceding and following each recombinant adenovirus infusion. The anti-CD4 antibody treatment resulted in a significantly diminished production of neutralizing antibody to the adenovirus (32). This permitted repeated injections and long-term correction, up to 6 months. These results suggest that the host immune system can be modulated to permit repeated injections of adenoviral vectors.

A final issue was whether transduction and expression following gene transfer would be sufficiently rapid to permit effective treatment of a hyperammonemic crisis. Our previous studies in human infants with urea cycle disorders had shown that hyperammonemic coma lasting more than 3 days was invariably associated with the subsequent development of mental handicap and other developmental disabilities (16). To test the rapidity of protection from hyperammonemia, we preinjected spf mice with a second-generation adenovirus containing the mouse OTC cDNA at 1, 2, 7, and 14 days before giving a fatal dose of ammonium chloride intraperitoneally. We found that by 24 hours, there was partial protection from the clinical symptoms as well as the biochemical abnormalities, and at 48 hours and afterward there was complete protection (33). These data suggest that transduction and expression following *in vivo* gene transfer are rapid and that this approach could theoretically be an effective adjunctive treatment for acute hyperammonemia in OTC-deficient humans.

Preclinical Toxicology Studies of Adenovirus Gene Transfer

After demonstrating the efficacy of adenovirus gene transfer in metabolic correction of OTC deficiency, we performed preclinical toxicity studies of adenovirus vectors on C57Bl/6 mice, C3H mice, Spf mice, rhesus monkeys, and baboons to determine the maximum tolerated dose (MTD) for human studies. We also studied intraarterial injection of adenovirus into the hepatic artery of baboons (the approach to be used in the human studies), and found that this technique was feasible and safe, and that spread of virus did not occur to the gonads at the doses proposed for the human trial (34).

Three basic forms of recombinant adenovirus constructs were studied: E1-deleted (with or without small deletions in the E3 region); E1-deleted, E2 temperature-sensitive mutations; and E1,E4-deleted (35). A general scoring system of hepatopathology was developed, based on published criteria (36). We found that there was a dose-dependent increase both in gene transfer (by X-gal histochemical analysis) and in portal and lobular hepatic inflammation. There was less inflammation with the later generation vectors (E1,E2 and E1,E4 deleted) than with the first-generation, E1-deleted vectors. In primates, dose-limiting toxicity was reached at 1×10^{13} particles/kg, with evidence of marked disturbances of liver function tests, hepatocellular necrosis, and disseminated intravascular coagulopathy that necessitated euthanasia in the affected monkey (37,38). In surviving animals, toxicity peaked at 3 days and returned to normal by 29 days after gene transfer. In these studies the no-observed-effect level (NOEL) was found to be 2.5×10^{11} particles/kg. These data informed our decision on dosage and toxicity criteria in our pilot human gene transfer trial.

A PILOT STUDY OF HUMAN GENE TRANSFER USING ADENOVIRUS IN OTC DEFICIENCY

On the basis of the above preclinical studies, which showed effective transduction and expression following gene transfer, correction of the metabolic defect in OTC-deficient spf/y mice, and an acceptable safety profile in mice and nonhuman primates—we proceeded with a pilot study of gene transfer in adult male patients with partial OTC deficiency and in clinically asymptomatic OTC-deficient, heterozygous adult women (39).

Design of the Clinical Trial

The basic design of this trial was a dose-escalation study for assessing toxicity. Subjects were tested at one of six escalating doses of the vector. The starting dose was 2×10^9 particles/kg. Gene transfer used an E1,E4-deleted adenovirus vector.

Each subject received a single dose of vector. The dose was increased in 1/2-log increments (2×10^9 to 6×10^{11} particles/kg) in cohorts of three or four subjects. Vector was infused into the right hepatic artery using a catheter placed under fluoroscopic guidance to decrease both the likelihood of extrahepatic dissemination and the risk of generalized liver inflammation. Subjects were monitored for evidence of toxicity, gene transfer, and changes in urea synthetic capacity (40). The clinical trial was initiated on April 7, 1997 with the dosing of the first subject, and was halted following the death of the 18th subject on September 17, 1999.

Study Results in the First 17 Subjects

Assessment of Safety

Several transient clinical signs and symptoms emerged in the first 17 subjects, none of which resulted in prolonged clinical sequelae (41). Most symptoms occurred during the first 24 to 48 hours after gene transfer. All subjects experienced fever, myalgia, nausea, and occasionally emesis during that period. The onset of fever was generally seen within 1 to 4 hours of completing the vector infusion, and the peak temperature was roughly dose-dependent.

Subjects were monitored for evidence of liver toxicity based on our preclinical models, which suggested that hepatitis could be dose-limiting. Evidence of biochemical liver injury was observed most significantly in cohort 4, with a rise of aspartate transaminase (AST) to a maximum of 286 U/l in one subject (normal female range, 14–36 U/l; normal male range, 17–59 U/l), which returned to baseline within 10 days. The subsequent subjects in cohort 5 and the first subject in cohort 6 had less hepatotoxicity, based on smaller rises in transaminases (AST ranging from 64 to 141 U/l) despite receiving higher doses of vector.

Percutaneous liver biopsy specimens were obtained on post–vector infusion day 7 in all subjects. Examination of the specimens showed glycogenated nuclei, steatosis, and fibrosis—histopathologic features consistent with the underlying OTC deficiency (42,43). Portal or lobular inflammation, Kupffer cell hyperplasia, and apoptotic cells were also reported and were thought to reflect either underlying liver disease or vector administration. Thus, hepatotoxicity did not appear to be dose-related, either biochemically or histologically.

There were various hematologic findings. A mild and transient decline in platelets without consistent abnormalities in basic coagulation indices was observed in virtually all subjects. Anemia was also noted, with a nadir in hemoglobin values ranging between 11.9 g/dl (highest) to 7.9 g/dl (lowest), which was not dose-related. There was no systematic or dose-related decline in the white blood cell count.

Vector-Specific Host Immune Responses

Evidence of preexisting immunity to adenovirus was variable in our subjects, but generally mirrored the experience in healthy individuals and patients with cystic fibrosis (44). All subjects showed an increase in neutralizing antibodies to adenovirus following vector infusion. Neutralizing antibody titers peaked between 14 and 60 days after vector administration at values ranging from 2^2 reciprocal dilutions (lowest) to 2^8 reciprocal dilutions (highest). Neutralizing antibody titers persisted, with some decline, over a period of 18 months.

An unanticipated finding was a decrease in preexisting adenovirus-specific lymphocyte proliferative response (LPR) in the six subjects who elicited strong to modest adenovirus-specific LPRs before vector administration. In contrast, subjects with low initial LPRs (stimulation index, ≤ 3.6) generally developed an increase in LPR within the first 2 weeks, consistent with a primary immune response. These enhanced responses eventually returned to baseline after approximately 6 months.

Evaluation of Transgene Expression

Gene transfer was assessed by detection of transgene expression using *in situ* hybridization in liver biopsy specimens. Transgene expressing cells were observed in eight liver biopsies of subjects in cohorts 1, 2, 3, and 5. Expression was seen in no more than 1% of hepatocytes in any of the subjects, however, and no dose–response effect was apparent. Functional gene expression was also assessed through metabolic measures of urea synthesis (stable isotope studies) and nitrogen accumulation (orotic acid excretion) (40). There were no statistically significant increases in urea synthetic capacity and no significant decline in urinary orotate for the group as a whole, comparing pre- and postdose values.

RESULTS IN SUBJECT 18

The 18th subject enrolled in this pilot study was the second in cohort 6. He had been diagnosed as having partial OTC deficiency at 30 months of age, following an episode of vomiting, seizures, and coma associated with hyperammonemia. He was subsequently found to have a unique mutation leading to mosaicism for OTC deficiency (45). Despite treatment with a low-protein diet and alternate-pathway therapy, he experienced multiple episodes of clinically symptomatic hyperammonemia, culminating in a hyperammonemic coma (peak plasma ammonium 382 μmol/l; normal <35 μmol/l) requiring ventilatory support in December 1998, 9 months before gene transfer. He recovered from this episode and graduated from high school. At the time of enrollment into the study, he was found to have 6% to 10% of normal urea synthetic capacity by studies of ^{15}N incorporation into urea. He remained on alternative-pathway therapy throughout the study.

On September 13, 1999, he received 6×10^{11} particles/kg (total dose of 3.8×10^{13} particles) of the adenovirus vector containing the human OTC cDNA (9). Approximately 18 hours after gene transfer he was noted to have altered mental status and jaundice. He then became progressively obtunded, with intermittent periods of agitation. Plasma ammonia rose to a maximum of 393 μmol/l. Hemodialysis was begun and plasma ammonia fell to 72 μmol/l at 40 hours after gene transfer. Disseminated intravascular coagulopathy was subsequently diagnosed on the basis of laboratory studies. At 48 hours, the patient developed severe and progressive hypoxemia, eventually requiring extracorporeal membrane oxygenation (ECMO). By 88 hours, he became anuric and his liver function tests deteriorated (alanine transaminase, 1,366 IU/l; aspartate transaminase, 4,998 IU/l), confirming progressive renal and hepatic failure. Somatosensory evoked potentials became absent, and transcranial Doppler showed no flow. With the family's consent, the subject was removed from ECMO and died 98 hours after gene transfer.

Although the precise cause of the death of this patient remains unclear, the timing of vector administration with the onset of clinical illness suggests that the two were linked. The proximate cause of death was acute respiratory distress syndrome (46). This developed in the context of apparent activation of a systemic inflammation complicated by multiorgan failure. Systemic inflammatory response syndrome (SIRS) has been described in various medical and surgical settings, including septic shock, severe burns, trauma, and cardiothoracic surgery (47). The cause of SIRS in these settings is unclear, although a substantial activation and dysregulation of the host immune response has been postulated (48). In particular, interleukin-6 (IL-6) has been implicated in the postinjury immune response (49,50). We retrospectively evaluated all subjects dosed with vector for evidence of raised serum cytokines. In fact, IL-6 was detected in the serum of subjects in the later cohorts, within hours of vector infusion (9).

WHAT CAN BE LEARNED FROM THIS GENE TRANSFER TRIAL?

Animal Models May Not Be Predictive of Human Responses

The first lesson learned is that studies of animal models may not be predictive of human responses. This variable response to gene transfer was first noted in our familial hypercholesterolemia trial. Familial hypercholesterolemia is an inherited disorder caused by a deficiency of low-density lipoprotein receptors. Preclinical data in an inbred strain of Watanabe rabbits that also has heritable hyperlipidemia (WHHL) suggested that biologically detectable gene transfer could be accomplished using liver-directed gene transfer. This conclusion was based on experiments involving transplantation of autologous hepatocytes that were genetically corrected *ex vivo* with recombinant retroviruses (51). WHHL rabbits uniformly showed a 30% to 50% decrease in total serum cholesterol that persisted for the duration of the experiment. In contrast, the lowering of cholesterol noted in a subsequent human pilot gene transfer study was more variable—prolonged reductions in low-density lipoprotein and cholesterol were demonstrated in only three of five subjects with familial hypercholesterolemia (52–54).

Similar variance was seen in the OTC deficiency trial. In spf mice, histochemical detection of transgene product was dramatic at doses that caused minimal liver injury. Using a third-generation, E4-deleted adenoviral vector expressing β-galactosidase at a dose of 1×10^{11} particles/mouse (5×10^{12} particles/kg, assuming a 25 g C57Bl/6 mouse) essentially every hepatocyte was transduced without evident toxicity (31). In baboons, a dose of 2.5×10^{12} particles/kg was infused selectively into the right lobe of the liver and approximately 40% of the parenchyma was positive for X-gal stain

(38), again without severe toxicity. In the human trial, no discernible transgene expression was seen at 6×10^{11} particles/kg and there was a fatality.

Relation Between Transgene Expression and Toxicity

As a corollary of the above observations, there appears to be a steep curve relating transgene expression to toxicity in nonhuman primates and humans. In rhesus macaques, a dose of 5×10^{12} particles/kg of a first-generation vector was associated with significant gene transfer and minimal toxicity, while a two-fold higher dose, 1×10^{13} particles/kg, was associated with transduction of essentially every hepatocyte, but with dose-limiting toxicity (37). An intermediate vector dose, 7.5×10^{12} particles/kg, was associated with high-level gene transfer but with severe hepatic biochemical toxicity and coagulopathy (55). In humans, a dose of 6×10^{11} particles/kg was associated with a lethal outcome in one of two subjects, despite the absence of detectable gene transfer. The same dose in baboons was associated with only mild transaminitis and no clinically detectable sequelae (38). A dose of 2×10^{11} particles/kg in humans was variably associated with low-dose transgene expression but minimal hepatic toxicity.

The cause of the steep curve is not known but it may be related to a sequestration of vector particles in Kupffer cells. In mice, cells of the reticuloendothelial system take up low doses of vector. Higher vector doses appear to saturate the reticuloendothelial system and lead to uptake in other cell types, such as hepatocytes (56). Vector particles capable of transgene transduction and inactive particles may both be capable of causing reticuloendothelial system saturation. Psoralen inactivated particles induce cytokines at the same level as intact vector particles (57). One important strategy for the production of future vectors, therefore, would be to decrease the ratio of particle to plaque forming unit to unity, as a means of ensuring that each infused vector particle is capable of transducing a target cell.

Role of IL-6 in Adenovirus Toxicity

An increase in IL-6 levels has been noted in the context of adenovirus vector administration since the original report of its use in CFTR gene transfer into the respiratory tract of subjects with cystic fibrosis (58,59). One possible explanation was that the vector abrogated the integrity of the epithelial barrier, allowing preformed IL-6 to leak into the systemic circulation. The increase in IL-6 was thought to be caused by the presence of inflammatory cells in the lungs of individuals with cystic fibrosis. IL-6 has been shown to have a $t^{1/2}$of 4.2 hours in humans; however, the increase following gene transfer lasted for 2 to 3 days, suggesting sustained release of IL-6 into serum after vector administration.

A similar time course of IL-6 elevation, as well as of IL-10, was noted in subjects in the OTC deficiency gene transfer trial. The presence of IL-6 and IL-10 in serum appeared to be dose-dependent, with the highest levels being seen in subjects in the last cohort, which received 6×10^{11} particles/kg (9). In the surviving member of that cohort, IL-6 levels followed the same time course as that of the subjects in the cystic fibrosis trial. In the subject who died, markedly raised IL-6 levels persisted until his death 4 days after gene transfer.

To model the role of IL-6 in vector-mediated gene transfer, additional mouse and rhesus monkey studies were performed after the OTC deficiency study was halted. The findings suggest that preferential activation of dendritic cells and macrophages may account for the adenovirus-triggered acute inflammatory response in mice. Elimination of dendritic cells and macrophages reduced the observed early release of IL-12, IL-6, and tumor necrosis factor-α (TNF-α) in response to vector administration (57). In primates, a sublethal dose of adenovirus vector was also associated with a substantial elevation of IL-6; and fluorescence-labeled vector particles were distributed into dendritic cells and macrophages (55). The OTC deficiency study data suggest that IL-6 levels are raised in response to systemically administered vector. In contrast to the cystic fibrosis trial, where airway neutrophils could have been the primary source of IL-6, dendritic cells and macrophages may play a preeminent role after systemic administration. Whether IL-6 plays a direct pathogenic role or serves as a surrogate marker for other pathogenic cytokines, possibly IL-12, has yet to be determined.

THE PROMISE OF GENE TRANSFER: HEMOPHILIA AND SEVERE COMBINED IMMUNE DEFICIENCY

In contrast to the tragic experience with adenovirus gene transfer in OTC deficiency, the use of adeno-associated virus (AAV) gene transfer in hemophilia and of retrovirus gene transfer in a form of severe combined immunodeficiency shows promise. In the case of hemophilia B, AAV was given into muscle at a dose of 2×10^{11} vector genomes (vg) per kilogram (10). Assessment of the first three subjects showed no evidence of toxicity, and vector sequences were present in muscle. Expression of factor IX was shown by immunohistochemistry, and modest clinical improvement occurred, including increased circulating levels of factor IX and a decreased requirement for factor IX protein infusion. These data suggest that AAV could convert severe hemophilia B into a milder form of the disease. Increased gene expression may be possible through administration of AAV to liver.

In severe combined immunodeficiency—X1 type there is an early block in T and natural killer (NK) lymphocyte differentiation. This block is caused by mutations of the gene encoding the γc cytokine subunit of interleukin-2, -4, -7, -9, and -15 receptors. A gene transfer trial was initiated using cDNA containing a defective γc Moloney retrovirus–derived vector and *ex vivo* infection of CD34 cells (11). After a 10-month follow-up period, γc transgene expressing T and NK cells were detected in the two subjects treated. T, B, and NK cell counts and function, including antigen-specific responses, were comparable to those of controls.

THE FUTURE OF GENE TRANSFER IN NUTRITIONALLY DEPENDENT DISORDERS

Despite the tragic and unanticipated death that ended the OTC pilot study, we believe that clinical gene transfer and gene transfer trials should proceed. As noted above, experience has taught us that preclinical studies can only provide a rough idea of how humans might react to vector administration. As an example, the SIRS seen in the human study was not predicted by preceding animal studies. However, the underlying concept that gene transfer may be a powerful tool to add to the armamentarium of treatment approaches to these disorders remains valid. New vectors and other delivery systems need to be explored, and an improved understanding of the underlying immunology needs to be gleaned from preclinical studies. Other issues that should be addressed include the risk of selective inactivation, necrosis, or apoptosis of transfected cells; concern about enzyme stability; the possible need for coregulation of proteins necessary for intracellular processing; and the mode of administration. There is reason for optimism that gene transfer will eventually become an effective treatment for nutritionally dependent inborn errors of metabolism.

REFERENCES

1. Treacy E, Childs B, Scriver CR. Response to treatment in hereditary metabolic disease: 1993 survey and 10-year comparison. *Am J Hum Genet* 1995; 56: 359–67.
2. Azen C, Koch R, Friedman EG, *et al.* Intellectual development in 12-year-old children treated for phenylketonuria. *Am J Dis Child* 1991; 145: 35–57.
3. Brusilow SW, Horwich AL. Urea cycle enzymes. In: Scriver CR, Beaudet AL, Sly WS, Valle D, eds. *The metabolic and molecular basis of inherited disease.* 8th ed. New York: McGraw-Hill, 2001:1909–64.
4. Whittington PF, Alonso EM, Boyle JT, *et al.* Liver transplantation for the treatment of urea cycle disorders. *J Inher Metab Dis* 1998; 21[Suppl 1]: 112–18.
5. Terheggen HG, Lowenthal A, Lavinha F, Colombo JP, Rogers S. Unsuccessful trial of gene replacement in arginase deficiency. *Z Kinderheilkd* 1975; 119: 1–3.
6. Friedmann T, Roblin R. Gene therapy for human genetic disease. *Science* 1972; 175: 949–55.
7. Rogers S. Gene therapy for human genetic disease. *Science* 1972; 178: 648.
8. Rogers S. Gene therapy: a potentially invaluable aid to medicine and mankind. *Res Commun Chem Pathol Pharmacol* 1971; 2: 587.
9. Raper SE, Chirmule N, Lee FS, *et al.* Lethal systemic inflammatory response syndrome in a patient with partial ornithine transcarbamylase deficiency following intravascular administration of recombinant adenoviral vector, *unpublished data.*
10. Kay MA, Manno CS, Ragni MV, *et al.* Evidence for gene transfer and expression of factor IX in haemophilia B patients treated with an AAV vector. *Nat Genet* 2000; 24: 257–61.
11. Cavazzana-Calvo M, Hacein-Bey S, de Saint Basile G, *et al.* Gene therapy of human severe combined immunodeficiency (SCID)–X1 disease. *Science* 2000; 288: 669–72.
12. Batshaw ML, Brusilow SW, Waber L, *et al.* Treatment of inborn errors of urea synthesis: activation of alternate pathways of waste nitrogen synthesis and excretion. *N Engl J Med* 1982; 306: 1387–92.
13. Rutledge SL, Havens PL, Haymond MW, McLean RH, Dan JS, Brusilow SW. Neonatal hemodialysis: effective therapy for the encephalopathy of inborn errors of metabolism. *J Pediatr* 1990; 116: 125–8.
14. Uchino T, Endo F, Matsuda I. Neurodevelopmental outcome of long-term therapy of urea cycle disorders in Japan. *J Inherit Metab Dis* 1998; 21: 151–9.
15. Matsuda I, Nagata N, Matsuura T, *et al.* Retrospective survey of urea cycle disorders: Part 1. Clinical and laboratory observations of thirty-two Japanese male patients with ornithine transcarbamylase deficiency. *Am J Med Genet* 1991; 38: 85–9.
16. Msall M, Batshaw ML, Suss R, Brusilow SW, Mellits ED. Neurologic outcome in children with inborn errors of urea synthesis. *N Engl J Med* 1984; 310: 1500–5.
17. Qureshi IA, Letarte J, Ouellet R. Ornithine transcarbamylase deficiency in mutant mice. I. Studies on the characterization of enzyme defect and suitability as animal model of human disease. *Pediatr Res* 1979; 13: 807–11.
18. Doolittle DP, Hulbert LL, Cordy C. A new allele of the sparse fur gene in the mouse. *J Hered* 1974; 65: 194–5.

19. Kozarsky K, McKinley D, Austin LL, Raper SE, Stratford-Perricaudet LD, Wilson JM. In vivo correction of low density lipoprotein receptor deficiency in the Watanabe heritable hyperlipidemic rabbit with recombinant adenoviruses. *J Biol Chem* 1994; 269: 13695–702.
20. Jaffe HA, Danel C, Longenecker G, *et al.* Adenovirus-mediated in vivo gene transfer and expression in normal rat liver. *Nat Genet* 1992; 1: 372.
21. Kay MA, Landen CN, Rothenberg SR, *et al.* In vivo hepatic gene therapy: complete albeit transient correction of factor IX deficiency in hemophilia B dogs. *Proc Natl Acad Sci USA* 1994; 91: 2353–7.
22. Yang Y, Nunes FA, Berencsi K, Gonczol E, Wilson JM. Cellular immunity to viral antigens limits E1-deleted adenoviruses for gene therapy. *Proc Natl Acad Sci USA* 1994; 91: 4407–11.
23. Yang Y, Ertl HCJ and Wilson JM. MHC class I restricted cytotoxic T lymphocytes to viral antigens destroy hepatocytes in mice infected with E1 deleted recombinant adenoviruses. *Immunity* 1994; 1: 433–42.
24. Yang Y, Xiang Z, Ertl HCJ, Wilson JM. Upregulation of class I MHC antigens by interferon-γ is necessary for the T cell-mediated elimination of recombinant adenovirus infected hepatocytes in vivo. *Proc Natl Acad Sci USA* 1995; 92: 7257–61.
25. Stratford-Perricaudet LD, Levrero M, Chasse J-F, Perricaudet M, Briand P. Evaluation of the transfer and expression in mice of an enzyme-encoding gene using a human adenovirus vector. *Hum Gene Ther* 1990; 1: 241–56.
26. Grompe M, Jones SN, Louseged H, Caskey T. Retroviral-mediated gene transfer of human ornithine transcarbamylase into primary hepatocytes of spf and spf-ash mice. *Hum Gene Ther* 1992; 3: 35–44.
27. Morsy MA, Alford EL, Bett A, Graham FL, Caskey TC. Efficient adenviral-mediated ornithine transcarbamylase expression in deficient mouse and human hepatocytes. *J Clin Invest* 1993; 92: 1580–6.
28. Morsy MA, Zhao JZ, Ngo TT, *et al.* Patient selection may affect gene therapy success. *J Clin Invest* 1996; 97: 826–32.
29. Ye X, Zimmer K-P, Brown R, *et al.* Differences in the human and mouse amino terminal leader peptides of ornithine transcarbamylase affect mitochondrial import and efficacy of adenoviral vectors. *Hum Gene Ther* 2001; 12: 1035–46.
30. Kiwaki K, Kanegae Y, Saito I, *et al.* Correction of ornithine transcarbamylase deficiency in adult spf-ash mice and in OTC-deficient human hepatocytes with recombinant adenoviruses bearing the CAG promoter. *Hum Gene Ther* 1996; 7: 821–30.
31. Ye X, Robinson MB, Batshaw ML, Furth EE, Smith I, Wilson JM. Prolonged metabolic correction in adult ornithine transcarbamylase-deficient mice with adenoviral vectors. *J Biol Chem* 1996; 271: 3639–46.
32. Ye X, Robinson MB, Pabin C, Batshaw ML, Wilson JM. Anti-CD4 antibody permits readministration of recombinant adenovirus to correct ornithine transcarbamylase deficiency in the sparse fur mice [abstract]. *Pediatr Res* 1997; 41: 109A.
33. Ye X, Robinson MB, Batshaw ML, Pabin CS, Quinn T, Wilson JM. Adenovirus-mediated in vivo gene transfer rapidly protects ornithine transcarbamylase-deficient mice from an ammonium challenge. *Pediatr Res* 1997; 41: 527–34.
34. Ye X, Gao GP, Pabin C, Raper SE, Wilson JM. Evaluating the potential of germ line transmission after intravenous administration of recombinant adenovirus in the C3H mouse. *Hum Gene Ther* 1998; 9: 2135–42.
35. Gao GP, Engdahl RK, Wilson JM. A cell line for high-yield production of E1-deleted adenovirus vectors without the emergence of replication-competent virus. *Hum Gene Ther* 2000; 11: 213–19.
36. O'Brien WE. Isolation and characterization of argininosuccinate synthetase from human liver. *Biochemistry* 1979; 18: 5353–6.
37. Nunes FA, Furth EE, Wilson JM, Raper SE. Gene transfer into the liver of nonhuman primates with E1-deleted recombinant adenoviral vectors: safety of readministration. *Hum Gene Ther* 1999; 10: 2515–26.
38. Raper SE, Haskal ZJ, Ye X, *et al.* Selective gene transfer into the liver of non–human primates with E1-deleted, E2A-defective, or E1-E4 deleted recombinant adenoviruses. *Hum Gene Ther* 1998; 9: 671–9.
39. Raper SE, Wilson JM, Yudkoff M, Robinson MB, Ye X, Batshaw ML. Developing adenoviral-mediated in vivo gene therapy for ornithine transcarbamylase deficiency. *J Inherit Metab Dis* 1998; 21[Suppl 1]: 119–37.
40. Yudkoff M, Daikhin Y, Ye X, Wilson JM, Batshaw ML. In vivo measurement of ureagenesis with stable isotopes. *J Inherit Metab Dis* 1998; 21: 21–9.
41. Raper SE, Yudkoff M, Chirmule N, *et al.* A pilot study of in vivo liver directed gene transfer with an adenoviral vector in partial ornithine transcarbamylase deficiency. *Human Gene Therapy* 2002; 13: 163–75.

42. Badizadegan K, Perez-Atayde AR. Focal glycogenosis of the liver in disorders of ureagenesis: its occurrence and diagnostic significance. *Hepatology* 1997; 26: 365–73.
43. Latham PS, LaBrecque DR, McReynolds JW, Klatskin G. Liver ultrastructure in mitochondrial urea cycle enzyme deficiencies and comparison with Reye's syndrome. *Hepatology* 1984; 4: 404–7.
44. Chirmule N, Propert K, Magosin S, Qian Y, Qian R, Wilson JM. Immune responses to adenovirus and adeno-associated virus in humans. *Gene Ther* 1999; 6: 1574–83.
45. Maddalena A, Sosnoski D, Berry G, Nussbaum R. Mosaicism for an intragenic deletion in a boy with mild ornithine transcarbamylase deficiency. *N Engl J Med* 1988; 319: 999–1003.
46. Ware L, Matthay M. The acute respiratory distress syndrome. *N Engl J Med* 2000; 342: 1334–48.
47. Moore FA, Sauaia A, Moore EE, Haenel JB, Burch JM, Lezotte DC. Postinjury multiple organ failure: a bimodal phenomenon. *J Trauma* 1996; 40: 510.
48. Faist E, Schinkel C, Zimmer S. Update on the mechanisms of immune suppression of injury and immune modulation. *World J Surg* 1996; 20: 454–9.
49. Biffl W, Moore E, Moore F, Peterson V. Interleukin-6 in the injured patient: marker of injury or mediator of inflammation? *Ann Surg* 1996; 224: 647–64.
50. Headley A, Tolley E, Meduri G. Infections and the inflammatory response in acute respiratory distress syndrome. *Chest* 1997; 111: 1306–21.
51. Chowdhury JR, Grossman M, Gupta S, Chowdhury NR, Baker JR, Wilson JM. Long-term improvement of hypercholesterolemia after ex vivo gene therapy in LDLR-deficient rabbits. *Science* 1991; 254: 1802–5.
52. Grossman M, Raper SE, Kozarsky K, *et al.* Successful ex vivo gene therapy directed to liver in a patient with familial hypercholesterolaemia (see comments). *Nat Genet* 1994; 6: 335–41.
53. Grossman M, Rader DJ, Muller DW, *et al.* A pilot study of ex vivo gene therapy for homozygous familial hypercholesterolaemia (see comments). *Nat Med* 1995; 1: 1148–54.
54. Raper SE, Grossman M, Rader DJ, *et al.* Safety and feasibility of liver-directed ex vivo gene therapy for homozygous familial hypercholesterolemia (see comments). *Ann Surg* 1996; 223: 116–26.
55. Schnell MA, Zhang Y, Tazelaar J, *et al.* Activation of innate immunity in nonhuman primates following intraportal administration of adenoviral vectors. *Mol Ther* 2001; 3: 708–22.
56. Tao N, Gao G-P, Parr M, *et al.* Sequestration of adenorival vector by Kupffer cells leads to a non-linear dose response of transduction in liver. *Mol Ther* 2001; 3: 28–35.
57. Zhang Y, Chirmule N, Gao G-P, *et al.* Acute cytokine response to systemic adenoviral vectors in mice is mediated by dendritic cells and macrophages. *Mol Ther* 2001; 3: 697–707.
58. Crystal RG, McElvaney NG, Rosenfeld MA, *et al.* Administration of an adenovirus containing the human CFTR cDNA to the respiratory tract of individuals with cystic fibrosis. *Nat Genet* 1994; 8: 42–51.
59. McElvaney NG, Crystal RG. IL 6 release and airway administration of human CFTR cDNA adenovirus vector. *Nat Med* 1995; 1: 182–4.

DISCUSSION

Dr. Clarke: One of the things I've been intrigued with is the possibility of using nutritional and regulated promoters, so that you can control the amount of expression by controlling the composition of the diet—a classic example would be a fatty acid control promoter. What do you think about the potential for that sort of strategy in the long term?

Dr. Batshaw: It might be very effective. I think it depends on where the enzyme is in the cycle and whether it is the rate-limiting enzyme for that cycle. In the case of OTC, you do not really have to control it because it is not the rate-limiting enzyme, but if it were CPS, the one above it, you would.

Dr. Endres: What is the oncogenic potential of the adenovirus? Is that problem solved or is it still an issue?

Dr. Batshaw: The adenovirus is not incorporated into the host DNA and is episomal, so it is my impression that there isn't an oncogene potential.

Dr. Bachmann: When you did your assessment of biodosage in animals, were those animals on a diet comparable with that of the patient, with respect to the capacity of CPS or other urea cycle enzymes? It takes a few days for urea cycle adaptation, as far as I remember from the old assays. Did you have a comparable situation in that respect?

Dr. Batshaw: Not exactly. These subjects were all clinically stable, and half of them were asymptomatic heterozygote mothers of affected children, so many of them were on unrestricted diets. In fact, to elicit the best response from the sparse fur mice, we tend to give them a high-protein diet, so those are not comparable from that point of view.

Dr. Bachmann: Does that include subject 18?

Dr. Batshaw: That was also true for subject 18. He was on about 1 g/kg/d of protein, and he was 18 years old, so his intake was quite substantial.

Dr. Rey: You said that half your patients were asymptomatic heterozygotes. What is the ethical justification for gene therapy in individuals not affected by a severe and untreatable genetic disease?

Dr. Batshaw: This was a phase 1 trial, which is a safety trial, not an efficacy trial, so it is basically a pharmacokinetic study. Initially we were thinking of doing this trial on newborn infants with OTC deficiency—that is, the most severely affected patients who could most benefit from it. Our bioethicist, however, said we shouldn't do that, for three reasons: first, one has to obtain safety data from a phase 1 trial, which is not designed to be therapeutic; second, newborn babies can't give informed consent for a study that will not be of benefit to them; and third, a parent whose child is near death is not in a position to give informed consent for such a study. Our ethicist advised that we should use the mothers of these children, who are asymptomatic heterozygotes and who could themselves make an informed decision on whether they wanted to participate in the trial. This study went through a very extensive series of regulatory decisions—including consideration by the Recombinant DNA Advisory Committee of the National Institutes of Health (NIH)—in trying to decide who should be the subjects involved in the initial treatment. This is a different question from your one about the mildly affected children with phenylketonuria (PKU). In that case, the risk/benefit ratio needed to be considered. If gene therapy turns out to be safe, and all it involves is an intravenous injection once in your life, then using it for PKU would be very appropriate. It is a question of balance.

Dr. Bachmann: Those who have dealt with urea cycle hyperammonemic defects will know that the situation is completely different from PKU. In the urea cycle defects, there is an ever-present danger from environmental changes; such dangers are simply not present in disorders like PKU, which evolve chronically into mental handicap. With the urea cycle disorders, we are always anxious about external factors, which is why the survival statistics are so inadequate—they are so dependent on the sudden crises that are common in classically treated patients, even in heterozygote mosaics, and so you are never sure that the patient will not suddenly die. There are even reports of patients in their 40s with OTC deficiencies who have had their first lethal attack during an infection (1). That is why it is important ethically to distinguish efforts to treat such disorders from treatment of disorders like PKU.

Dr. Sollid: I understand that the problem here is the genetic variation in the host response. You presumably need to test a sufficient number of individuals to make sure that you will never see such a host response. Would one possibility be not to use any vector that is immunogenic, whether it stimulates the innate or the adaptive immune system? Then you would have a delivery that totally bypasses that possibility.

Dr. Batshaw: When we did these studies, we looked first for neutralizing antibody against adenovirus. Any individual who had a high neutralizing antibody response to adenovirus before gene transfer was excluded from the study, so we did have some concern about that, and we had to exclude two subjects for that reason. We also started to look at whether you could use an anti-CD4 antibody—such as are now being used for rheumatoid arthritis—to try to suppress any immune reactions. However, I agree that using a virus like the AAV, which has much less of an immune response, may be one approach. I am not sure it has to be an all-or-nothing decision, but it is clear that one has to deal with the immune response better than we have done in the past.

Dr. Böhles: Can we learn anything from cystic fibrosis treatment, also using adenoviruses (2)?

Dr. Batshaw: What we learned from the cystic fibrosis study was that when you gave the virus through the nares, or directly into the lungs, you can get inflammation of the lung. We assumed therefore that if we gave the virus into the liver of OTC-deficient patients, we might get hepatitis. The reason why a systemic inflammatory response, as opposed to a localized immune response, occurred is still unclear. One of our questions is whether subject 18 had something unique about him that placed him at greater risk for this response. I did not show the data, but his bone marrow showed marked suppression, which could not have occurred in the 4-day period after gene transfer. So we wondered whether he had some sort of problem beforehand that suppressed his immune system and put him had greater risk. This points out the importance of developing a better means for screening individuals for immunogenetic response.

Dr. Pencharz: C gene therapy is very exciting, so the experience you had is valuable, but what struck me—both with the cystic fibrosis work and with what you have described—is that you're not really creating a permanent cure. The problem with cystic fibrosis was getting it into the stem cells. In some of the bone marrow disorders, you can take bone marrow, identify stem cells, do your gene therapy, and then retransplant. That seems to me to be about the only situation where you could get a permanent cure. Maybe this is just my lack of understanding. Can you help with this?

Dr. Batshaw: It depends on what you mean by permanent. If you mean life-long, then I think you are probably right. Our study was the first of a two-part study. Our plan was that you would use the adenovirus to get the infants out of coma in the newborn period, because it works so rapidly and lasts for 40 to 60 days. You would then follow that up with an AAV injection, which would not have cross-reactivity and which takes about 3 to 4 weeks to work. The effects of adeno-associated viral gene transfer last for 3 years or so, and because it does not elicit an immune response, it was thought that you could give repeated injections. That was the rationale behind the treatment. I agree, it would not have been permanent.

Dr. Rey: Can you say a few words about vectors other than viruses?

Dr. Batshaw: There are many other approaches to delivery, from naked DNA to using artificial chromosomes. To my knowledge, none of these has gone far enough for us to know whether they are going to work or not. The reason people have used viruses up to this point has been that the adenovirus, for example, has an affinity for the liver. The idea of having a "zip code" is very appealing, and that is why the focus has been on that mode of delivery. Obviously the problems that have developed with the adenovirus have called into question the use of high doses of this virus. There is more work being done in other areas, but I don't think any of it is near the stage of use in human therapy yet.

REFERENCES

1. Nishiyori A, Yoshino M, Kato H, *et al.* The R40H mutation in a late onset type of human ornithine transcarbamylase deficiency in male patients. *Human Genet* 1997; 99: 171–6.
2. Liu X, Jiang Q, Mansfield SG, *et al.* Partial corrections of endogenous Delta F508 CFTR in human cystic fibrosis airway epithelia by spliceosome-mediated RNA transsplicing. *Nat Biotechnol* 2002; 20: 47–52.

Genetic Expression and Nutrition, edited by Claude Bachmann and Berthold Koletzko. Nestlé Nutrition Workshop Series, Pediatric Program, Vol. 50. Nestec Ltd., Vevey/Lippincott Williams & Wilkins, Philadelphia, © 2003.

Concluding Discussion

Berthold Koletzko

Division of Metabolic Diseases and Nutritional Medicine, Dr. von Hauner Children's Hospital, Kinderklinik and Kinderpoliklinik, University of Muenchen, Muenchen, Bavaria, Germany

In my concluding remarks, I will round off this workshop by recalling the various papers in chronologic order. I will give a subjective selection of some of the high points in the presentations, without trying to summarize them. The concept of the program was to look at various aspects of metabolic disease and at what we can learn about general physiology from these diseases. Such lessons might then be transferable to other applications. We looked at genetic variation in the general population, the impact on nutrition, and the impact of nutrition. We looked at the range from molecular to functional and clinical aspects. I'm happy to say that I found no weak spots in the program.

We started with Dr. John Scott's discussion of homocysteine metabolism in relation to arteriosclerosis—a very stimulating paper in which he discussed whether homocysteine should be regarded as a risk marker or as a risk factor for cardiovascular disease. He is inclined to think that it is a risk factor, although he clearly indicated to us that we need evidence from properly conducted intervention trials on which we can base appropriate preventive strategies.

Next we had Dr. Hélène Ogier de Baulny, who reviewed the extensive experience of the Paris group in the treatment of phenylketonuria (PKU) and maple syrup urine disease (MSUD). She estimated the intakes of phenylalanine and leucine in adequately growing children with these disorders, and reported that with an intake of about 250 mg phenylalanine a day and 350 to 450 mg of leucine, these children appear to grow normally. In the discussion, some questions were raised about potential limitations in extrapolating the data to healthy populations, questions on differences in metabolic turnover of these amino acids in the disease state, the role of amino acid supplements, and functions and outcomes other than purely growth.

Dr. Paul Pencharz introduced us to a novel approach to looking at amino acid needs—the measurement of tracer oxidation of a marker amino acid in response to a variation in amino acid intake—and then to a mathematical break point analysis. He came up with an estimate for tyrosine and phenylalanine needs in PKU of the order of 20 mg/kg/d. We wondered whether there were potential limitations such as the role of protein quality, adaptation to habitual intake, and the question of whether a 24-hour measurement could lead to different conclusions.

I will not comment on my own presentation—you can make up your mind for yourselves!

Dr. Steve Clarke's presentation was one of the highlights of the meeting. He looked at the question of whether nutrients are powerful modulators of genes and metabolism—something that has been referred to repeatedly here as nutrigenomics. I was particularly impressed by the coordinated up- and down-regulation of metabolic steps that he showed. He gave us some examples, such as the supply of ω-3 fatty acids, which appears to have effects that would be particularly relevant in relation to the metabolic syndrome and long-term health. Here we had one of the sayings of the week: "Nutrients were the first hormones." I will certainly take that home with me and remember it.

Dr. Ron Wanders then gave us a brilliant overview of fatty acid oxidation metabolism, its defects, and the role of peroxisome proliferator-activated receptors (PPARs). He showed us a newly described defect, 2-methylacyl-CoA racemase deficiency, with adult onset sensory-motor neuropathy and other symptoms. From a diagnostic point of view, their long-chain fatty acids can be normal. He also gave some very impressive indications of crosstalk between fatty acid oxidation systems and the regulator of PPARα in the oxidation of fatty acids.

Then Dr. Helmut Sies provided a very broad overview of oxidant/antioxidant systems and their relation to degenerative neurologic disease, macular degeneration, and other disorders such as Down syndrome. He told us about promising interventions, both of a preventive and a therapeutic nature. I think we all agree that this is an area that deserves much more work and holds great promise. Intervention studies appear desirable.

Dr. Thorsten Marquardt gave us a brilliant guided tour through the emerging and complex field of glycosylation defects, where we are continually exposed to new disorders at short intervals. It is now apparent that for two defects there is effective treatment available, although we need to know a bit more about the effects of therapy, particularly in the case of fucose. The challenges obviously are to learn more about the physiology of these conditions and the variation in presentation between patients. Dr. Marquardt's message was that we need to increase our diagnostic awareness about such patients, so that they aren't missed. We need improvements in diagnostic methods and therapeutic approaches.

Dr. Clemens Kunz discussed the oligosaccharides. He introduced us to a class of more than 100 newly identified compounds in human milk, showed us details of their characterization, and showed us that they are absorbed and metabolized in the recipient infant. He indicated their potential role as prebiotics, providing food for microbiotic colon flora, their receptor binding anti-infective properties, their potential effects on immune regulation, and obviously their potential for producing systemic effects as they are absorbed into the circulation. The challenge here is to characterize their functional effects.

Dr. Claude Bachmann gave another brilliant talk, from arginine to creatinine, with the message that different brain cell types vary in expression of genes that are involved in arginine and creatine transport metabolism. It appears that brain synthesis of creatine is more important than its uptake from the blood, in contrast to the metabolism of arginine. Of particular interest are the intercellular interactions, which have probably relevance in hyperammonemia. His saying of the week: "The brain is not a homogenate."

Dr. Ludvig Sollid gave a fascinating description of the molecular and cellular basis of celiac disease. He showed us that gluten peptides, after deamidation by transglutaminase, are bound to the HLA DQ2 hydrodimer and are presented to T cells. Obviously this knowledge could form the basis for further advances in treatment. He showed us some examples—the inhibition of transglutaminase, tolerance induction to deamidated gliadin, and the approach with soluble dimers of HLA—peptide complexes.

Dr. Dennis Bier then tried to make a good case for the beneficial aspects of the common modifications in food production. He told us that biotechnology might contribute to an increase in food production, a decrease in food spoilage, an increased quality of foods, and ultimately to improvements in terms of health outcomes of the recipient consumers. He clearly indicated that the public demand for a guarantee of absolute safety is not realistic, but as in other areas, what we need is a careful benefit/risk analysis. His saying of the week, "No evidence of risk is not evidence of no risk."

Dr. Hannelore Daniel then returned to the questions that were raised by Dr. Clarke on nutrigenomics. She showed us that cell cycle growth in apoptosis could be modulated by dietary components, by total energy intake, by micronutrient intake, and by nonnutrients. She emphasized that these effects are tissue- and cell-specific, and she showed us examples of how dietary components may influence both the initiation of tumors and the later stages of tumor development. I have a problem with interpreting her key messages. She gave us two messages that were, for me, somewhat conflicting. One was that energy restriction is good for you. The other, after dinner, was that life is a pleasure at a Nestlé Nutrition Workshop! How you bring those two together is a challenge. Her memorable saying: "Not all tomatoes are equal and not all potatoes are equal."

I don't need to summarize the final presentation that you have just heard, by Dr. Mark Batshaw, describing the unanticipated complications of gene therapy, emphasizing that animal studies may not be predictive of human disease. In view of their great therapeutic potential, it is clear to me that gene transfer trials should proceed.

At the end of this workshop, I would like to thank everyone: the speakers for excellent presentations and for submitting their manuscripts on time; the stimulating discussions; the audience for your interest; the people from the audiovisual service for their fantastic work, which provided us with great communication conditions; and the people in the organizing committee and from Nestlé who have enabled this excellent workshop to take place.

In the first presentation, Dr. Scott proposed that there might not be any normal people. When I look around this room, I fully agree with him! I have met an outstanding collection of great people, great scientists, and great personalities, and I am truly grateful for this. Thank you all.

Subject Index

A

ABO blood group
  function, 149–150
  and human milk composition, 139–152
Aboriginal peoples of Australia, diet, 262
Acetyl-CoA carboxylase, and unsaturated fatty acids, 70
N-acetyltransferases, 241
Acyl-CoA dehydrogenases
  in β-oxidation system, 92–93
  genetic deficiency in, 96
Acyl-CoA oxidase 1 deficiency, 101
Acyl-CoA oxidases, in oxidation, 94
Adeno-associated virus vectors
  in hemophilia, 272
  in ornithine transcarbamylase deficiency, 277
S-Adenosylhomocysteine, and homocysteine formation, 1–3
S-Adenosylmethionine
  and homocysteine formation, 1–3, 5
  and methionine catabolism, 5
Adenovirus 12, celiac disease, 170
Adenovirus vectors, 264–277
  immune response to, 265, 268–269, 276–277
  and interleukin-6 levels, 271–272
  in ornithine transcarbamylase deficiency, 264–277
    animal model, 265–267, 270–271
  toxicity, 267, 271–272
    and transgene expression, 271
Adrenoleukodystrophy
  infrafamilial variability, 105
  overview, 98–99
  treatment, 99, 105
Adrenomyeloneuropathy, 98–99, 105
  adrenoleukodystrophy comparison, 105
  overview, 98–99
Age-related macular degeneration, 113–114
Aging, and peroxisomal oxidation, 104
Agricultural biotechnology, 201–217
  and child nutrition, 209–211
  in drug production, 217
  safety concerns, 205–207
Agrobacterium tumefaciens, 207
Akt protein, overexpression, 249
Alcoholism, glycosylation defects, 134
Alpha-oxidation, mechanism, 86–87
Alpha-tocopherol transfer protein abnormalities, 113
Alzheimer disease
  antioxidants role, 111–112, 119
    causative effect question, 119, 120–121
  oxidative stress role, 109, 120
Amino acid requirements
  antenatal steroid effects, 47
  assessment methods, overview, 37–42
  factorial analysis, 27–31
  human milk model calculation, 26–31
  inborn errors of metabolism models, 25–49
  indicator assessment method, 38–50
Amlodipine, grapefruit juice interactions, 224–225
Ammonium chloride
  arginine and creatine effects, brain, 157–165
  axon development effects, 160–161, 164
Amyotrophic lateral sclerosis
  antioxidant effects on, 113
  oxidative stress role, 111
Anoikis, 248
Antenatal steroids, 47
Antihistamines, grapefruit juice interactions, 224

Antihypertensive drugs, and grapefruit juice, 223–224, 226
Antilipemic agents, grapefruit juice interactions, 223–224
Antioxidants, 107–122
  Alzheimer disease benefit, 111–112, 119, 120–121
  and apoptosis, 251, 262
  cataract benefits of, 114–115
  flavonoids as, 262
  natural versus synthetic, 121
  in neurodegenerative diseases, 111–113
  and omega-3 fatty acids, 80
Antiretroviral drugs, and grapefruit juice, 228
Apaf-2, and cancer, 248
Apoptosis, 239–262
  and diet, cancer prevention, 239–262
  flavonoid effects, 251–254
  and macronutrients, 249–250
  major pathways, 246–248
  and micronutrients, 250–251
  p53 induction of, 244
  triggers, 246
Arachidonic acid supplementation, in phenylketonuria, 61
Arginase deficiency, gene therapy, 263
Arginase II
  ammonia effects, astrocytes, 159
  mRNA analysis, brain, 155–156
Arginine, 153–166
  fetal rat brain, hyperammonemia, 157–161
  hyperammonemia effects on, 153–165
  metabolic pathways, 154–155
  serum monitoring, preterm babies, 165
Arginine deficiency, 165
Arginine:glycine amidino transferase
  ammonia effects, brain, 161
  mRNA analysis, brain, 155–157
Argininosuccinate lyase
  ammonia effects, astrocytes, 159
  mRNA analysis, brain, 155–156
Argininosuccinate synthetase
  ammonia effects, astrocytes, 159
  arginine synthesis, 154–155
  mRNA analysis, brain, 155–156
Artemether, grapefruit juice interactions, 227
Ascorbic acid, and apoptosis, 251
Astemizole, grapefruit juice interactions, 224–225
Astrocytes
  ammonia effects, fetal rat, 157–161
  arginine and creatine metabolism, 155–165
Atherosclerosis, homocysteine role, 1–24
Atorvastatin, grapefruit juice interactions, 226
Autoantibodies
  celiac lesion role, 177
  and gluten-specific T cells, 175–176
Axon growth
  ammonium chloride effects, 160–161, 164
  and creatine, 160–161
  pH versus ammonium chloride influence, 164

B

Barley
  gliadin-like proteins, 190
  HMW-glutenin gene transformation, 196
  prolamins, 188, 191–192
  taxonomy, 186
Bcl proteins
  and apoptosis, 248–249, 252
  flavone effect on, 252–253
  and nutrients, 249
Bergamottin, 229–230
Beta-carotene
  dementia effect of, 112
  high dose negative effects, 121

Beta-oxidation
enzymology, 92–95
genetic disorders, 95–105
mechanism, 86–92
peroxisomal versus mitochondrial, 89–92
and peroxisome proliferator activated receptor-α, 100–101
polyunsaturated fatty acids, 81-82
Betaine
and homocysteine levels, 9
homocysteinuria intervention, 22
D-Bifunctional protein deficiency, 99, 101–102
L-Bifunctional protein deficiency, 101
Bifunctional proteins, in peroxisome oxidation, 94–95
Biotechnology. *See* Agricultural biotechnology
Blood groups
diet relationship, 216
function, 149–150
and human milk oligosaccharides, 139–152
Brain
breast milk oligosaccharides, 152
and oxidative stress, 108–109
Breast cancer, and diet, 244–245
Breastfeeding. *See also* Human milk
celiac disease prevention, 33, 181–182
gluten introduction, 181–182
Burkitt lymphoma, 244
Butyrate, and colon cancer, 245, 249

C

C22: 6, in adrenoleukodystrophy, 105
Calcium antagonists, and grapefruit juice, 223–224, 226
Calcium intake, and apoptosis, 251
Camptothecin, 252, 254
Cancer
apoptosis role, 246–248
and diet, 239–262
Canola oil, 201–202
Carbohydrate response element, 75
Carbohydrate therapy, 123–136
congenital disorders of glycosylation, 123–136
CDG-Ia, 126–130, 134–135
CDG-Ib, 129, 131, 134–136
CDG-If, 131–132
CDG-IIc, 131, 133–134, 135
and human milk oligosaccharides, 141–152
Carbon, and unsaturated fatty acid activity, 76
Carbon oxidation method, 38–41
versus nitrogen balance method, 40
strengths and weaknesses, 40–41
Cardiac presentation, β-oxidation deficiency, 96
Cardiovascular disease, homocysteine role, 1–24
Carnitine
genetic deficiency effect, 96
β-oxidation role, 89–90
Carnitine/acylcarnitine translocase deficiency, 95–96
Carnitine palmitoyltransferases
genetic deficiency effect, 95–97
and β-oxidation, 89–90
unsaturated fatty acids effect, 69–70, 82
L-Carnitine supplementation, 97–98
β-Carotene
dementia effect of, 112
high dose negative effects, 121
Caspases, in apoptosis, 246–249, 251
Cataract, and oxidative stress, 114–115
Cationic amino acid transporters
ammonia effects, astrocytes, 159
arginine metabolism, 155
brain mRNA, 155–157
$CD4^+$ T cells
celiac disease, 169–176
treatment implications, 177
gluten peptide recognition, 172–175
CD28 gene, celiac disease, 169

Celiac disease, 167–199
  and autoimmunity, 34, 175–176, 183
  deamidated gluten recognition, T cells, 172–175
  environmental factors, 170
  genetically-modified gliadins in, 185–198
  genetics, 167–169, 171–175
  gluten intake effects, 32–33, 182
  and HLA genes, 168–169, 171–175
  lesion characteristics, 170, 176–177
  molecular and cellular basis, 167–183
  non-HLA genes, 169, 181
  oat tolerance, 32, 177, 181
Cereals
  and genetic engineering, 192–198
  prolamins in, 187–192
  taxonomy, 186
Cerebrovascular disease, and homocysteine, 8
Childhood cerebral adrenoleukodystrophy, 98–99
Childhood nutrition, and recombinant foods, 209–211
Cholesterol
  and colon cancer, 259
  and plasma homocysteine, 23–24
Cholesterol-lowering drugs, and grapefruit juice, 223–226
Chromosomal translocations, 244–245
Cisapride, grapefruit juice interactions, 224–225
Citrulline, arginine synthesis, 153–156, 164
Clarithromycin, grapefruit juice interactions, 225
Coagulation system
  and congenital disorders of glycosylation, 136
  and plasma homocysteine, 24
Coenzyme Q_{10}, in neurodegenerative diseases, 112–113
Cognitive effects, and docosahexaenoic acid, 57–58
Colon cancer
  and cholesterol intake, 259
  and dietary fiber, 245
  flavonoid effects, 252–254
Colostrum, immune effects, 151
Complementary DNA expression arrays, 254–255
Complex I, in Parkinson disease, 110
Congenital disorders of glycosylation, 123–136
  carbohydrate therapy, 126–136
    CDG-Ia, 126–130, 134–136
    CDG-Ib, 129, 131, 134–136
    CDG-If, 131–132
    CDG-IIc, 131, 133–135
    CDG-IIb, 135
  and N-glycosylation principles, 123–126
Copper deficiency, and neurologic disorders, 113
Cow's milk, oligosaccharides, 139, 152
COX-2 expression
  and dietary fat, 249
  flavone effect on, 252–253
  and flavonoids, 246, 252–253
Creatine, 153–165
  and axon development, 160–161
  clinical perspective, 164
  fetal rat brain, hyperammonemia, 157–161
  hyperammonemia effects, 153–165
  metabolic pathways, 154–155
Creatine kinase, brain isoenzyme
  creatine synthesis, 154
  in mitochondria, 165
Creatine transporter 1
  ammonia effects, rat brain, 159–161
  genetic defects, 154
  mRNA analysis, brain, 155–157
CTLA4/CD28 genes, celiac disease, 169
Cyclosporine-ketoconazole interactions, 226–227
Cyclosporine-Seville orange juice interactions, 230

CYP3A4
  in drug metabolism, 222
  grapefruit juice effect on, 222–225
Cystathionine-β-synthase
  genetic disease effects, 9, 11–12, 15–16
  homocysteine formation, 3
Cysteine, methionine ratio, 20
Cystic fibrosis, gene therapy, 277
Cytochrome c, and apoptosis, 248, 251

D

Deamidated gluten
  celiac disease, 172–175, 177
  mechanism, 182–183
  T-cell recognition, 172–174
  tissue transglutamidase role, 173–175
  treatment implications, 177
Dementia. *See* Alzheimer disease
Δ-6 Desaturase pathway
  activation induction, 73–74
  and docosahexaenoic acid, 63
  fuel repartitioning function, 68–70
  and n-6 versus n-3 fatty acids, 76
Diet-drug interactions, 219–237
  adverse clinical consequences, 225–226
  beneficial clinical consequences, 226–231
  CYP3A4 mechanism, 222–231
  grapefruit juice in, 219–237
Dietary fiber, and colon cancer, 245
Dihydropyridine calcium antagonists, 223–224, 226
6',7'-Dihydroxybergamottin, 229–230
DISC (fas-associated death domain protein), 247
DNA adducts, 241
DNA repair
  macronutrient effects, apoptosis, 249
  and monosaturated fatty acids, 82–83
  p53 protein role, 242
Docosahexaenoic acid
  and breast cancer, 245
  gene expression regulation, 52
  peroxisomal β-oxidation, 91–92
  in phenylketonuria, 51–64
  plasma and blood cell levels, 60—61, 64
  in vegan women, 63
  visual evoked potentials effect, 52–64
Dolichol-linked oligosaccharide, in glycosylation, 124–125
Donepezil, 112
Dough functionality. *See* Wheat dough
Down syndrome
  antioxidants in, 120–121
  oxidative stress role, 111
DQ2 haplotype. *See* HLA-DQ2
Drinking water, and agricultural biotechnology, 215
Drosophila, nutrition and apoptosis, 250
Drug response-diet link, 219–237
  CYP3A4 mechanism, 222–231
  grapefruit juice role, 219–237
  and oral bioavailability, 219–221, 228–231

E

Eicosanoids, 76–77
Energy intake effects, 249, 259
Enoyl-CoA hydratases, and oxidation, 92–94
Epigallocatechin-3-gallate, 252–254
Erythromycin, grapefruit juice interactions, 225
Escherichia coli, and oligosaccharide receptors, 144
Essential amino acids. *See* Amino acid requirements
Estrogens, and breast cancer, 245
Ethics, gene therapy, 276
Etomoxir, 100–101
Evidence-based medicine, and antioxidant benefits, 120–121
Evolutionary factors
  fish oil effectiveness, 80, 262
  and recombinant agriculture, 208–209
Expressed sequence tag sequencing, 197
Expression profiling method, 261

F

Factorial method, protein requirements, 27–34
Familial hypercholesterolemia, 270
Fas ligand, in apoptosis, 247
Fat intake, and breast cancer, 245
Fatty acid synthase gene, 72–75
Fatty acids. *See also* Omega-3 fatty acids
- fuel repartitioning function, 68–70
- gene expression regulation, 67–83
- oxidation, 85–92
  - genetic disorders, 95–105
  - overview, 85–92
- peroxisome proliferator activated receptor role, oxidation, 68–71, 100–105
- in red blood cells versus plasma, 60–61, 64
- visual evoked potentials effect, 51–64

Felodipine-grapefruit juice interaction
- discovery, 220–221
- mechanism and relevance, 221–223, 229–230
- oral bioavailability studies, 229–230

Fish oil. *See also* Omega-3 fatty acids
- antioxidant effects, 80
- cancer prevention, 249
- evolutionary considerations, 80
- and muscle gene expression, 68, 69
- phenylalanine effect of, 64
- and visual evoked potentials, 52–64

Flavone
- apoptosis regulation, 252–254, 261
- molecular effects, cancer cells, 253

Flavonoids
- antioxidant properties, 262
- apoptosis regulation, 251–254
- and cancer, 242, 246, 251–254

Folate
- chemotherapy effects, 241
- genetic deficiencies effect, 10–12, 21
- and homocysteine, 3, 5, 10–12, 21–23
- and neural tube defects, 21–22
- structure, 4

Folic acid
- and DNA methylation 240–241
- intervention trials, 13–15, 21–23
- structure, 4

Food production
- future needs, 202–204
- recombinant modifications in, 201–217
  - and child nutrition, 209–211

Friedreich ataxia, 111, 113
Fructans, cancer prevention, 249
Fructosylated oligosaccharides, 151
Fucose, 131, 133, 135
Fucosylated oligosaccharides
- antiadhesive properties, 141–152
- fructosylated oligosaccharides comparison, 151
- human milk concentrations, 137–152

Furanocoumarins, 229–230

G

GAMT expression, 155–161
Garlic, anticancer activity, 242
GDP-mannose, in glycosylation, 123–125
Gene amplification, breast cancer, 244
Gene suppression, gliadins, 193—194
Gene therapy/transfer, 263–277
- animal models, predictive value, 270–271
- and diet, 275
- ethics, 276
- in hemophilia, 272
- ornithine transcarbamylase deficiency, 265–277
  - toxicity, 267, 271–272
  - vector choice, 264–267
- in severe combined immune deficiency, 272

Genetic engineering
- celiac-related peptides, 192–198
- drug production, 217
- food production, 201–217
  - and childhood nutrition, 209–211
  - risks, 205–206

Genistein, 252–253
Gingko biloba, in Alzheimer disease, 112
α-Gliadin
  in cereals, 187–189
  genetic engineering, strategies, 192–196
  repeat domain modification, 195
  structure, 189–192
Gliadins, 185–198
  and celiac disease, 185–198
  in cereals, 189–192
  genetic engineering, 192–198
  repetitive domains, 189, 195
    modification, 195
  structure, 189–192
  T-cell recognition, 172, 174–175
  tissue transglutaminase effect on, 173–174, 186
Glucose intake, and fatty acid oxidation disorders, 97
Glucose-6-phosphate dehydrogenase mRNA, 75
Glutamate carboxypeptidase polymorphisms, 12
Glutamine
  apoptosis role, 249
  enterocyte effects, 249, 260–261
Glutathione, and homocysteine, 19–21
Gluten. *See also* Gliadins
  and breastfeeding, 181–182
  deamidated form, celiac disease, 172–175
  oral tolerance, 173, 177
  structure, 187
Gluten-free diet
  alternatives, 177–178
  limitations, 185–186
Gluten intake, and celiac disease, 32–33, 182
Glutenins
  in cereal, 187
  gene suppression, 193
  gene transformation, 196–197
Glycine-*N*-methyltransferase, methionine catabolism, 5
N-Glycosylation
  and congenital disorders of glycosylation, 123–136, 280
  principles, 123–126
Golden rice, 210–211, 215–216
Grapefruit juice-drug interactions, 219–237
  active ingredient studies, 229–230
  adverse clinical consequences, 225–226
  beneficial clinical consequences, 226–231
  discovery, 220–221
  mechanism and relevance, 221–223, 229–230
Guanidinoacetic acid, 159–160

H

Helicobacter pylori, antiadhesive treatment, 144, 150
Hemophilia, gene transfer, 272
Hepatic effects, mitochondrial β-oxidation deficiency, 96
Hepatic nuclear factor 4, fatty acid regulation, 75
Hexacosanoic acid, 91
High-throughput screening arrays, 254–255
Highly unsaturated fatty acid
  fuel repartitioning function, 68–70, 74, 76
  gene expression regulation, 67–83
  lipid metabolism, genomic mechanism, 70–75
HLA-DG8
  celiac disease, 168–169, 171–175, 177
  deamidated gluten recognition, 172–175
HLA-DQ2
  celiac disease, 32, 168–169, 171–177, 181, 281
  deamidated gluten recognition, 172–175
  peptide binding, 171
  T cell binding, 174
  treatment implications, celiac disease, 177

HLA genes, celiac disease, 168–169, 171–177, 181
HMG-CoA reductase inhibitors, and grapefruit juice, 223–226
Homocysteine, 1–24
atherosclerosis role, 1–24
catabolism, 6
genetic studies, 9–12
intervention trials, 13–17, 22–24
metabolism, 1–6
reactivity, 19–21
renal disease effect, 13
as risk factor versus risk marker, 7–8, 22–23, 279
structure, 2
toxicity, 6–8
Homocysteinuria
genetics, 11–12, 21
intervention, 22
Human milk
minimum protein requirements model, 26–31
oligosaccharides genetic variability, 137–152
and antiadhesive properties, 141–152
Hunter-gatherer period, fatty acid diet, 262
Huntington disease
antioxidants effect, 112–113
oxidative stress role, 110–111
3-Hydroxyacyl-CoA dehydrogenase, 93–94
Hyperammonemia
arginine and creatine metabolism, 153–165
fetal rat brain, creatine pathway, 157–161
in ornithine transcarbamylase deficiency, 264, 266–267, 276
Hypotensive reactions, and drug-grapefruit interactions, 226

I

Idebenone, 112–113
Inborn errors of metabolism. *See also specific disorders*
amino acid requirements assessment, 28–50
gene therapy, 263–281
and homocysteine toxicity, 9–10
Indicator amino acid balance method, 38–50
adaptation effects in, 38–39, 45
maple syrup urine disease application, 44–45
phenylalanine application, PKU, 42–44
strengths and weaknesses, 40–41
Indicator oxidation model, 39–41
Information processing
docosahexaenoic acid effect, PKU, 52–58
and visual evoked potentials, 62–63
Insulin response element, 72–74
Interferon-γ, celiac disease lesion, 176
Interleukin-6, adenovirus toxicity, 271–272
Intraepithelial lymphocytes, celiac disease, 170, 176
Inulin, 151
Iron overload, 121
Isothiocyanates, anticancer activity, 242
Itraconazole, grapefruit juice interactions, 225

K

Keratinocyte growth factor, celiac disease, 176
3-Kethothiolases, 93–94
Ketoconazole-cyclosporine interactions, 227

L

L-type pyruvate kinase gene, 75
Lactase, and human milk oligosaccharides, 149
Lactation, oligosaccharide differences, 150
Lacto-N-fucopentaose, 137–143
Lacto-N-octaoses, 137–138

Lacto-N-tetraose
fucosylation effect, 139
human milk concentrations, 137–143
Lactose, human milk concentrations, 149
Leucine
factorial analysis, 27–28
as indicator amino acid, 39
minimum requirements, 27–28
Leukocyte adhesion, and selectins, 146–148
Leukocyte adhesion deficiency syndrome type II, 131, 133
Lewis blood groups
function, 149–150
and human milk oligosaccharides, 139–152
Linoleic acid, and breast cancer, 245
Lipogenic gene transcription, 72–75
Long-chain fatty acids. *See* Omega-3 fatty acids
Lorenzo's oil, 99–100
Lovastatin, grapefruit juice interactions, 225
Low-molecular-weight gliadins, 191
Lutein, macular protection, 113–114
LXR receptor, 81
Lysine, as indicator amino acid, 43, 47–48

M

Macronutrients, and apoptosis, 249–250
Macular degeneration, 113–114
Magnetic resonance imaging, fatty acid utilization, 62
Maize, prolamins, 188, 192
Malonyl-CoA, 69–70, 82
Mannose treatment
congenital disorders of glycosylation, 126–131, 135–136
side effects, 136
Maple syrup disease, 25–34
Matrix metalloproteinases, celiac disease, 170
Medium-chain acyl-CoA dehydrogenase deficiency, 96
Metabolic syndrome, 67–83, 280
Metalloproteinases, celiac disease, 170
Methionine
DNA methylation, 240–241
and homocysteine formation, 1–6, 20, 22–23
metabolism, 1–6
structure, 2
Methionine load, 20, 22
Methionine synthase
genetic deficiencies effect, 11–12
homocysteine formation, 3
Methionine synthase reductase, 11–12, 15–16
2-Methylacyl-CoA racemase deficiency, 99, 280
Methylated DNA
cancer mechanism, 239–241
dietary factors, 240–241
Methylation ratio, homocysteine formation, 2–3
5,10-Methylene-THF-reductase gene, 21
5,10-Methylenetetrahydrofolate reductase
genetic deficiencies effect, 9–12, 15–17
in homocysteine formation, 3, 5
structure, 4
Methylmalonicaciduria, 26
Methylmalonyl-CoA mutase, 9
Methyltransferase reactions, homocysteine formation, 1–6
Micronutrients, and apoptosis, 250–251
Minimum protein requirements
assessment methods overview, 37–42
factorial analysis, 27–31
inborn error of metabolism models, 25–49
indicator amino acid oxidation method, 38–50
Mitochondrial creatine kinase, 154
Mitochondrial fatty acid oxidation
carnitine role, 89–90
enzymes, 92–93
genetic disorders, 95–98
diagnosis, 104
treatment, 97–98, 105
mechanism, 88
versus peroxisomal oxidation, 88–92

Mitochondrial pathway
in apoptosis, 247–248, 251
and macronutrients, 251
Mitochondrial trifunctional protein, 93
Monosaturated fatty acids, and DNA repair, 82–83
MPTP, complex inhibition, 110

N

Naringin, 229–230
NeuAc milk levels, 137–152
biologic consequences, 137–152
carboxyl group function, 148
and leukocyte adhesion, infants, 146–148
Neural tube effects, folate-related genes, 21
Neurodegenerative diseases
antioxidative role, 111–113, 119, 120–121
oxidation stress role, 108–113, 120
Neurons
ammonia effects, fetal rat, 157–161
arginine and creatine metabolism, 155–165
NF-κB, flavone effects, 252–253
Nitrates, grapefruit juice interactions, 226
Nitric oxide, creatine synthesis, brain, 154–156
Nitric oxide synthetase, brain isoenzyme, 154–155
Nitrogen balance method, 37–38, 40–41
S-Nitroso-homocysteine, and cardiovascular disease, 23
Nitrotyrosine, measurement, 120
Non-HLA genes, celiac disease, 169, 181
Nuclear factor Y, 72–74
Nutrigenomics, 280, 281

O

Oat tolerance
celiac disease, 32, 181
molecular evidence, 181
Oats
celiac disease safety, 186
HMW glutenin gene transformation, 196
prolamins, 191–192
Oligodendrocytes
ammonia effects, fetal rat, 157–161
arginine and creatine metabolism, 155–169
creatine synthesis locus, 163
Oligosaccharides
antiadhesive properties, 141–152
cow's milk, 139, 152
human milk, 137–152
blood group influence, 139–142
immune function, infants, 146–148
and leukocyte adhesion, 146–149
microbe receptor relationship, 144
Olive oil
ammonia effects, fetal rat, 157–161
antioxidant effects, and apoptosis, 80
arginine and creatine metabolism, 155–169
muscle gene expression effect, 68–69
Omega-3 fatty acids. *See also* Docosahexaenoic acid
antioxidant effects, 80
and breast cancer, 245
and colon cancer, 249
evolutionary considerations, 80, 262
fuel partitioning activity, 76
gene expression regulation, 67–83
hunter-gatherer period, 262
n-6 fatty acid comparison, 76
in phenylketonuria, 51–64
plasma and blood cell levels, 60–61, 64
visual evoked potentials effect, PKU, 52–64
Ω-oxidation, mechanism, 86–87
Ophthalmological disease, oxidative stress, 113–15
Oral drug bioavailability
and drug response, 219–220
grapefruit juice effect, 219–237

Oral tolerance
celiac disease, 173, 177
transgenic plants induction of, 211, 216–217
Ornithine transcarbamylase deficiency
current treatment and outcome, 264
gene therapy, 263–277
animal model, 265–267, 270, 271
toxicity, 267, 271–272
Otitis media, 145
Oxidation. *See* Beta-oxidation; Fatty acid oxidation
Oxidative stress, 107–122, 280
and Alzheimer disease, 109, 120
homocysteine role, 19
and Huntington disease, 110–111
in neurodegenerative diseases, 108–113
ophthalmological diseases, 113–115
quantification, 119–120
8-Oxo-dG, in neurodegenerative diseases, 109–111

P

P53 protein
apoptosis induction, 244, 249
DNA repair, 242–244
P100 latencies, omega-3 fatty acids effect, 54–56
Parkinson disease, 109–110, 112
Pauling, Linus, 118–119
Peroxisomal acyl-CoA oxidase, 70
Peroxisomal fatty acid oxidation, 88–105
age effects, 104
carnitine role, 89–90
enzymes, 94–95
genetic disorders, 98–100
frequency, 104
treatment, 99–100
mechanism, 88
versus mitochondrial oxidation, 88–92
substrate specificities, 89–92
Peroxisomal thiolase 1 deficiency, 99
Peroxisome proliferator activated receptor α, 85–105
desaturase activation, 73–74
fatty acid oxidation and synthesis role, 68–72
nomenclature, 104
and β-oxidation genetic disorders, 100–105
pH effect, axon growth, 164
Phase I/phase II enzymes, 241–242
Phenylalanine
blood level values, 49
factorial analysis, 27–28
human milk model application, 27–28
indicator amino acid oxidation application, 42–50
minimum requirements, 27–30
phenylketonuria requirements, 29–30, 42–44, 48–50
Phenylketonuria, 25–65
amino acid requirements, 25–49
gene therapy ethics, 276
indicator amino acid oxidation application, 40–49
omega-3 fatty acids in, 51–65
phenylalanine requirements, 29–30, 40–42, 47–48
Phosphomannomutase 2, 127–130
Photooxidative damage, macular degeneration, 113–114
Phytanic acid, 86–87, 102
Phytoestrogens, and breast cancer, 245
Polyunsaturated fatty acids. *See also* Omega-3 fatty acids
antioxidant effects, 80
fuel repartitioning function, 68–70, 74
gene expression regulation, 67–83
hunter-gatherer period, 262
in lipid metabolism, 68–75
plasma membrane transport, 81
in red blood cells and plasma, 60–61, 64
visual evoked potentials effect, PKU, 61–64
Pravastatin, grapefruit juice interactions, 224–226

Pregnancy, and phenylketonuria, 50
Pristanic acid, 91, 95–96, 102
Prolamins. *See also* Gliadins
 celiac disease trigger, 186
 in cereals, 187–189
 classes and repetitive domains, 189
 genetic engineering, 192–198
 structure, 189–192
Prooxidants, 120
Propafenone, and grapefruit juice, 230
Propionic aciduria, 26
Protease inhibitors, and grapefruit juice, 228
Protein intake
 minimum requirements calculation, 26–28
 and plasma homocysteine, 23–24
Protein requirements. *See* Minimum protein requirements
Protein-restricted diets, 25–34

Q

QT interval, and grapefruit juice-drug interactions, 225
Quercetin, and apoptosis, 252–253

R

Rapid acetylator genotype, 241
Ras proto-oncogene, 242–244
Recombinant agriculture. *See* Agricultural biotechnology
Red blood cells, docosahexaenoic acid, 60–61, 64
Red meat, and colon cancer, 241
Reduced folate carrier, 12
Refsum disease, 86
Renal disease, and plasma homocysteine, 13
Retinoic acid, and apoptosis, 250–251
Retinoid X receptor, phytanic acid activation, 102
Rhabdomyolysis, grapefruit juice-drug interactions, 225–226
Riboflavin
 fatty acid oxidation disorders treatment, 98
 and homocysteine formation, 3
 and plasma homocysteine, 7, 10–11
Risk factors/markers, and homocysteine, 7–8, 22–23, 279
Ritonavir, grapefruit juice interactions, 228
mRNA expression
 in arginine and creatine metabolism, 154–165
 enzyme activity relationship, 163
 highly unsaturated fat regulation of, 75
Rye
 gliadin-like proteins, 190
 prolamines, 188, 191–192

S

S14 gene, 72–74
Saquinavir, grapefruit juice interactions, 228
Saturated fatty acid
 and colon cancer, 249
 and Neolithic revolution, 262
SCPx, oxidation, 94–95
Selenium, and apoptosis, 251
Severe combined immune deficiency, gene transfer, 272
Seville oranges, 229–230
Sialyllactoses
 antiadhesive properties, 141–152
 Helicobacter pylori treatment, 144, 150
 human milk concentrations, 137–152
 leukocyte adhesion effect, 146–148
Sildenafil, grapefruit juice interactions, 226
Simvastatin, grapefruit juice interactions, 225
Single nucleotide polymorphisms, and cancer, 254–255

Skeletal muscle, oxidation, 68–70, 82, 96–97
Soya products, and breast cancer, 245
Sparse fur mice, 265–267, 276
Specific transcription factor 1, 72–74
Statins, and colon cancer, 259
Stem cells, and colorectal cancer, 250
Steroids, antenatal administration, 48
Sterol carrier protein X deficiency, 101–102
Sterol regulatory binding protein-1, 72–74
Streptococcus pneumoniae, 144–145
Synthetic antioxidants, 121

T

T-cell receptor, celiac disease, 169–176
T cells
  celiac disease, 169–177
    treatment implications, 177
  gluten peptide recognition, 172–175
Terfenadine, grapefruit juice interactions, 224–225
Tetracosahexaenoic acid, 92
Theophylline, and grapefruit juice, 230
Thiols, and homocysteine toxicity, 6, 20–21
Tissue transglutaminase
  autoantibody trigger, 175–177, 182
  gluten peptide deamidation role, 173–175
  physiologic role, 33–34, 182–183, 281
  treatment implications, celiac disease, 177
α-Tocopherol transfer protein abnormalities, 113
Torsade de pointes, and drug interactions, 225
TRAIL ligand
  apoptosis mechanism, 247
  retinoic acid effect, 250–251
Transferrin, isoelectric focusing, 134
Transgenics, 201–217
  celiac-related peptides, 192–198
  in drug production, 217
  expression profiling, plants, 261
  in food production, 201–217
    and childhood nutrition, 209–211
    vaccine delivery, 211
Transglutaminase. *See* Tissue transglutaminase
Transport proteins, and fatty acids, 81, 92
Triticale, taxonomy, 186
Triticeae grasses, 186, 188–192
Tumor necrosis factor-α, 176
Tumor suppressor genes, 242–244
Twenty-four hour balance method, 37–41
Tyrosemia type I, 26
Tyrosine requirements, phenylketonuria, 42–44, 47

U

Ubiquitin-proteasome pathway, cataracts, 114–115
Upstream stimulatory factor, 72–74

V

Vaccine delivery, transgenic plants, 211, 216
Vegan diet, docosahexaenoic acid levels, 63
Viagra, grapefruit juice interactions, 226
Visual evoked potentials
  age dependency, 62
  information processing relationship, 62–63
  omega-3 fatty acids effect on, 52–64
Vitamin A deficiency, and golden rice, 215–216
Vitamin B_2. *See* Riboflavin
Vitamin B_6
  in homocysteine formation, 3
  and plasma homocysteine levels, 7

Vitamin B_{12}
DNA methylation, 240–241
in homocysteine formation, 3, 5
inborn errors effect, homocysteine, 9–10
and plasma homocysteine levels, 7, 16
Vitamin C
cataract benefits, 114–115
in neurodegenerative diseases, 111–112
Vitamin E
apoptosis induction, 251
in neurodegenerative diseases, 111–113

W

Warfarin, and grapefruit juice, 230
Wheat
genetic engineering, 192–198
gliadin structure, 190
prolamins, 187–189, 191–192
taxonomy, 186
Wheat dough
and gliadin genetic engineering, 193–195
and glutinin gene transformation, 196
Wnt (Wingless-type) pathway, 250
WY 14,643, 73

X

X-linked adrenoleukodystrophy, 98–100
intrafamilial variability, 105
overview, 98–99
treatment, 99–100, 105

Z

Zeaxanthin, 113–114
Zellweger syndrome, 91